Springer Handbook of Auditory Research

Volume 61

More information about this series at http://www.springer.com/series/2506

The ASA Press

The ASA Press imprint represents a collaboration between the Acoustical Society of America and Springer dedicated to encouraging the publication of important new books in acoustics. Published titles are intended to reflect the full range of research in acoustics. ASA Press books can include all types of books published by Springer and may appear in any appropriate Springer book series.

ASA Press

Nina Kraus · Samira Anderson
Travis White-Schwoch · Richard R. Fay
Arthur N. Popper
Editors

The Frequency-Following Response

A Window into Human Communication

With 66 Illustrations

ASA Press

Springer

Editors
Nina Kraus
Auditory Neuroscience Laboratory,
Communication Sciences, Neurobiology
and Otolaryngology
Northwestern University
Evanston, IL
USA

Samira Anderson
Department of Hearing and Speech Sciences
University of Maryland
College Park, MD
USA

Travis White-Schwoch
Auditory Neuroscience Laboratory,
Communication Sciences
Northwestern University
Evanston, IL
USA

Richard R. Fay
Marine Biological Laboratory
Woods Hole, MA
USA

Arthur N. Popper
Department of Biology
University of Maryland
College Park, MD
USA

ISSN 0947-2657 ISSN 2197-1897 (electronic)
Springer Handbook of Auditory Research
ISBN 978-3-319-47942-2 ISBN 978-3-319-47944-6 (eBook)
DOI 10.1007/978-3-319-47944-6

Library of Congress Control Number: 2016958502

Printed on acid-free paper

This Springer imprint is published by Springer Nature
The registered company is Springer International Publishing AG
The registered company address is: Gewerbestrasse 11, 6330 Cham, Switzerland

Acoustical Society of America

The purpose of the **Acoustical Society of America** (www.acousticalsociety.org) is to generate, disseminate, and promote the knowledge of acoustics. The Acoustical Society of America (ASA) is recognized as the world's premier international scientific society in acoustics, and counts among its more than 7,000 members, professionals in the fields of bioacoustics, engineering, architecture, speech, music, oceanography, signal processing, sound and vibration, and noise control.

Since its first meeting in 1929, the ASA has enjoyed a healthy growth in membership and in stature. The present membership of approximately 7,000 includes leaders in acoustics in the United States of America and around the world. The ASA has attracted members from various fields related to sound including engineering, physics, oceanography, life sciences, noise and noise control, architectural acoustics; psychological and physiological acoustics; applied acoustics; music and musical instruments; speech communication; ultrasonics, radiation, and scattering; mechanical vibrations and shock; underwater sound; aeroacoustics; macrosonics; acoustical signal processing; bioacoustics; and many more topics.

To assure adequate attention to these separate fields and to new ones that may develop, the Society establishes technical committees and technical groups charged with keeping abreast of developments and needs of the membership in their specialized fields. This diversity and the opportunity it provides for interchange of knowledge and points of view has become one of the strengths of the Society.

The ASA's publishing program has historically included the *The Journal of the Acoustical Society of America, JASA-Express Letters, Proceedings of Meetings on Acoustics*, the magazine *Acoustics Today*, and various books authored by its members across the many topical areas of acoustics. In addition, ASA members are involved in the development of acoustical standards concerned with terminology, measurement procedures, and criteria for determining the effects of noise and vibration.

The editors dedicate this volume to Gary C. Galbraith (left), whose FFR experiments remain unrivaled in their ingenuity and imagination and to the late Norman M. Weinberger (right), whose deeply thoughtful explorations of hearing, learning, and the brain continue to resonate.

Series Preface

The following preface is the one that we published in volume 1 of the Springer Handbook of Auditory Research back in 1992. As anyone reading the original preface, or the many users of the series, will note, we have far exceeded our original expectation of eight volumes. Indeed, with books published to date and those in the pipeline, we are now set for over 60 volumes in SHAR, and we are still open to new and exciting ideas for additional books.

We are very proud that there seems to be consensus, at least among our friends and colleagues, that SHAR has become an important and influential part of the auditory literature. While we have worked hard to develop and maintain the quality and value of SHAR, the real value of the books is very much because of the numerous authors who have given their time to write outstanding chapters and to our many co-editors who have provided the intellectual leadership to the individual volumes. We have worked with a remarkable and wonderful group of people, many of whom have become great personal friends of both of us. We also continue to work with a spectacular group of editors at Springer. Indeed, several of our past editors have moved on in the publishing world to become senior executives. To our delight, this includes the current president of Springer US, Dr. William Curtis.

But the truth is that the series would and could not be possible without the support of our families, and we want to take this opportunity to dedicate all of the SHAR books, past and future, to them. Our wives, Catherine Fay and Helen Popper, and our children, Michelle Popper Levit, Melissa Popper Levinsohn, Christian Fay, and Amanda Fay Seirra, have been immensely patient as we developed and worked on this series. We thank them and state, without doubt, that this series could not have happened without them. We also dedicate the future of SHAR to our next generation of (potential) auditory researchers—our grandchildren—Ethan and Sophie Levinsohn, Emma Levit, and Nathaniel, Evan, and Stella Fay.

Preface 1992

The Springer Handbook of Auditory Research presents a series of comprehensive and synthetic reviews of the fundamental topics in modern auditory research. The volumes are aimed at all individuals with interests in hearing research including advanced graduate students, post-doctoral researchers, and clinical investigators. The volumes are intended to introduce new investigators to important aspects of hearing science and to help established investigators to better understand the fundamental theories and data in fields of hearing that they may not normally follow closely.

Each volume presents a particular topic comprehensively, and each serves as a synthetic overview and guide to the literature. As such, the chapters present neither exhaustive data reviews nor original research that has not yet appeared in peer-reviewed journals. The volumes focus on topics that have developed a solid data and conceptual foundation rather than on those for which a literature is only beginning to develop. New research areas will be covered on a timely basis in the series as they begin to mature.

Each volume in the series consists of a few substantial chapters on a particular topic. In some cases, the topics will be ones of traditional interest for which there is a substantial body of data and theory, such as auditory neuroanatomy (Vol. 1) and neurophysiology (Vol. 2). Other volumes in the series deal with topics that have begun to mature more recently, such as development, plasticity, and computational models of neural processing. In many cases, the series editors are joined by a co-editor having special expertise in the topic of the volume.

Arthur N. Popper College Park, MD, USA
Richard R.Fay Woods Hole, MA, USA

SHAR logo by Mark B. Weinberg, Bethesda, Maryland, used with permission.

Volume Preface

The frequency-following response (FFR) is a measure of synchronous sound-evoked neural activity that reveals the integrity of sound processing in the brain. Studies of the FFR are organized around two intertwining themes that are reviewed in this volume: learning and everyday communication. These studies tie into a conceptual framework whereby making sense of sound is fundamental to everyday life and is at the intersection of cognitive, sensorimotor, and reward networks in the brain. Understanding how well an individual listener processes sound provides a snapshot of auditory function and its impact on everyday communication skills.

Chapter 1, by Nina Kraus, Samira Anderson, and Travis White-Schwoch, provides an overview of FFR research and contends that the FFR reflects an individual's past and potential in sound. The part of this volume devoted to learning starts in Chap. 2 where Jeng reviews how auditory experience early in life shapes auditory brain development, intertwined with a discussion of theories of early speech and language development. In Chap. 3, Krishnan and Gandour discuss how everyday linguistic experience shapes auditory processing, with an emphasis on the neural coding of pitch-bearing information. Following this, Carcagno and Plack (Chap. 4) provide a comprehensive review of FFR studies of short-term training and perceptual learning. Escera (Chap. 5) considers a different form of auditory plasticity—the ability to rapidly adapt to a sensory environment *online.* White-Schwoch and Kraus (Chap. 6) end the section on learning and bridge into the section on everyday communication. They review principles of auditory learning, emphasizing the enduring biological legacy that everyday experiences impart.

Shinn-Cunningham, Varghese, Wang, and Bharadwaj (Chap. 7) open the section on everyday communication by reviewing cutting-edge work that unravels biological processes that facilitate and constrain sound-directed attention. Following this, Bidelman (Chap. 8) discusses two insidious constraints on everyday communication: noise and reverberation. Schochat, Nunes Rocha-Muniz, and Filippini (Chap. 9) then tackle auditory processing disorder—poor auditory function despite normal audiograms—a clinical condition that continues to vex audiologists and scientists. Following this, Reetzke, Xie, and Chandrasekaran (Chap. 10) review the extensive literature using the FFR to study reading impairments such as dyslexia. The volume ends with a discussion of clinical translation in the context of aging, hearing loss, and amplification by Anderson (Chap. 11).

Together, these chapters illustrate the diversity of research applying the FFR and the rich granularity of biological insight into auditory function that the FFR provides. The core theme that emerges is that human communication is intimately tied to experience with sound. These experiences range from in-the-moment adaptation to lifelong experience with language or music. These communicative skills extend into everyday life, including listening in noise, spatial hearing, and literacy. Interest in capitalizing in the communication–experience link motivates an eventual goal of using the FFR in clinical settings, to evaluate listening skills, predict future listening challenges, and reveal outcomes from interventions.

Nina Kraus, Evanston, IL, USA
Samira Anderson, College Park, MD, USA
Travis White-Schwoch, Evanston, IL, USA
Richard R. Fay, Falmouth, MA, USA
Arthur N. Popper, College Park, MD, USA

Contents

1 **The Frequency-Following Response: A Window into Human Communication** ... 1
Nina Kraus, Samira Anderson, and Travis White-Schwoch

2 **Infant and Childhood Development: Intersections Between Development and Language Experience** ... 17
Fuh-Cherng Jeng

3 **Shaping Brainstem Representation of Pitch-Relevant Information by Language Experience** ... 45
Ananthanarayan Krishnan and Jackson T. Gandour

4 **Short-Term Learning and Memory: Training and Perceptual Learning** ... 75
Samuele Carcagno and Christopher J. Plack

5 **The Role of the Auditory Brainstem in Regularity Encoding and Deviance Detection** ... 101
Carles Escera

6 **The Janus Face of Auditory Learning: How Life in Sound Shapes Everyday Communication** ... 121
Travis White-Schwoch and Nina Kraus

7 **Individual Differences in Temporal Perception and Their Implications for Everyday Listening** ... 159
Barbara Shinn-Cunningham, Leonard Varghese, Le Wang, and Hari Bharadwaj

8 **Communicating in Challenging Environments: Noise and Reverberation** ... 193
Gavin M. Bidelman

9 Understanding Auditory Processing Disorder Through the FFR 225
Eliane Schochat, Caroline Nunes Rocha-Muniz, and Renata Filippini

10 Neurobiology of Literacy and Reading Disorders 251
Rachel Reetzke, Zilong Xie, and Bharath Chandrasekaran

11 Clinical Translation: Aging, Hearing Loss, and Amplification 267
Samira Anderson

Contributors

Samira Anderson Department of Hearing and Speech Sciences, University of Maryland, College Park, MD, USA

Hari Bharadwaj Martinos Center for Biomedical Imaging, Massachusetts General Hospital, Boston, MA, USA

Gavin M. Bidelman School of Communication Sciences and Disorders, University of Memphis, Memphis, TN, USA

Samuele Carcagno Department of Psychology, Lancaster University, Lancaster, UK

Bharath Chandrasekaran Department of Communication Sciences and Disorders and Psychology, Institute for Neuroscience, Center for Perceptual Systems, The University of Texas at Austin, Austin, TX, USA

Carles Escera Brainlab-Cognitive Neuroscience Research Group, Department of Clinical Psychology and Psychobiology, Faculty of Psychology, University of Barcelona, Catalonia, Spain

Renata Filippini Department of Physiotherapy, Audiology and Speech Therapy and Occupational Therapy, LIM 34, Neuroaudiology Lab, Universidade de São Paulo, São Paulo, Brazil

Jackson T. Gandour Department of Speech Language Hearing Sciences, Lyles-Porter Hall, Purdue University, West Lafayette, IN, USA

Fuh-Cherng Jeng Communication Sciences and Disorders, Ohio University, Athens, OH, USA

Nina Kraus Auditory Neuroscience Laboratory, Departments of Communication Sciences, Neurobiology and Otolaryngology, Northwestern University, Evanston, IL, USA

Ananthanarayan Krishnan Department of Speech Language Hearing Sciences, Lyles-Porter Hall, Purdue University, West Lafayette, IN, USA

Christopher J. Plack Department of Psychology, Lancaster University, Lancaster, UK; Manchester Centre for Audiology and Deafness, University of Manchester, Manchester Academic Health Science Centre, Manchester, UK

Rachel Reetzke Department of Communication Sciences and Disorders, The University of Texas at Austin, Austin, TX, USA

Caroline Nunes Rocha-Muniz Department of Physiotherapy, Audiology and Speech Therapy and Occupational Therapy, LIM 34, Neuroaudiology Lab, Universidade de São Paulo, São Paulo, Brazil

Eliane Schochat Department of Physiotherapy, Audiology and Speech Therapy and Occupational Therapy, LIM 34, Neuroaudiology Lab, Universidade de São Paulo, São Paulo, Brazil

Barbara Shinn-Cunningham Biomedical Engineering, Boston University, Boston, MA, USA

Leonard Varghese Biomedical Engineering, Boston University, Boston, MA, USA

Le Wang Biomedical Engineering, Boston University, Boston, MA, USA

Travis White-Schwoch Auditory Neuroscience Laboratory, Department of Communication Sciences, Northwestern University, Evanston, IL, USA

Zilong Xie Department of Communication Sciences and Disorders, The University of Texas at Austin, Austin, TX, USA

Chapter 1
The Frequency-Following Response: A Window into Human Communication

Nina Kraus, Samira Anderson, and Travis White-Schwoch

Abstract The frequency-following response (FFR) is a measure of synchronous sound-evoked neural activity that reveals the integrity of sound processing in the brain. Studies of the FFR are organized around two intertwining themes: learning and everyday communication. These studies tie into a conceptual framework wherein making sense of sound is fundamental to everyday life and is at the intersection of cognitive, sensorimotor, and reward networks. Understanding how well an individual listener processes sound provides a snapshot of auditory function and its impact on everyday communication skills. This chapter provides an overview of FFR research and contends that the FFR is a measure that reflects an individual's past and potential in sound. Despite diverse terminology in the field, it is argued that FFR provides a good umbrella term for these biological approaches. A brief historical perspective illustrates how FFR has a longstanding history in auditory neuroscience and has addressed many basic and clinical questions in hearing. The FFR is on its way to becoming a mainstream tool in neuroscience. Perhaps most exciting is the potential for use in brain screening to assess hearing in newborns to evaluate risk for communication impairments, setting the stage for early interventions that offset a life spent struggling to learn and communicate.

N. Kraus (✉)
Auditory Neuroscience Laboratory, Departments of Communication Sciences, Neurobiology and Otolaryngology, Northwestern University, 2240 Campus Drive, Evanston, IL 60208, USA
e-mail: nkraus@northwestern.edu

S. Anderson
Department of Hearing and Speech Sciences, University of Maryland, 0100 Lefrak Hall, College Park, MD 20742, USA
e-mail: sander22@umd.edu

T. White-Schwoch
Auditory Neuroscience Laboratory, Department of Communication Sciences, Northwestern University, 2240 Campus Drive, Evanston, IL 60208, USA
e-mail: whiteschwoch@northwestern.edu

N. Kraus et al. (eds.), *The Frequency-Following Response*,
Springer Handbook of Auditory Research 61,
DOI 10.1007/978-3-319-47944-6_1

Keywords Auditory brainstem response · Auditory evoked potentials · Auditory learning · Auditory midbrain · Auditory processing · cABR · Envelope-following response · EFR · History of neuroscience · Inferior colliculus · Learning and memory · Neuroplasticity · Speech perception

1.1 Introduction

Making sense of sound is fundamental to everyday life. Sound is an invisible but powerful force that provides a critical medium for learning about the world. Much of this learning is tangible, such as a child's prodigious ability to soak up speech and, eventually, learn to talk. But sound also provides a channel for phenomena that are less concrete, such as making friends, building relationships, and learning how to navigate the social world.

The ability to make sense of sound relies on the remarkable spectrotemporal precision in the auditory system. Listeners can detect auditory events that are shorter in duration than an action potential, and neurons in the auditory system can respond to sound more than 1,000 times more quickly than photoreceptors in the visual system. This temporal precocity is intimately tied to everyday communication. Thus, enriched auditory milieus that facilitate the coordination of cognitive, sensorimotor, and reward systems also enhance the integrity with which the brain processes sound with concomitant gains in communication skills. In contrast, disruptions to any chain in this system cascade to communication impairments that are coupled to poor auditory coding (Kraus and White-Schwoch 2015).

The frequency-following response (FFR) is a measure of synchronous sound-evoked brain activity that reveals the integrity of sound processing in the brain and reflects auditory-neurophysiological processes with granularity and precision rarely offered by other tools in human neuroscience. FFR provides a snapshot of the hearing brain and reflects the confluence of cognitive, sensorimotor, and reward systems on auditory processing, reliably showing individual differences that align with everyday communication skills.

1.2 Why Measure Sound Processing in the Brain?

A longstanding goal in auditory neuroscience has been to understand the relationship between hearing and everyday life and to elucidate the biological mechanisms underlying this link in humans. This goal has translational consequences because understanding how sound processing and communication are disrupted can pave a way toward strategies to evaluate and manage communication impairments, spanning listening, language, and literacy.

A first step in achieving this overarching goal is to understand the biological mechanisms that underlie auditory processing and its impairments. This theme pervades the chapters in this volume and shows how measuring neurophysiological responses to complex sounds illuminates the role of auditory processing in communication, language development, literacy, and other important functions of everyday life. Additionally, this approach documents the disruption of auditory processing in clinical populations. Yet this processing is not static; rather, it is sculpted by a life in sound. Thus, auditory neurophysiology reveals the imprint of learning. As reviewed throughout this book, the same neurophysiological markers implicated in communication impairments are amenable to explicit training, motivating the use of targeted interventions to boost communication skills and their underlying biological mechanisms. The FFR reveals biological hearing health in individual humans with unprecedented granularity.

1.3 What Is the Frequency-Following Response?

The FFR is a reflection of sound-evoked synchronous neural activity that is distinguished from other evoked potentials by its *transparency*. Whereas other potentials are abstract representations of sound that are identified by "neural waves," the FFR reflects phase-locked activity that codes sound; thus, it physically resembles the eliciting stimulus, as illustrated in Fig. 1.1.

What distinguishes the FFR from other types of sound-evoked neuroelectric responses is that an individual's FFR offers a wealth of information about sound processing in the brain—a biological mosaic that goes far beyond the timing and amplitude measures gleaned from most types of sound-evoked electrical activity. Because the FFR reconstructs most properties of the eliciting stimulus (Fig. 1.1), the response is as complex as the sound that elicits it. Thus, the integrity of an individual's neural coding of discrete cues, such as those that convey a speech

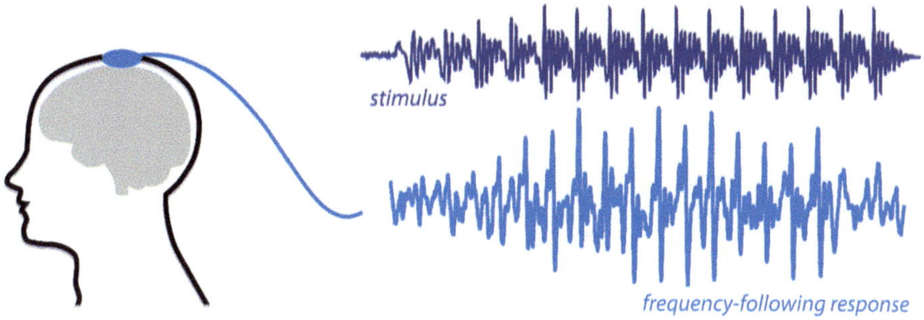

Fig. 1.1 The FFR is a scalp-recorded auditory evoked potential. Unlike most measures of biological activity that provide abstract measures, the FFR is transparent—it recreates many physical features of the evoking sound. As may be seen, the stimulus and response are similar with respect to duration, periodicity, rise, and more. Thus, the FFR is an avenue to evaluate the neural coding of multiple features in sound

sound's identity, may be teased apart. The diversity of FFR-derived measures is reflected in the diverse chapters in this book. What will become clear when thinking about these chapters as a whole is that each FFR component is somewhat independent from the others: a large response is not necessarily a stable response, and a stable response is not necessarily a large response.

The chapters presented in this book cover a vast territory, but all of them review work focused on the FFR. Together, these chapters illustrate how the FFR meets two high-reaching goals. For one, FFR studies encapsulate how making sense of sound is coupled to communication and shed light on basic principles of sound processing in the brain, its malleability, and its stability. Unlike other approaches, though, the FFR reliably reveals *individual differences*. A single FFR can reflect the past (White-Schwoch et al. 2013) and predict the future (White-Schwoch et al. 2015). Thus, the second goal: the FFR is an approach that may be applied clinically to understand communication disorders. The FFR, therefore, is a candidate clinical tool because it moves beyond asking *whether* an individual makes sense of sound to shed light on *how well* an individual makes sense of sound and *which* of the biological processes that are important for making sense of sound are enhanced or diminished.

1.4 The FFR: Nothing New

The last decade has borne a surge of interest in the FFR. As shown in Fig. 1.2, a Google Scholar search shows over 700 references to "frequency-following response" between 2010 and 2016, compared to 436 references between 2000 and 2010. It is, however, interesting to trace the early history of the FFR because it illustrates how the FFR has always played a role in auditory neuroscience and neurophysiology in general. For over 50 years, the FFR has been used by scientists interested in hearing assessment, pitch perception, diagnostics, cochlear transduction, and attention.

The first FFR-like recordings even predate Lord Adrian, often considered one of the founding fathers of neurophysiology, who would win the Noble prize for (among many other discoveries) conducting some of the first single neuron

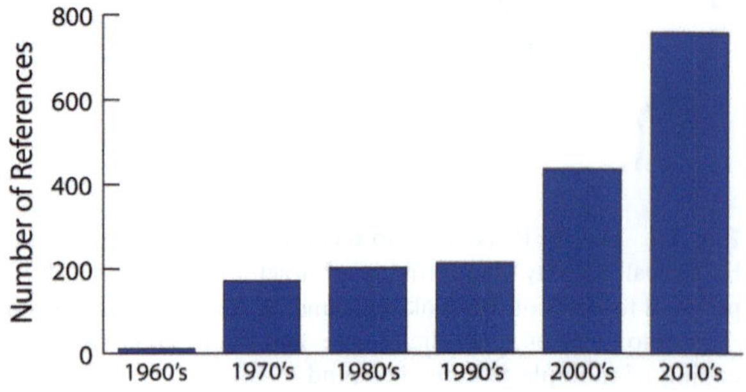

Fig. 1.2 A Google Scholar search (May 2016) revealed the surge in papers referencing the "frequency-following response" in the past two decades, but also illustrated its long history

recordings and establishing the all-or-none principle of action potentials (Adrian 1926). But almost two decades earlier, Buytendijk (1910) recorded sound-evoked electric activity in guinea pigs, rabbits, and frogs (although he refered to similar, unpublished observations made in 1904). While his recordings were likely dominated by eighth nerve activity, he did record an FFR-like dipole in the rabbit with an active electrode near the internal auditory meatus and a reference electrode "on an indifferent spot of the hindmost skull-cavity." He also noted differences in response properties between anesthetized and deceased animals, portending discovery of active cochlear mechanics by more than a half century (for review see Dallos 1992).

Derbyshire and Davis (1935) conducted one of the first comprehensive studies of sound-evoked electrical activity and, like neurophysiologists to follow, were struck by the similarity between periodicities in the stimulus and the response. Like many FFR scientists to come, they were interested in the neural basis of pitch perception, a topic that was to recur periodically through the FFR's history. As early as 1965, the FFR was used to arbitrate between place and volley theories of pitch perception (Boudreau 1965). This was to become a focus of FFR research in the 1970s and 1980s (Hall 1979; Greenberg et al. 1987). It was clear that the FFR lent itself to experiments aiming to understand pitch processing in humans (e.g., Galbraith 1994). This remains a topic of intense scrutiny (Gockel et al. 2011), especially given new evidence that pitch coding, vis-à-vis the FFR, is subject to experience (Krishnan and Gandour, Chap. 3; Carcagno and Plack, Chap. 4; White-Schwoch and Kraus, Chap. 6).

During the 1970s much attention turned to the origins and basic properties of the FFR in an effort to develop it as an objective measure of hearing thresholds. Worden, Marsh, and their colleagues made the first FFR recordings in humans and dedicated energy to understanding its origins (Worden and Marsh 1968; Marsh et al. 1970). Complementary studies in animal models worked to distinguish the FFR from the cochlear microphonic and to rule out stimulus artifact—a challenge for the FFR researcher that remains to this day (Faingold and Caspary 1979; Snyder and Schreiner 1984). It was perhaps fate that pioneering scientists in the field of auditory neuroplasticity, such as Michael Merzenich (Gardi et al. 1979) and Norman M. Weinberger (Weinberger et al. 1970), briefly forayed into this biological approach that has now become a powerful approach to study auditory learning in humans (Sect. 1.6).

Eventually, FFR researchers felt adventuresome. Rather than just measuring responses to pure tones, they sought to test the limits of just how much an FFR could resemble the stimulus. It soon became apparent that the rich spectrotemporal details contained in natural sounds, such as speech and music, were beautifully recreated by the FFR (Fig. 1.1). In fact, if a computer was tricked into playing an FFR, listeners could identify the evoking stimulus (Galbraith et al. 1995). Soon, stimuli combining transient and sustained features (e.g., consonant-vowel syllables) and complex listening situations (e.g., speech in background noise) were getting closer and closer to approximating everyday listening environments, revealing biological bottlenecks in everyday sound processing (Cunningham et al. 2001). Parallel experiments in an animal model elucidated the specific biological mechanisms underlying these phenomena (Cunningham et al. 2002). This crucial discovery opened a door to the use of complex stimuli to understand complex auditory phenomena, which is the focus of this volume. Additionally, this approach showed that a single FFR offered a wealth of information about discrete aspects of sound processing in the brain, which has motivated FFR work since, and allows for a more thorough evaluation of sound processing than periodicity tracking or response amplitude (Anderson et al. 2012). FFR technology has now advanced to the point where it is no longer constrained to the laboratory, facilitating clinical and community-based studies of auditory processing and learning (Kraus et al. 2014a, b).

1.5 Call It "FFR"

Following the resurgence in interest, the FFR has entered the throes of a terminology identity crisis. At times it seems there are as many terms to refer to the FFR as there are papers using it! While this might be seen as a point of consternation, the editors hold that it is a sign of maturity: the FFR is on its way to becoming a mainstream approach in neuroscience. "FFR" provides an excellent umbrella term that ties together diverse approaches, populations, and questions. The FFR has come a long way and accomplished a lot and, like other approaches to evaluating sound processing, it can encompass many offshoots. FFR can be thought of as a suite of methods that can be tailored to the population and stimuli of interest.

Common terms, aside from FFR, include: cABR, auditory brainstem response to complex sounds (Skoe and Kraus 2010); EFR, envelope-following response (Dolphin and Mountain 1992; Aiken and Picton 2008); AMFR, amplitude-modulation following response (Kuwada et al. 2002); sABR, speech-evoked auditory brainstem response (Russo et al. 2004); and SSSR, subcortical steady-state response (Bharadwaj and Shinn-Cunningham 2014). Weinberger et al. (1970) poetically termed it the "auditory neurophonic," a term that caught on for a brief period to distinguish it from the cochlear microphonic; however this name also was used to refer to auditory nerve activity (Snyder and Schreiner 1984). Sometimes the FFR is simply called an "auditory brainstem response" (ABR)—woe to the reader imagining hearing thresholds! In addition, a single evoked potential is sometimes dichotomized into its "ABR" and "FFR" portions (Cunningham et al. 2001). Even more confusing is that all of these terms are often hybridized, such as "speech-evoked-envelope-following response" (Easwar et al. 2015), even though it is unlikely there is a one-to-one mapping between *acoustic envelope* and temporal fine structure and *neural envelope* and temporal fine structure (Shamma and Lorenzi 2013). In other words, the FFR envelope may not solely reflect coding the stimulus envelope, and the FFR fine structure may not solely reflect coding the stimulus fine structure.

The advantage of the term "ABR" and its derivatives is that it provides a good technical description. FFRs are similar to ABRs in many ways, including with respect to technique, such as the collection parameters (electrode montage, filtering, averaging, and more). Additionally, ABRs are classically thought of as responses to transients and when FFRs are elicited to complex sounds, they contain these rich transient cues, such as those found in consonants (Fig. 1.1). Unfortunately, the term ABR undermines the rich biological information offered by the FFR because only latencies and amplitudes are classically analyzed in ABRs. Moreover, the term "brainstem" is something of a misnomer. First, the FFR is thought to have a strong contribution from the inferior colliculus of the auditory midbrain, or at least from synchronized inputs to the midbrain (for review see Chandrasekaran and Kraus 2010). Second, an emerging view characterizes the auditory system as a distributed, but integrated, experience-dependent network, and it has been argued that the FFR reflects this interactivity (Kraus and White-Schwoch 2015, 2016).

In fact, recent evidence from Zatorre and colleagues suggests a cortical contribution to the FFR (Coffey et al. 2016), and recent work from Shinn-Cunningham and colleagues suggests a contribution from eighth nerve fibers (Shinn-Cunningham, Varhgese, Wang, and Bharadwaj, Chap. 7). Thus, terminology that implies anatomic generators can be misleading, especially "ABR." In fact, the editors of this volume and their colleagues introduced the term "cABR" (Skoe and Kraus 2010; Anderson et al. 2013) but in retrospect regret the localization implied. "Brainstem" may be especially problematic by implying low-level afferent processes when, in fact, the activity revealed by the FFR is exquisitely tuned and retuned by the convergence of afferent and efferent influences. Moreover, in many cases, click-evoked ABRs appear normal in listeners with an abnormal FFR that reveals a communication disorder (King et al. 2002; Banai et al. 2009).

The advantage of the term "FFR" is that the response does just that—it follows the frequencies of the stimulus, thus offering its wonderful transparency and richness as an evoked potential. Unfortunately, this term is not without its problems either. Traditionally the FFR referred solely to phase-locked activity to pure tones, intended to measure low-frequency hearing sensitivity (see Sect. 1.4). For many in the field, then, it does not imply the rich information across frequencies offered by the FFR to a complex sound. In addition, real-world sounds, such as speech, contain transients. Although these are technically brief, broadband bursts of acoustic information, they are rarely thought of on a frequency-specific basis and are instead thought of in terms of timing. Thus, "FFR" risks eliding important aspects of the technique.

Although no term is perfect, it is the view of this volume's editors that "FFR" is the best compromise. FFR can be thought of as an umbrella term that encapsulates all of the others. FFR stimuli and recording parameters can be tailored to the specific population and question of interest.

1.6 A Window into Human Communication

This volume is organized around two themes: the neurobiology of (1) learning and (2) everyday communication. What should be clear upon reading any chapter—and especially when considering the book as a whole—is that these themes are connected and interactive. That is, the ability to communicate is shaped by experience, and experience is shaped by everyday communication.

Fuh-Cherng Jeng (Chap. 2) reviews how auditory experience early in life shapes brain development. This is intertwined with a discussion of theories of early speech and language development and how language experience during the first year of life shapes auditory neurophysiology. He then reviews FFR studies during infancy and early childhood that illustrate both the rapid developmental plasticity incumbent in young children and how this maturational course intersects with everyday linguistic experience. FFR studies during infancy and early childhood have translational implications. As reviewed later in this volume, in older children and adults FFR measures indicate communication impairments. Jeng makes a convincing case that the FFR is a robust and reliable measure during infancy, opening up an avenue for early identification of communication disorders to facilitate early interventions.

Ananthanarayan Krishnan and Jackson Gandour (Chap. 3) discuss how everyday linguistic experience shapes auditory processing, with an emphasis on the neural coding of pitch-bearing information. Different linguistic systems employ distinct acoustic cues to convey lexical information. Tone languages, such as Mandarin, use pitch contours to convey meaning, and Krishnan and Gandour highlight their seminal work using the FFR to examine how this experience shapes automatic auditory response properties. They couch this in a discussion of models of language and pitch processing through the auditory system. In addition to revealing the profound influence of everyday experience on the auditory system, Krishnan and Gandour show how elegant FFR experiments shed light on the biological legacy of experience, the organization of pitch processing in the auditory system, and the fundamental link between language and hearing.

Samuele Carcagno and Christopher Plack (Chap. 4) provide a comprehensive review of FFR studies of short-term training and perceptual learning. Auditory abilities are not static, and short courses of intensive auditory training shape perceptual skills. The FFR is increasingly used as an outcome measure in these experiments. Following a brief review of perceptual learning and some of the major questions facing the field, Carcagno and Plack lucidly cover each FFR training experiment, including those in children, young adults, and older adults, with a critical assessment of each experiment's strengths and weaknesses. They connect the dots to the broader literature on auditory neuroplasticity, pulling on work in animal models to evaluate several frameworks for learning that have been posited in light of these FFR experiments. While auditory training is often recommended for listeners with communication impairments, Carcagno and Plack lay out what must be accomplished in future work to strengthen this clinical potential.

Carles Escera (Chap. 5) considers a different form of auditory plasticity—the ability to rapidly adapt to a sensory environment "online". Listeners must navigate constantly changing auditory worlds, and sensory systems need to be dynamic enough to accommodate this diversity. Escera considers experiments in humans and animal models of context-dependent adaptation observed throughout the auditory system. This leads to a discussion of work in humans that examines how FFR response properties are shaped by stimulus context, which is couched in a discussion of context-dependent modulation in the auditory midbrain. As Escera argues, this work provides insights into the fundamental organization of the auditory system, and he rejects the view that context-dependent modulation is a strictly cortical phenomenon in favor of a more integrated model for auditory processing.

Travis White-Schwoch and Nina Kraus (Chap. 6) complete the section on learning and provide a bridge to the section on everyday communication. They review principles of auditory learning, emphasizing the enduring biological legacy that everyday experiences impart. A central argument is that experiences—good or bad—shape automatic processing, and they argue that both may do so through congruous pathways. Thus, the auditory brain's default state is in a constant push-and-pull between stability and plasticity. After reviewing FFR studies of communication abilities and disabilities, they juxtapose lifelong music training (a case of enrichment) to growing up in poverty (a case of deprivation). Against the backdrop of understanding how different FFR measures indicate communication impairments, they argue that enrichment activities, such as music training, language experience, and auditory training, can be targeted to strengthen the neurobiological bottlenecks endemic to specific populations.

Barbara Shinn-Cunningham, Leonard Varghese, Le Wang, and Hari Bharadwaj (Chap. 7) open the section on everyday communication by reviewing cutting-edge work that unravels biological processes that facilitate and constrain sound-directed attention. The work they review illustrates how the auditory system operates as a distributed but integrated circuit, highlighting complex interactions between the integrity of the auditory periphery, fine-grained temporal coding, and guided attention. Shinn-Cunningham and colleagues review the challenges and opportunities offered by the FFR in the study of these individual differences. Next, they highlight one candidate mechanism for individual differences, namely, a noise-induced deafferentation of synapses at the inner hair cells. While they emphasize what the FFR has contributed to the study of everyday listening skills and individual differences in those abilities, they clearly outline its limitations, identifying important avenues for future experiments and highlighting how everyday communication skills rely on many interactive auditory and non-auditory processes.

Gavin Bidelman (Chap. 8) discusses two insidious constraints on everyday communication: noise and reverberation. Few everyday listening situations are pristine, and for too long the field imagined that auditory performance in the sound booth was a good predictor of listening skills in a restaurant. Bidelman cogently discusses how both noise and reverberation constrain the intelligibility of a signal and how these constraints can be seen in the FFR. Next, he discusses several experiments that show links between the integrity of the FFR in adverse listening conditions and a listener's auditory performance. He ties this back to the question of auditory experience and shows how different experiences shape the contingency between the FFR and listening skills. One of the fundamental questions for auditory neuroscience is how listeners manage to understand speech in noisy, everyday environments such as the cocktail party. Bidelman emphasizes just how much the field has learned through FFR experiments.

Eliane Schochat, Caroline Nunes Rocha-Muniz, and Renata Filippini (Chap. 9) tackle auditory processing disorder—poor auditory function despite a normal audiogram—a clinical condition that continues to vex audiologists and scientists. They emphasize the importance of objective biological approaches in evaluating listening skills, especially in children with related cognitive and language impairments, and how the FFR has contributed in the context of auditory neurophysiology. What Schochat et al.'s summary highlights is that the FFR is that rare tool in translational science that both teaches basic lessons about the mechanisms underlying communication skills and offers clinicians a strategy to improve diagnosis and management of their patients. In short, the FFR inherently is a biological index of auditory processing and its disorders.

Rachel Reetzke, Zilong Xie, and Bharath Chandrasekaran (Chap. 10) review the extensive literature using the FFR to study reading impairments such as dyslexia. If it is at first surprising that a book on the auditory system includes a discussion of reading and dyslexia, Reetzke et al. quickly make clear that auditory processing is fundamental to literacy development, and they highlight the contributions from the FFR in this lesson. As they review, literacy is coupled to listening, and many children with poor reading skills have poor neural coding reflected in the FFR. Research in dyslexia and reading remains fraught with controversy; the FFR provides a stabilizing view. Thus, from a pragmatic standpoint, Reetzke et al. make a convincing case that FFR work dovetails with several models of reading impairment and that irrespective of the underlying causes of poor reading, one can still appreciate the FFR's contributions as an experimental—and potentially clinical—tool.

Samira Anderson (Chap. 11) closes the volume with a discussion of clinical translation in the context of aging and hearing loss. The communication problems that older adults face are of strong interest in the hearing sciences and audiology, and they are exacerbated by age-related hearing loss. Anderson reviews studies that show the FFR reveals distinct bottlenecks in sound processing associated with aging and hearing loss. Next, she discusses how the FFR serves as a research tool in studies of auditory training, and how it is beginning to emerge as a technique in the study of amplification. Finally, she lays out the directions necessary to translate the FFR from the lab to the clinic.

Together, these chapters illustrate the diversity of research applying the FFR. The core theme that emerges is that human communication is intimately tied to experience with sound. These experiences range from in-the-moment adaptation to lifelong experience with language or music. These communicative skills extend into everyday life, including listening in noise, spatial hearing, and literacy. Interest in capitalizing in the communication-experience link motivates an eventual goal of using the FFR in clinical settings to evaluate listening skills, predict future listening challenges, and reveal outcomes from interventions.

1.7 Next Steps: A Mainstream Role for the FFR in Neuroscience

A few broad conclusions can be drawn from these chapters; these highlight some of the future directions for this field. From a technical standpoint, the FFR has reached a reasonable level of sophistication. Although ongoing work is dedicated to refining its collection and analysis, how to go about measuring and interpreting an FFR is now basically understood (Skoe and Kraus 2010). This opens the door to applying the FFR to new and diverse areas in the study of communication, listening, and experience. Particularly exciting is the potential for the systematic study of individual differences in the FFR, including in clinical populations. This can eventually lead to a better understanding of how listening skills can be disrupted. If the FFR continues to pattern in distinct ways in different clinical populations, it may prove to be a sensitive and specific biological marker for communication impairments.

At the same time, there are areas in need of technical refinement. A persistent challenge to FFR work is the signal-to-noise problem: an FFR needs to be the averaged response of many repetitions of sound. As techniques develop to reduce collection time and, perhaps, make sense of the response to just a few stimuli, the FFR can provide a stronger measure of real-time listening and adaptation. Additionally, use of the FFR will become more practical, especially in difficult-to-test clinical populations.

Tracing the FFR's history (Sect. 1.4) shows that it was first predominantly applied to animal models. The past few decades, however, have seen it turn almost exclusively to a technique used in humans. While it is a robust marker of auditory processing in humans, several authors in this volume and elsewhere outline questions that are best answered in animal models. An exciting new avenue is to study humans and animal models in conjunction to understand the neural mechanisms underlying auditory phenomena and their consequences for every day, real-world listening (Warrier et al. 2011; White-Schwoch et al. 2016). It is clear that the midbrain is subject to experience-dependent modulation (Gao and Suga 2000; Bajo et al. 2010); using the FFR can provide a deeper understanding of the mechanisms underlying experience-dependent plasticity.

The FFR has proven its worth in studying groups of listeners. However, as many authors here note, it is unusual among evoked potentials in its reliability and interpretability in an individual. Systematically studying individual differences with the FFR represents a new frontier for auditory neuroscience, which can take the field to a point where it considers how auditory function is shaped by an individual's life in sound. The evidence reviewed here, particularly in the second half of this volume, shows how individual differences in the FFR reveal an individual's strengths and weaknesses in sound processing. These individual differences are stable across stimuli and test sessions, motivating longitudinal studies employing the FFR. It remains to be seen whether this information can be harnessed to be clinically useful, but early evidence is promising.

FFR studies span a wide scope. This volume focuses on learning and communication and emphasizes the more longstanding spheres of FFR research. However, new domains are rapidly being applied to the FFR, including mental health (Tarasenko et al. 2014), amplification/auditory prostheses (Easwar et al. 2015; Anderson, Chap. 11), attention deficit/hyperactivity disorder (Jafari et al. 2015), amusia (Lehmann et al. 2015), concussion (Kraus et al. 2016), and more. Perhaps it is not a surprise that the FFR has so many future avenues: it evaluates the incredibly fast and challenging auditory brain computations that are hypothesized to be easily disrupted by acute and long-lasting insults (Kraus and White-Schwoch 2015).

With increasing knowledge of how FFR signatures distinguish listeners and indicate an individual's strengths and weaknesses in making sense of sound, the FFR can inform medicine, education, and social policy. Early studies show that the FFR is an effective field-based technique for conducting neurophysiological studies outside of the traditional laboratory (Kraus et al. 2014a, b). As FFR technology continues to be refined, the FFR can be placed in schools and clinics to provide an adjunct for the evaluation of listening skills. Finally, an exciting potential for the FFR is to predict future communication skills (e.g., White-Schwoch et al. 2015). The FFR is a robust measure of auditory-neurophysiological processing in infants (Anderson et al. 2015; Jeng, Chap. 2), suggesting that it could be used as a measure to screen for communication disorders in newborns and, given the plasticity of these neurophysiological processes, identify approaches for early intervention (Carcagno and Plack, Chap. 4; White-Schwoch and Kraus, Chap. 6). One of hearing science's largest contributions to public health has been the introduction of universal newborn hearing screening. Research in the FFR, as reviewed in this book's diverse chapters, has the potential to make a second such contribution by screening the newborn brain, opening a door to early interventions that prevent a life spent struggling to make sense of sound.

1.8 Summary

This volume spans diverse work employing the FFR, revolving around intertwining themes of learning and everyday communication. The editors are grateful to each of the authors for contributing to this volume, and would like to emphasize the following points in closing:

- The FFR is a biological snapshot of the integrity of sound processing in the brain. This sound processing is shaped by experience, predicts the future, and reflects the integration of cognitive, sensorimotor, and reward networks.
- "FFR" provides the best terminology to refer to this biological approach and should be adopted as a standard.
- FFR has always had a place in auditory neuroscience. As it becomes a more mainstream technique, FFR has the potential to inform basic and applied questions in learning and communication.

- Perhaps one of the most exciting clinical outlets for the FFR is to screen the newborn brain. This could expedite early and targeted interventions for a myriad of communication disorders.

Acknowledgments The editors thank each of the authors in this volume for their contributions and are deeply grateful to series editors Richard R. Fay and Arthur N. Popper for their guidance and support. Thank you also to colleagues, past and present, in the Auditory Neuroscience Laboratory and Hearing Brain Laboratory. NK and TW-S are supported by NIH (R01 DC010016 and R01 HD069414), NSF (BCS 1430400), the National Association of Music Merchants, the Dana Foundation, and the Knowles Hearing Center. SA is supported by the Hearing Health Foundation, the American Hearing Research Foundation, and NIH (R21DC015843A).

Compliance with Ethics Requirements Nina Kraus is Chief Scientific Officer of Synaural, a company working to develop a user-friendly measure of auditory processing. Samira Anderson declared that she had no conflict of interest. Travis White-Schwoch declared that he had no conflict of interest.

References

Adrian, E. D. (1926). The impulses produced by sensory nerve endings. *The Journal of Physiology, 61*(1), 49–72.

Aiken, S. J., & Picton, T. W. (2008). Envelope and spectral frequency-following responses to vowel sounds. *Hearing Research, 245*(1), 35–47.

Anderson, S., Parbery-Clark, A., White-Schwoch, T., & Kraus, N. (2012). Aging affects neural precision of speech encoding. *The Journal of Neuroscience, 32*(41), 14156–14164.

Anderson, S., Parbery-Clark, A., White-Schwoch, T., & Kraus, N. (2013). Auditory brain stem response to complex sounds predicts self-reported speech-in-noise performance. *Journal of Speech, Language, and Hearing Research, 56*, 31–43.

Anderson, S., Parbery-Clark, A., White-Schwoch, T., & Kraus, N. (2015). Development of subcortical speech representation in human infants. *The Journal of the Acoustical Society of America, 137*(6), 3346–3355.

Bajo, V. M., Nodal, F. R., Moore, D. R., & King, A. J. (2010). The descending corticocollicular pathway mediates learning-induced auditory plasticity. *Nature Neuroscience, 13*(2), 253–260.

Banai, K., Hornickel, J., Skoe, E., Nicol, T., et al. (2009). Reading and subcortical auditory function. *Cerebral Cortex, 19*, 2699–2707.

Bharadwaj, H. M., & Shinn-Cunningham, B. G. (2014). Rapid acquisition of auditory subcortical steady state responses using multichannel recordings. *Clinical Neurophysiology, 125*(9), 1878–1888.

Boudreau, J. C. (1965). Neural volleying: Upper frequency limits detectable in the auditory system. *Nature, 208*, 1237–1238.

Buytendijk, F. (1910). On the negative variation of the nervus acusticus caused by a sound. *Proceedings of the Royal Netherlands Academy of Arts and Sciences (KNAW), 13*, 1910–1911.

Chandrasekaran, B., & Kraus, N. (2010). The scalp-recorded brain stem response to speech: Neural origins and plasticity. *Psychophysiology, 47*(2), 236–246. Doi:10.1111/j.1469-8986.2009.00928.x

Coffey, E. B., Herholz, S. C., Chepesiuk, A. M., Baillet, S., & Zatorre, R. J. (2016). Cortical contributions to the auditory frequency-following response revealed by MEG. *Nature Communications, 7*, 11070.

Cunningham, J., Nicol, T., King, C., Zecker, S. G., & Kraus, N. (2002). Effects of noise and cue enhancement on neural responses to speech in auditory midbrain, thalamus and cortex. *Hearing Research, 169*(1), 97–111.

Cunningham, J., Nicol, T., Zecker, S. G., Bradlow, A., & Kraus, N. (2001). Neurobiologic responses to speech in noise in children with learning problems: Deficits and strategies for improvement. *Clinical Neurophysiology, 112*(5), 758–767.

Dallos, P. (1992). The active cochlea. *The Journal of Neuroscience, 2*(12), 4575–4585.

Derbyshire, A. J., & Davis, H. (1935). The action potentials of the auditory nerve. *American Journal of Physiology-Legacy Content, 113*(2), 476–504.

Dolphin, W., & Mountain, D. (1992). The envelope following response: Scalp potentials elicited in the Mongolian gerbil using sinusoidally AM acoustic signals. *Hearing Research, 58*(1), 70–78.

Easwar, V., Purcell, D. W., Aiken, S. J., Parsa, V., & Scollie, S. D. (2015). Effect of stimulus level and bandwidth on speech-evoked envelope following responses in adults with normal hearing. *Ear and Hearing, 36*(6), 619–634.

Faingold, C. L., & Caspary, D. M. (1979). Frequency-following responses in primary auditory and reticular formation structures. *Electroencephalography and Clinical Neurophysiology, 47*(1), 12–20.

Galbraith, G. C. (1994). Two-channel brain-stem frequency-following responses to pure tone and missing fundamental stimuli. *Electroencephalography and Clinical Neurophysiology: Evoked Potentials Section, 92*(4), 321–330.

Galbraith, G. C., Arbagey, P. W., Branski, R., Comerci, N., & Rector, P. M. (1995). Intelligible speech encoded in the human brain stem frequency-following response. *NeuroReport, 6*(17), 2363–2367.

Gao, E., & Suga, N. (2000). Experience-dependent plasticity in the auditory cortex and the inferior colliculus of bats: Role of the corticofugal system. *Proceedings of the National Academy of Sciences of the USA, 97*(14), 8081.

Gardi, J., Merzenich, M., & McKean, C. (1979). Origins of the scalp-recorded frequency-following response in the cat. *Audiology, 18*(5), 353–380.

Gockel, H. E., Carlyon, R. P., Mehta, A., & Plack, C. J. (2011). The frequency following response (FFR) may reflect pitch-bearing information but is not a direct representation of pitch. *Journal of the Association for Research in Otolaryngology, 12*(6), 767–782.

Greenberg, S., Marsh, J. T., Brown, W. S., & Smith, J. C. (1987). Neural temporal coding of low pitch. I. Human frequency-following responses to complex tones. *Hearing Research, 25*(2), 91–114.

Hall, J. W. (1979). Auditory brainstem frequency following responses to waveform envelope periodicity. *Science, 205*(4412), 1297–1299.

Jafari, Z., Malayeri, S., & Rostami, R. (2015). Subcortical encoding of speech cues in children with attention deficit hyperactivity disorder. *Clinical Neurophysiology, 126*(2), 325–332.

King, C., Warrier, C. M., Hayes, E., & Kraus, N. (2002). Deficits in auditory brainstem pathway encoding of speech sounds in children with learning problems. *Neuroscience Letters, 319*(2), 111–115.

Kraus, N., & White-Schwoch, T. (2015). Unraveling the biology of auditory learning: A cognitive-sensorimotor-reward framework. *Trends in Cognitive Sciences, 19*(11), 642–654.

Kraus, N., & White-Schwoch, T. (2016). Neurobiology of everyday communication: What have we learned from music? *The Neuroscientist*. Doi:10.1177/1073858416653593

Kraus, N., Slater, J., Thompson, E. C., Hornickel, J., et al. (2014a). Auditory learning through active engagement with sound: Biological impact of community music lessons in at-risk children. *Frontiers in Neuroscience*, *8*, 351. Doi:10.3389/fnins.2014.00351

Kraus, N., Slater, J., Thompson, E. C., Hornickel, J., et al. (2014b). Music enrichment programs improve the neural encoding of speech in at-risk children. *The Journal of Neuroscience, 34*(36), 11913–11918.

Kraus, N., Thompson, E. C., Krizman, J., Cook, K., White-Schwoch, T., & LaBella, C. R. (2016). Auditory biological marker of concussion in children. *Nature, Scientific Reports, 6*(39009).

Kuwada, S., Anderson, J. S., Batra, R., Fitzpatrick, D. C., et al. (2002). Sources of the scalp-recorded amplitude-modulation following response. *Journal of the American Academy of Audiology, 13*(4), 188–204.

Lehmann, A., Skoe, E., Moreau, P., Peretz, I., & Kraus, N. (2015). Impairments in musical abilities reflected in the auditory brainstem: Evidence from congenital amusia. *European The Journal of Neuroscience, 42*(1), 1644–1650.

Marsh, J. T., Worden, F. G., & Smith, J. C. (1970). Auditory frequency-following response: Neural or artifact? *Science, 169*(3951), 1222–1223.

Russo, N., Nicol, T., Musacchia, G., & Kraus, N. (2004). Brainstem responses to speech syllables. *Clinical Neurophysiology, 115*(9), 2021–2030.

Shamma, S., & Lorenzi, C. (2013). On the balance of envelope and temporal fine structure in the encoding of speech in the early auditory system. *The Journal of the Acoustical Society of America, 133*(5), 2818–2833. Doi:10.1121/1.4795783

Skoe, E., & Kraus, N. (2010). Auditory brainstem response to complex sounds: A tutorial. *Ear and Hearing, 31*(3), 302–324.

Snyder, R., & Schreiner, C. (1984). The auditory neurophonic: Basic properties. *Hearing Research, 15*(3), 261–280.

Tarasenko, M. A., Swerdlow, N. R., Makeig, S., Braff, D. L., & Light, G. A. (2014). The auditory brainstem response to complex sounds: A potential biomarker for guiding treatment of psychosis. *Frontiers in Psychiatry, 5*, 142. Doi:10.3389/fpsyt.2014.00142

Warrier, C. M., Abrams, D. A., Nicol, T. G., & Kraus, N. (2011). Inferior colliculus contributions to phase encoding of stop consonants in an animal model. *Hearing Research, 282*(1–2), 108–118.

Weinberger, N., Kitzes, L., & Goodman, D. (1970). Some characteristics of the "auditory neurophonic". *Experientia, 26*(1), 46–48.

White-Schwoch, T., Woodruff Carr, K., Anderson, S., Strait, D. L., & Kraus, N. (2013). Older adults benefit from music training early in life: Biological evidence for long-term training-driven plasticity. *The Journal of Neuroscience, 33*(45), 17667–17674.

White-Schwoch, T., Nicol, T., Warrier, C. M., Abrams, D. A., & Kraus, N. (2016). Individual differences in human auditory processing: Insights from single-trial auditory midbrain activity in an animal model. *Cerebral Cortex*. doi:10.1093/cercor/bhw293

White-Schwoch, T., Woodruff Carr, K., Thompson, E. C., Anderson, S., et al. (2015). Auditory processing in noise: A preschool biomarker for literacy. *PLoS Biology, 13*(7), e1002196.

Worden, F., & Marsh, J. (1968). Frequency-following (microphonic-like) neural responses evoked by sound. *Electroencephalography and Clinical Neurophysiology, 25*(1), 42–52.

Chapter 2
Infant and Childhood Development: Intersections Between Development and Language Experience

Fuh-Cherng Jeng

Abstract This chapter provides an overview of the development and plasticity of the neural encoding of speech and non-speech stimuli at the subcortical level with an emphasis on the influence of an individual's language experience during infancy and childhood. Sections of this chapter are developed based upon a theoretical framework to embrace all possible sources and interactions that may have a significant effect on the development of the auditory system during the early stages of life. The discussion begins with the acoustic environment of a human fetus and the possible influence of prenatal listening experience on the development of the auditory system at the subcortical level. Next, the development during an infant's immediate postnatal days and first year of life are presented. Developmental trajectories and possible influences of linguistic experience on speech processing, as illustrated by the many aspects of neural encoding (e.g., tracking acuity, pitch strength, and the spectral and timing accuracy at the fundamental frequency, harmonics, and speech formants) are discussed. The presentation continues through childhood, which manifests itself as exponential growth of the developmental trajectory and adaptation of the auditory system. Neural encoding of the various aspects of human speech is described as it pertains to children situated in a quiet or noisy acoustic environment. The effects of acquiring more than one language, sequentially or simultaneously, are also discussed.

Keywords Auditory system · Frequency-following response · Fundamental frequency · Harmonics · Innate ability · Linguistic experience · Neural circuitry · Neonate · Neuroplasticity · Newborn · Pitch · Speech formant · Subcortical

---

F.-C. Jeng (✉)
Communication Sciences and Disorders, Ohio University, Athens, OH 45701, USA
e-mail: jeng@ohio.edu

N. Kraus et al. (eds.), *The Frequency-Following Response*,
Springer Handbook of Auditory Research 61,
DOI 10.1007/978-3-319-47944-6_2

2.1 Introduction

After a gestational period, human infants are delivered to the "outside" world. Immediately, infants encounter novel sensory inputs. Interactions between the infants and their native environments take place promptly in multiple forms (Hollich et al. 2000). Human infants, equipped with astonishing anatomical structures and functional capacities (Eggermont and Moore 2012) have the ability to interact with their environment through modalities such as vision, touch, smell, and, clearly, sound. The sensory inputs that are available in the infant's native environment, together with the infant's interactions with caregivers, facilitate brain development, which includes the perception and production of prosody, music (Trainor and Unrau 2012), gesture, and later on, the mastery of a language (Panneton and Newman 2012).

Immediately after birth infants are exposed to the rich acoustic stimuli that are available in their native environments. Development of the auditory system thus intersects with the infant's exposure to their acoustic environment during the early stages of life. However, the pathway to acquire language is a long process. While it takes approximately three years for most children to master the perception and production skills necessary for communicating in their native languages (Jusczyk 1997), some children may take a longer time to reach the same speech and language proficiencies than their peers (see Schochat, Rocha-Muniz, and Filippini, Chap. 9). Acquiring language can be challenging when infants and children are placed in adverse acoustic environments, such as background noise and reverberation (see Bidelman, Chap. 8). When a person encounters difficulty processing specific aspects of speech, individualized training and rehabilitative protocols may be considered (see Carcagno and Plack, Chap. 4). With or without difficulty processing human speech during the early stages of life, the brain continues to develop during adolescence and adulthood (see Krishnan and Gandour, Chap. 3) and starts to decline during the senior years (see Anderson, Chap. 11).

This chapter provides an overview of the development and plasticity of the neural encoding of speech and non-speech stimuli at the subcortical level with an emphasis on the influence of an individual's language experience during infancy and childhood. The chapter begins with a brief description of the measurements that are utilized to examine the various aspects of speech processing at the subcortical level. The following sections are then developed based upon a theoretical framework to embrace all possible sources and interactions that may have a significant effect on the development of the auditory system during the early stages of life. The discussion begins with the acoustic environment of a human fetus and the possible influence of prenatal listening experience on the development of the auditory system at the subcortical level. Next, the development during an infant's immediate postnatal days and first year of life are presented, together with the recent documentations of the FFR (frequency-following response) literature, including longitudinal follow-up of speech encoding during the first year of life and cross-sectional studies that have included infants with various age groups. Developmental trajectories and possible influences of linguistic experience on speech processing as illustrated by the various aspects of neural encoding will be discussed (e.g., tracking acuity, pitch strength, and the spectral and timing accuracy at the fundamental frequency (F_0), harmonics, and speech formants). Of course, the importance of the stability and precision of neural firing at the subcortical level cannot be omitted. The presentation continues through childhood—additional exponential growth of the developmental trajectory and adaptation of the auditory system. Neural encoding of the various aspects of human speech is described as it pertains to children who are situated in a quiet or noisy acoustic environment. Effects of acquiring more than one human language, sequentially or simultaneously, are also discussed. Lastly, current issues concerning the possible influence of prenatal listening experience, lack of age-appropriate normative databases, and the urgency of developing functional and computational models suitable for the different age groups during infancy and childhood are discussed.

2.2 Frequency-Following Response

Interactions between the development of a human brain and the listener's linguistic experience take place both cortically and subcortically. While many studies have focused on the development and neuroplasticity of the neural structures at the cortical level (He et al. 2007; Näätänen et al. 2007; Brauer et al. 2011; Butler and Trainor 2013; Partanen et al. 2013), few have emphasized the neuroplasticity and functional organization of the neural circuitries at the subcortical level.

An electrophysiological measurement that provides insight into the subcortical pitch processing mechanisms of the auditory system is known as the frequency-following response (FFR): a term first used by Worden and Marsh (1968) to describe a response that is phase locked to the frequency components of the stimulus in cats. Phase locking is a phenomenon indicating that neural structures are firing action potentials that are synchronized with the waveform morphology of a stimulus. The FFR was later used to define an electrophysiological measure of the human auditory system's ability to track the frequencies of 500 and 1,000 Hz pure tones (Moushegian et al. 1973). Generally speaking, FFR is a collective response from a group of neurons that responds to the periodicity of the stimulus waveform. The F_0 of the acoustic stimulus presented to a listener, if encoded accurately, appears as periodic peaks on the recorded waveform. The inter-peak intervals of the stimulus and corresponding response waveform are a reflection of the neural phase-locking abilities of the auditory system. Investigations showing the neural phase-locking properties of the FFR reflect the *temporal* and *place theories*. The temporal and place theories are thought to contribute to the brain's ability to encode spectral information from speech stimuli. Because the FFR is a scalp-recorded auditory evoked potential and requires no active participation of the listener, it serves as a noninvasive and objective measurement of the capacity that human subcortical neurons have to receive and track changes in the frequency content embedded in an acoustic stimulus (Skoe and Kraus 2010).

Through the use of complex sounds (e.g., speech), various aspects of neural encoding at the subcortical level can be examined. Complex sounds elicit equally complex responses, offering a rich set of ingredients the brain has to process. Due to this nature, the FFR to complex sounds contains enriched information that may be useful to help decipher the complex, and yet dynamic, neural networks inside the brain. It is likely that the neural elements involved in the processing of complex speech sounds are not limited to one area of the brain but instead are distributed across various subcortical neural structures. These neural structures, although dispersed among the nuclei, respond concertedly with the various features of human speech. Examination of such complex but integrated neural circuitry requires a method that allows the investigation of the neural elements that respond synchronously with the incoming signal. The FFR provides such a method and opens a door to examine the various and yet distinct biological processing of speech sounds at the subcortical level. Details about the methodology and procedures of recording an FFR in response to complex sounds can be found in Chap. 1 by Kraus, Anderson, and White-Schwoch.

Developmental trajectories of subcortical pitch representation, as reflected by the scalp-recorded FFR, undergo stages of progression across a human lifespan. During the immediate postnatal days, newborns are able to process tones (Gardi et al. 1979) and the changes in voice pitch (Jeng et al. 2011b). Through infancy, pitch representations at the subcortical level undergo a phase of rapid maturational changes (Jeng et al. 2010; Anderson et al. 2015). During childhood, subcortical pitch representation continues to improve and sometimes is accompanied by an overshoot in the response latency and amplitude: shorter response latency and larger response amplitude than those observed in adults (Skoe et al. 2015). This overshoot is followed by a gradual increase in the response latency and a gradual decrease in the response amplitude that continues through adolescence. Pitch representation at the subcortical level remains relatively stable during adulthood but starts to decline when aging-related changes come into effect (see Anderson, Chap. 11). The following sections emphasize the possible influence of prenatal listening experience on subcortical pitch processing and the influence of postnatal linguistic experience on the development of neural responses to speech during infancy and childhood.

2.3 Gestational Age: Possible Influence of Prenatal Listening Experience

Before an infant is born, its mother's womb allows transmission of low-frequency sounds to the fetus (Gerhardt and Abrams 2000). Approximately three weeks after conception, the inner ear and cochlea start to take shape, and they become fully functional by five months of gestation (Frenz et al. 2001); the fetus is then equipped with the sensory and neural elements necessary to receive acoustic information from the outside world. Thus, during the gestational period, the fetus is able to receive frequency-specific information from acoustic signals that are generated either by its mother or from others in its native environment. Because the fetus is submerged in an abundance of amniotic fluid, transmission of the acoustic signals likely will propagate to the fetus through the bone conduction pathway of the auditory system.

When a person vocalizes, vibration of his or her vocal folds creates sound waves. These sound waves then travel through the air in all directions. When the sound waves hit the mother's abdomen, most of the sound gets reflected because of the impedance mismatch between the air and the mother's abdominal tissue (Griffiths et al. 1994). However, if the sound producer is the fetus's mother herself, the sound waves generated by the mother can travel directly through her body to the fetus. In this case, acoustic sounds initiated by the mother can be transmitted, although attenuated, to the fetus (Gerhardt and Abrams 2000). Playback of intrauterine acoustic recordings demonstrates that the sounds available to the fetus are rich but somewhat muffled (Abrams et al. 1998a).

When acoustic vibrations travel through the mother's womb, not all frequencies become attenuated by the same amount. In other words, the loss of acoustic energy of the incoming signal is frequency specific (Abrams et al. 1998a). The mother's womb functions as a low-pass filter that allows frequencies below 300 Hz to reach the head of the fetus with very little or no attenuation (Abrams et al. 1998b); that is, for acoustic energies below 300 Hz, the intrauterine sound pressure is nearly identical to that generated by the mother's vocal folds. For acoustic energies above 300 Hz, the intrauterine sound pressure decreases with a slope of about 5 dB per octave. A decrease of 5 dB means that the amplitude of the intrauterine sound pressure becomes a little bit larger than half of its original amplitude. Thus, recordings of intrauterine sounds demonstrated relatively high intelligibility of externally generated human speech and music (Querleu et al. 1989; Griffiths et al. 1994). Although muffled, intrauterine sounds are still enriched with frequency-specific information, particularly at the low frequencies.

Toward the end of the second trimester and during the third trimester of a pregnancy, the mother often experiences the sensation of the fetus' movement in response to environmental sounds—consistent with results of scientific experiments where pure tones with various frequencies were delivered to fetuses while they were still in their mother's womb (Gelman et al. 1982; Shahidullah and Hepper 1994). Many experiments began at about the middle of the second trimester of a pregnancy, and the researchers were looking at the fetus' bodily reactions to sounds through an ultrasound. As time advanced and hearing progressed, fetuses responded to a greater range of frequencies and at a lower sound intensity. In Shahidullah and Hepper's (1994) study, they also examined the fetuses' responses to human speech. By observing the fetuses' habituation responses through ultrasounds, they found that fetuses at 35 weeks of gestational age were able to differentiate the pre-recorded human speech /baba/ versus /bibi/. This finding indicated that human fetuses at 35 weeks' gestation already possessed the ability to discriminate between different phonemes. These informational cues may be helpful while the fetus is beginning to listen to sounds that are universal to all human languages and music. Some frequency components may even contain information that is specific to the native language of the fetus' external acoustic environment.

FFR recordings obtained in postnatal infants also provide evidence supporting the possible influence of prenatal listening experience on the development of the neural circuitry at the subcortical level. Anderson and colleagues (2015) reported that amplitudes of the FFRs at the first formant (F_1) and high harmonics (HH) increased significantly with increasing age, whereas the FFR amplitudes at the F_0 were clearly discernible in young infants and did not increase significantly in older infants. One possible explanation is that the low frequency vibrations (e.g., acoustic energies at F_0) are readily available during the prenatal stage of life, whereas the high frequency vibrations (e.g., acoustic energies at F_1 and HH) are unavailable to the fetus. The acoustic energies at the high frequencies (e.g., F_1 and HH) will become available to the fetus after birth. Exposure to these high frequency sounds in the listening environment after birth may have an effect on the growth of the F_1 and HH amplitudes during infancy. Importantly, the fact that the F_0 encoding is evident during early infancy corroborates the idea that the prenatal listening experience, at least for the low frequency components, may play an important role in facilitating normal development of the neural circuitry at the subcortical level during the early stages of life. This is also consistent with physiological evidence that low-frequency hearing sensitivity develops prior to high-frequency sensitivity in avian (Rubel and Ryals 1983) and mammalian (Echteler et al. 1989) systems.

2.4 Infancy

Immediately after infants are born they can detect almost all phonetic distinctions found in speech (Eimas et al. 1971; Kuhl et al. 2006). Interestingly, newborns and young infants exhibit a similar pattern of sound perception regardless of the language environment into which they are born (Kuhl 2010). This evidence indicates that the perception of speech is strongly influenced by innate factors (i.e., the *biological capacity model*). It is important to understand that the specific language environment to which an infant is exposed also effects perception of speech sounds (i.e., the *linguistic experience model*). Exposure to a specific language environment during the early stages of life results in a reduction in the ability to perceive differences among speech sounds of other languages (Kuhl et al. 1992; Kuhl 2004). For example, Kuhl and colleagues (1992) analyzed 6-month-old American and Swedish infants' perception of both native-language and foreign-language vowel sounds. They reported that the ability to hear differences among many of the sounds not used in the infant's language was lost by six months of age. During this same 6-month time frame, the infants' developmental speech perception of native sounds showed substantial enhancement and continued to do so until 12 months of age. For example, American infants showed significant improvements in the discrimination of the English /r/-/l/ contrast in comparison to age-matched Japanese infants (Kuhl et al. 2006). Additionally, both the Chinese-learning and English-learning infants showed improvement on affricate-fricative contrasts between 6 and 12 months of age (Tsao et al. 2006). These linguistic experiences during the early stages of life, along with innate factors, play an important role in speech and language development.

2.4.1 Theories and Evidence About the Early Acquisition of Speech and Language

Universal traits and language-specific experiences both influence the acquisition of pitch perception. For theories related to the "biological capacity" model, Jakobson (1968) introduced the *law of irreversible solidarity*. This theory proposed that early acquisition of sound could be explained by the frequency distribution of that sound among the world's languages. Acoustic features that were more basic and central to all human languages, such as intonation, voice pitch, and rhythm, would be acquired earlier than the other aspects of speech sounds. Dinnsen (1992) suggested that there might be a *universal hierarchical structure* with a limited set of ordered acoustic features that were applicable to the inventories of all languages. Each feature in the hierarchy had a default (or unmarked) value. Therefore, acquisition of any feature of a specific language would involve a process of replacing a default value with a language-specific value. Dinnsen's (1992) model indicated that the order in which an infant acquires a specific feature of a language would depend on the dominant and default values of that feature. That is, acoustic features ranked high in the hierarchy would be acquired earlier than features ranked low. Jakobson's "law of irreversible solidarity" and Dinnsen's "universal hierarchical structure" emphasized the innate ability of humans to acquire language and were consistent with the "biological capacity" model.

In contrast, there are a number of theories that emphasize the role of perceptual importance on language acquisition. Locke (1980) proposed three mechanisms for language acquisition: maintenance, learning, and loss. Once an infant starts to acquire an account of targeted features of a language, certain sounds will be solidified within the infant's inventory. Sounds or specific voice patterns not present in the infant's early inventory are then learned through interactions within the postnatal linguistic experience. The infant will abandon and lose the sounds or certain voice patterns not present in their targeted language system. According to Locke, the interaction of these three mechanisms results in the acquisition of the targeted language to which the infant is exposed.

Kuhl (1994) proposed a *native language magnet theory* and then a revised version in 2008. The expanded version divides an infant's language development into four phases. *Phase 1* indicates that the infant's initial state is universally the same. At birth, infants perceive sounds by their natural auditory processing mechanisms. During this phase, the infants' abilities do not depend on linguistic environment. *Phase 2* shows that early exposure to a specific language may cause physical changes in neural structure and circuitry, which become committed to recognizing acoustic patterns of native languages. For example, acoustic features that occur frequently in the infant's native language will stimulate certain neural structures repeatedly and may result in changes of specific neural structures and circuitry. Social interactions may also play a role in this phase by increasing the infant's attention and its awareness to specific acoustic patterns, and thus may have facilitated the functional reorganization of the infant's brain. *Phase 3* indicates how early linguistic experience repeatedly alters the initial state of the infant's perception of speech (i.e., magnet effects of speech perception). By six months of age, the infant's perception of speech not only deviates from the innate boundaries but also follows the distribution properties of sounds specific to its native language. *Phase 4* takes place when the neural commitment becomes stable. As infants come in contact with language-specific sounds, some form of this information is stored in their memory. A good example of this phase can be seen in Swedish, American, and Japanese infants. Behavioral responses of the infants were measured and showed that they produced distinctive representations that mirrored the distribution properties of ambient speech input. Over time, such magnet effects functionally erase certain boundaries that are irrelevant to the infant's native language (Kuhl et al. 2008).

One amazing feature of the human brain is its ability to adapt to the features of surroundings. For example, the acoustic and linguistic features of the listener's native language have substantial influences on the development of his/her processing of human speech. When infants are just born, they are capable of detecting subtle differences in speech sounds. That is, newborns can differentiate essentially all features of human speech (Eimas et al. 1971; Carral et al. 2005). Throughout the early stages of a human life, the brain develops and adapts to acoustic signals found in its environment. Such linguistic experiences initiate anatomical and functional refinement of the neural circuitry of the human brain. Over time, neural pathways that respond to the specific features of a language will be enhanced (Kuhl et al. 2008). For example, neonates who are born and raised in a tonal language environment will have substantial exposure to the distinctive intonation patterns (i.e., pitch contours) that are important in their native languages. Thus, neural circuits may be fine tuned to best respond to the pitch contours of the infant's linguistic environment. For example, Mandarin Chinese is a tonal language that utilizes distinctive pitch contours to deliver the different meanings of the same words. In Mandarin Chinese, there are four lexical pitch contours: *Tone 1*, *Tone 2*, *Tone 3*, and *Tone 4*. *Tone 1* has a flat pitch contour that remains relatively stable over its production. *Tone 2* starts from a low pitch utterance and gradually rises to a higher pitch. *Tone 3* has a falling and rising pitch contour with a reflection point around the mid portion of the utterance. *Tone 4* begins with a relatively high pitch utterance and gradually descends to a lower pitch. Each of the four pitch contours can carry a different meaning of the same word. For example, when the Mandarin syllable /yi/

is pronounced in Tone 1, it means "壹 [one]"; with Tone 2, it means "姨 [aunt]"; with Tone 3, it means "椅 [chair]"; and with Tone 4, it means "易 [easy]."

Behavioral studies have shown that infants learning tonal languages respond to changes in voice pitch in a categorical manner (Yip 2002; Panneton and Newman 2012). It was observed that four-months-old Chinese infants could discriminate the four lexical pitch contours with accuracy, and their ability in differentiating the four pitch contours persisted through infancy (Mattock and Burnham 2006). However, when American and French infants were tested with low versus rising lexical pitch contours in the Thai language, infants between four and six months old were able to discriminate the two lexical pitch contours with accuracy, but their performance dropped in infants who were nine months of age (Mattock et al. 2008).

2.4.2 Electrophysiological Measurements During Infancy

Electrophysiological measurements do not require behavioral feedback from the infant, enabling researchers to draw conclusions from an infant during the early stages of life. By about 4 to 5 months of age, an infant's brain is already sensitive to language-specific acoustic patterns and contrasts (Friederici et al. 2007). Electrophysiological studies that recorded cortical responses in infants indicated that early language exposure facilitates functional reorganization of brain networks (Grossmann et al. 2007; Friedrich and Friederici 2010). Although recordings of the cortical responses are useful in helping us to understand how the brain processes speech sounds at the cortical level, they have one drawback: they are affected by the state of the subject (Hall 2006). Thus, infants are required to remain awake throughout recordings (Friederici et al. 2007). This can be difficult to accomplish, particularly in young infants who have an attention span that is fairly short. One way to counterbalance this drawback is to record responses from neural structures at the subcortical level.

Although there is an abundance of literature documenting the behavioral and cortical responses in infants, few have focused on the functional organization at the subcortical level. Thus, the focus of the remainder of this chapter will be on the pitch processing and functional organization at the subcortical level for human infants. Electrophysiological studies that record responses from the subcortical neural structures have a major advantage over those that record responses from the auditory cortex. When recording responses from neural structures at the subcortical level, the infant does not have to stay awake or alert during data collection. Instead, the infant is encouraged to rest and fall asleep during research because alertness of the infant does not affect the results. An additional advantage of the FFR to complex sounds is that the FFR provides a precise representation of the stimulus features (Skoe and Kraus 2010), whereas the cortical response is an abstract representation (Hall 2006).

The FFR literature for neonates was first documented in 1979. During that time, Gardi and colleagues (1979) recorded the FFRs to low-frequency tone bursts in full-term, healthy neonates. The FFRs recorded in neonates shared common characteristics in terms of the response morphology with those recorded in adults. That is, the response waveform followed the periodicity of the stimulus waveform. Latency of the first peak in the FFR decreased with increasing frequency. This was consistent with the general theories of hearing, the biomechanical properties of the basilar membrane, and the anatomy of the auditory system. The amplitude and threshold values of neonatal FFRs in response to low-frequency tone bursts are also similar to those obtained in normal-hearing adults. All together, these findings indicate that the integrity of the neural elements, particularly for those that are sensitive to low frequencies at the subcortical level, can be assessed starting from the first day of life.

2.4.3 Encoding of the Fundamental Frequency

After a gap of more than 30 years, characteristics of neonatal FFRs in response to speech stimuli were investigated. Jeng and colleagues (2011b) utilized a monosyllabic Mandarin stimulus that mimicked the English vowel /i/ and elicited FFRs in American and Chinese neonates during their immediate postnatal days. The FFRs were visualized by plotting spectral energies of the recordings as a function of time (Fig. 2.1). Spectrograms of the stimulus (Fig. 2.1, left column) clearly showed energy at the F_0 and its harmonics. Spectrograms of the recordings taken from 12 American and 12 Chinese neonates (Fig. 2.1, *middle column*) showed FFR energy that followed the fundamental frequency of the stimulus. The FFR energy following the harmonics was not as apparent. The FFRs recorded from both groups of neonates exhibited clear energy that followed the periodicity, such as the pitch contours, of the speech stimuli. Importantly, the FFRs obtained from American and Chinese neonates resembled each other and showed little differentiation. This finding provides evidence for the "biological capacity model," indicating that the neonates are born with similar innate abilities of pitch encoding at the subcortical level.

While the influence of the infant's linguistic environment has been shown through behavioral studies (Mattock and Burnham 2006; Mattock et al. 2008), little is known about how the various pitch contours are processed at the subcortical level and the developmental course during the first year of life. The influence of a person's linguistic experience on the development of the auditory circuitry at the subcortical level can be examined by recording FFRs in both neonates and adults who are born and raised in two different language environments. Cross-linguistic comparisons with the additional data that were obtained from 12 American and 12 Chinese adults (Fig. 2.1, *right column*) revealed the influence of a person's linguistic experience on the spectral encoding of a speech stimulus at the subcortical level. A study design including the four groups of participants (American neonates, American adults, Chinese neonates, and Chinese adults) allowed examination of the influence across and between maturity (neonate versus adult) and the listeners'

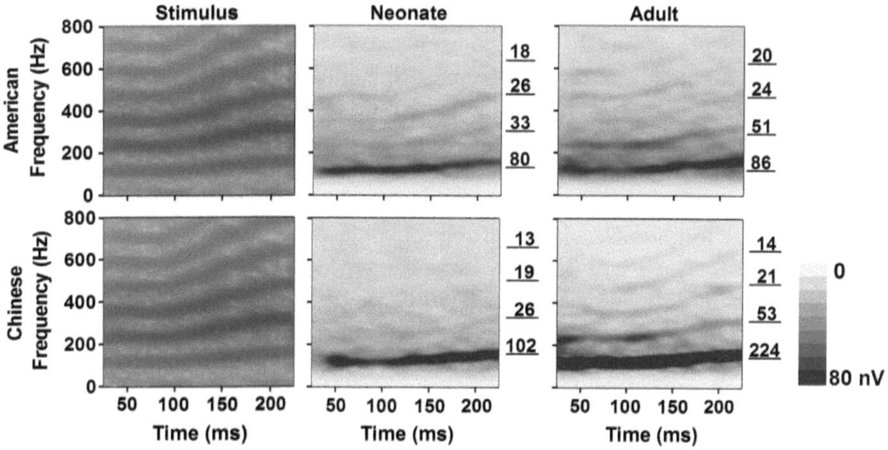

Fig. 2.1 Spectrograms of the stimulus (*left column*) and grand-averaged spectrograms obtained from 12 American neonates, 12 Chinese neonates (*middle column*), 12 American adults, and 12 Chinese adults (*right column*). The stimulus is a pre-recorded, monosyllabic, Mandarin speech stimulus that mimics the English vowel /i/ with a rising pitch. A *gray gradient scale* on the right of the spectrograms indicates the spectral amplitudes (nV) for the recordings obtained from neonate and adult participants. Underlined *numeric symbols* on the right of each spectrogram represent the means of the spectral amplitudes (nV) at the fundamental frequency, second, third, and fourth harmonics of the grand-averaged spectrograms. Spectral amplitudes of the harmonics were determined by finding the spectral peaks closest to those of the stimulus. The spectrograms of the stimulus are plotted on a normalized scale. All spectrograms were obtained using a Hanning window of 50 ms in length, overlap of 47.5 ms in length, and a frequency step of 1 Hz. (Reproduced with permission of publisher from Jeng et al. 2011b. © Ear and Hearing)

language experience (American versus Chinese) factors. The FFR data demonstrated a significant difference for the language factor, but not for the age factor nor for the interaction between the two factors. Furthermore, the FFRs obtained from the Chinese adults were significantly larger than Chinese neonates, whereas the FFRs obtained from American neonates and American adults were not significantly different from each other. These findings, together with the fact that American neonates and American adults had comparable FFRs, provide evidence supporting the "linguistic experience model."

Characteristics and maturational trends of the FFRs recorded in infants demonstrate the feasibility of studying the structural and functional reorganization of neural circuitry at the subcortical level during the first year of life. Jeng and colleagues (2010) recruited nine American infants, ranging from 1 to 11 months old. All infants were born and raised in native English-speaking households. The FFRs recorded from these infants showed discernible energy at the F_0 that followed the pitch contour of a speech stimulus. Four objective measures were applied to quantify the various aspects of pitch processing: *frequency error*, *slope error*, *tracking accuracy*, and *pitch strength*. Results obtained from the American infants were compared to those obtained in a group of American adults who were native speakers of English. The four objective measurements were all focused on different aspects of pitch processing in the brainstem, and yet similar maturational trends were observed across all four measurements. Specifically, pitch-tracking acuity and phase-locking magnitude in infants appeared very similar to those in the adult population. This finding indicates that the neural circuitry needed to respond to the speech stimulus and formulate a discernible FFR morphology is readily available in infants.

Longitudinal follow-ups with each participant would potentially reveal a maturational trend for each individual. A prospective, longitudinal study with regular follow-ups was first made in a case study. Through a special opportunity, among the nine infants that were recruited in a previous study (Jeng et al. 2010), one infant was brought back for FFR evaluations at 1, 3, 5, 7, and 10 months of age. Spectrograms of the recordings obtained from this infant (Fig. 2.2) are arranged according to the age of the infant. As expected, FFRs recorded from individual listeners were not as robust as the grand-mean averages across all participants. Specifically, recordings obtained from individual listeners showed a relatively lower signal-to-noise ratio in the spectrogram, which resulted in occasional disruptions of the FFR in response to the F_0 contour of the stimulus (e.g., recordings obtained when this infant was 1 and 3 months old). This infant showed a weak response at 1 month old, but her responses became more visible at 3, 5, 7, and 10 months of age. This improvement suggested maturation in FFRs for infants, who do not show robust responses when they are young (e.g., 1 month old) but have developed strong pitch representation in the early stages of life (e.g., 3 months old). Objective indices regarding the tracking acuity and response magnitude of the FFRs obtained from this infant (Fig. 2.3) demonstrate a developmental trend of pitch encoding during the first year of life.

A cross-sectional study was performed and corroborated the idea of early maturation of neural encoding of F_0. Anderson and colleagues (2015) recorded FFRs to a synthesized /da/ syllable in 28 American infants 3–10 months of age. Results demonstrated that the F_0 amplitude in the FFR to the speech stimulus remained

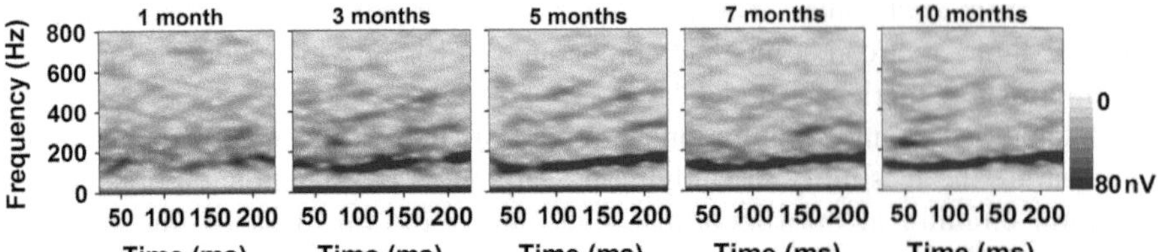

Fig. 2.2 Longitudinal follow-ups of an American infant revealed a maturational trend during the first year of life. This infant did not have a clearly identifiable FFR at 1 month old but showed a clearly identifiable FFR at 3 months old. After 3 months of age, this infant's FFR remained relatively stable. A gradient scale on the right indicates the spectral amplitudes (nV). All spectrograms were derived using a Hanning window of 50 ms in length, overlap of 47.5 ms in length, and a frequency resolution of 1 Hz. (Reproduced with permission of publisher from Jeng et al. 2010. © Perceptual and Motor Skills)

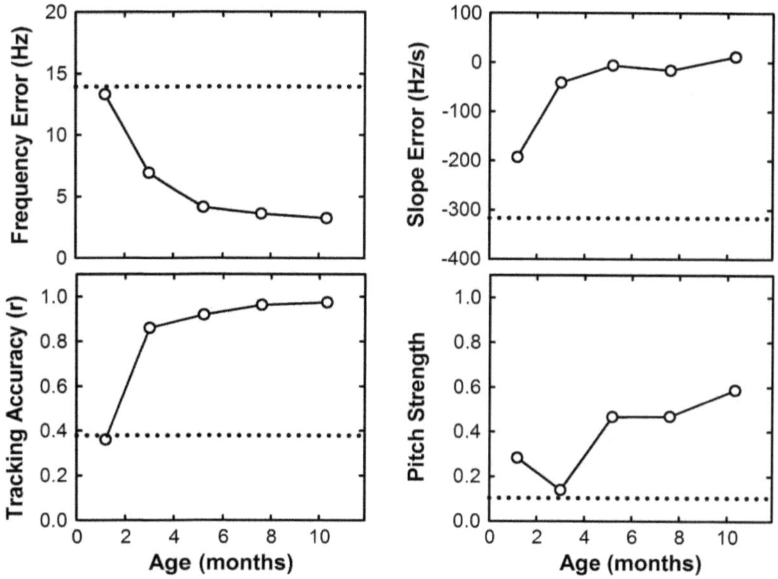

Fig. 2.3 Maturational trends of the FFRs obtained in an American infant at 1, 3, 5, 7, and 10 months of age. Frequency error, slope error, tracking accuracy, and pitch strength were computed from recordings obtained in this infant (see Fig. 2.2). Tracking acuity and phase-locking magnitude of the auditory system increased as this infant advanced in age. A *horizontal dotted line* within each panel indicates the mean of the control recordings, where the sound tube was occluded and moved away from the infant participant. (Reproduced from Jeng et al. 2010, with permission of publisher. © Perceptual and Motor Skills)

relatively stable for infants between 3 and 10 months of age. The growth of F_0 amplitude in the FFR occurs primarily during the early stages of life. This cross-sectional study demonstrated similar findings to those obtained in a longitudinal follow-up study of an infant. These findings signify that the development of neural circuitries responsible for encoding the F_0 information of a speech stimulus takes place early in time, likely sometime between 1 and 3 months old or even earlier.

2.4.4 Encoding Harmonics and Speech Formants

Frequency resolution is immature at birth for both animals and humans. Although the cochlea is fully functional at birth, neural elements at the subcortical level do not mature until later (Rubel and Ryals 1983). Specifically, neural elements sensitive to low frequencies emerge and mature earlier than those sensitive to high frequencies. Romand and Ehret (1990) studied the electrophysiological mapping in mice and reported that after birth the first recordable neural responses from the inferior colliculus are at low frequencies. As the mice matured, frequency responsiveness of the neurons extended into the high frequencies. This postnatal development and maturation of the neural elements is likely driven by listening activities after birth and may have implications for the development of the encoding of harmonics and speech formants for human infants.

Neural structures at the subcortical level play an important role in deciphering harmonic and formant information for infants. Harmonics and speech formants are at higher frequencies than the F_0, and neural phase locking is clearer and more robust at low frequencies than high frequencies. For example, frequencies beyond 5000 Hz are too fast for any neuron to follow. Thus, when examining the harmonic or formant responses in an FFR, people direct their attention to frequencies below 5000 Hz. Although the characteristics of the harmonics and speech formants in the FFR have been reported in normal-hearing adults (Aiken and Picton 2006, 2008), few studies have examined the characteristics and implications of the harmonics and formants in the FFR for infants.

To date, only one paper reports the characteristics and development of the harmonics and formants in the FFR for infants. Anderson and colleagues (2015) recorded FFRs in American infants 3–10 months old. They reported that the amplitude of the F_0 in the FFR remained relatively stable, while the amplitudes of the first formant and high harmonics in the FFR increased as age increased (Fig. 2.4). Furthermore, when these infants were divided into two groups, younger and older (the younger group: 3–5 months old; the older group: 6–10 months old), the older infants demonstrated larger harmonic and formant amplitudes than the younger infants. These results not only provide evidence supporting improved neural encoding of speech features with age, but also highlight the importance of auditory neurodevelopment at the subcortical level for human infants.

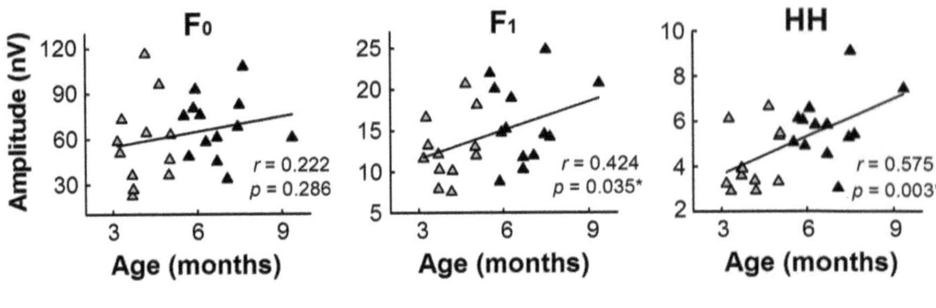

Fig. 2.4 Scatterplots of the F_0, F_1, and HH amplitudes of the FFRs recorded to a 40-ms /da/ syllable in 25 American infants, 3–10 months of age. No correlation was found between F_0 amplitudes of the FFRs and the age of the participating infants, but positive correlations were observed between the age of the infants and the F_1 and HH amplitudes of the FFRs. Younger infants (3–5 months old, *gray triangles*) have significantly smaller F_1 and HH amplitudes than older infants (6–10 months old, *black triangles*). *Solid lines* are the linear regressions of the FFR amplitudes as a function of age. *Asterisks* indicate $p < 0.05$; r, correlation coefficient; F_0, fundamental frequency; F_1, first speech formant; HH, high harmonics. (Reproduced from Anderson et al. 2015 with permission. © The Journal of the Acoustical Society of America)

Results of infant FFR studies are consistent with behavioral and other electrophysiological studies that have shown that frequency encoding of a human brain is immature at birth but improves during the first 6 months of age. For example, behavioral studies employing frequency discrimination tasks have reported that infants can detect frequency changes as small as 2–3% (Olsho et al. 1987). Importantly, infants at 3 months of age demonstrate a significantly larger frequency difference limen than those who are 6 months of age (Abdala and Folsom 1995). These findings indicate that the infant's ability to detect changes in frequency improves as they mature. Electrophysiological studies that record mismatch negativity responses from cortical neural structures confirmed the immature frequency resolution at birth followed by a significant improvement during the first few months.

2.4.5 Timing Aspects of the FFR

Timing aspects of neural encoding can be studied by examining the temporal waveform of the FFR. Gardi et al. (1979) utilized tone bursts (250, 500, and 1,000 Hz) and successfully recorded FFRs in neonates during their immediate postnatal days. Latency of the neonatal FFRs shared the characteristics of the physiological properties of the basilar membrane. Specifically, FFR latency decreased as a function of increasing frequencies of the tone bursts as in normal-hearing adults. Despite the similar characteristics shared by the neonates and adults, several differences were observed. For example, the neonatal FFRs elicited by 250 Hz and 500 Hz have longer latencies than those of normal-hearing adults. This is likely due to the fact that the neonate's brain remains under development, suggesting they require more input from their linguistic environment to further define the neural circuits of their auditory system.

In addition to examining the spectral components of the response, Anderson and colleagues (2015) investigated timing of onsets and offsets in response to a /da/ syllable in infants. It was discovered that the latency of the onset (*A*) peak, latency of the offset (*O*) peak, the inter-peak latency between the *A* and *O* peaks, and the onset slope from the *V* peak to the *A* peak were negatively correlated with the age of the infants (Fig. 2.5). Additionally, younger infants (3–5 months old) had longer peak latencies, shorter inter-peak *A–O* latency, and less abrupt onset of the *V–A* slope than older infants (6–10 months old). In other words, the younger the infant, the longer is the latency of the FFR and the less synchronous is the neural firing. These findings were consistent with the development of the human brain during infancy.

Although the cochlea is largely functional at birth, neural myelination and synaptic organization are still developing. For example, through the use of magnetic resonance imaging in infants, it has been reported that myelination of neural elements at the subcortical level takes place gradually during the first few months of life. For instance, the cochlear nucleus, superior olivary complex, and lateral

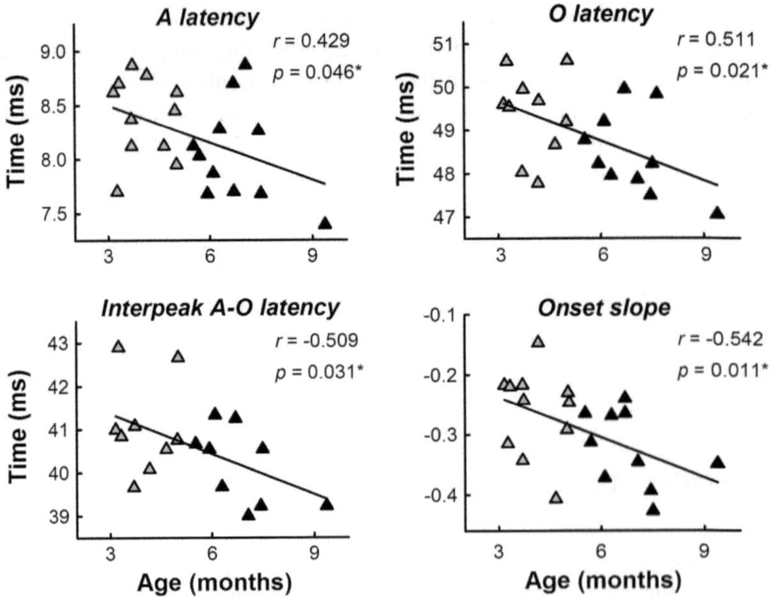

Fig. 2.5 Scatterplots demonstrating negative correlations between age and the latency of the *A peak* (onset), latency of the *O peak* (offset), interpeak latency between the *A* and *O peaks*, and the onset slope from the *peak V* to *peak A*. The FFR data were derived from recordings obtained in 25 American infants who were between 3 and 10 months old. Younger infants (3–5 months old, *gray triangles*) have significantly longer latencies than older infants (6–10 months old, *black triangles*). *Solid lines* are the linear regressions of the FFR measurements as a function of age. *Asterisks* indicate $p < 0.05$, r = correlation coefficient. (Reproduced from Anderson et al. 2015. © The Journal of Acoustical Society of America)

lemniscus show increased myelination density for up to 13 weeks of age, whereas the inferior colliculus shows improved myelination density for up to 39 weeks of age (Sano et al. 2007). The continued myelination of neural structures at the subcortical level may lead to the decreased latency of the FFRs recorded during the first year of life.

2.5 Childhood

Development and refinement of neural structures continue throughout childhood (Eggermont and Moore 2012). Specific speech characteristics and linguistic features that exist in the child's native language environment stimulate and enhance the functionalities of the neural elements and auditory circuitry on both the cortical (Huttenlocher and Dabholkar 1997) and subcortical (Song et al. 2008) levels. For the purpose of this chapter, discussion will be focused on subcortical changes. One major advantage of using a complex stimulus, such as the speech token /da/, is that it allows a detailed examination of the timing and spectral properties of neural processing related to the abrupt onset of the consonant /d/, its transition to the vowel /a/, and the responses to the steady-state vowel portion of the stimulus. This approach provides enriched information about the neural processing of speech. As such, the FFR has been used as a neurophysiological marker that associates and predicts reading readiness, literacy capability, and academic performance for school-aged children and before they enter elementary school (White-Schwoch and Kraus 2013; White-Schwoch et al. 2015b). The FFR in response to complex stimuli also has been reported to be a viable method in assessing the impact of the child's linguistic environment on the functional and structural changes of the auditory neural circuitry for toddlers and school-aged children (Skoe et al. 2013; Krizman et al. 2015).

Precision of neural encoding at the subcortical level is correlated with a child's reading readiness and literacy (see Reetzke, Xie, and Chanderasekaran, Chap. 10). Additionally, the precision of neural encoding is decreased in children who have auditory processing disorders, dyslexia, or autistic spectrum disorders (see Schochat, Rocha-Muniz, and Filippini, Chap. 9). A more thorough discussion on these topics is available in the other chapters of the book (e.g., short-term learning and memory by Carcagno and Plack, Chap. 4; auditory experience and communication by White-Schwoch and Kraus, Chap. 6; and communicating in challenging environments by Bidelman, Chap. 8). The following sections of this chapter will emphasize normal development and the influences of linguistic experience on the subcortical neural encoding of speech during childhood.

2.5.1 Stability and Precision of Neural Firing

A reliable pattern of neural firing plays a pivotal role as a child continues to develop and adapt to the various, and sometimes adverse, listening environments they may encounter throughout childhood (White-Schwoch et al. 2015b). A consistent firing pattern among the distributed, but integrated, neural elements at the subcortical level is a physiological prerequisite for accurate encoding of the various linguistic and paralinguistic information embedded in human speech (Bidelman, Chap. 8). A child will be unable to process and perceive speech information accurately if the firing patterns of the involved neural elements fail to reflect the necessary and specific features of the speech sounds important in the child's native language.

Stability of neural firing can be quantified by measuring the *test-retest reliability* or *trial-by-trial variability* among the various sweeps of FFR recordings. While the test-retest reliability can be examined through recordings from the same participants within one test session, it can also be evaluated through analyzing the FFR recordings from the same participants but in different test sessions. An alternative approach is to examine the trial-by-trial variability by randomly selecting a fixed number of sweeps from a pool of all the recording sweeps obtained within a test session. The FFR literature from normally developing children reveals a reliable and consistent firing pattern in a quiet listening environment (Russo et al. 2004; White-Schwoch et al. 2015a). It is worth mentioning that the various aspects of neural recording, as reflected by the different portions of FFR (e.g., the transient onset response versus the steady-state response) may have different levels of consistency in neural firing. This differentiation is important because speech perception is affected not only by how the brain processes a steady-state vowel but also by how a transient consonant is encoded by the brain. Russo and colleagues (2004) measured the test-retest reliability in eight normally developing children and found that FFR measurements derived from the sustained responses are more stable than those derived from the transient onset responses. Similar findings were discovered when comparing the trial-by-trial variability within a single FFR recording (White-Schwoch et al. 2015a). For example, Hornickel and Kraus (2013) reported a systematic relationship between the stability of the FFR and literacy skills, with reading-impaired children showing more variable responses to speech than their age-matched peers. Together, these results verify the notion that a consistent pattern of neural firing is a positive indicator regarding the maturity and readiness of the subcortical neural elements to receive and decipher the various specific features of human speech.

Efficiency and consistency of reliable firing patterns may be compromised when a child is listening in an adverse acoustic environment, such as in a reverberant room or in the presence of a substantial amount of background noise. White-Schwoch and colleagues (2015b) examined the FFRs recorded from a group of children 3–4 years old and reported that the precision and stability of the neural encoding of consonants in noise were strongly correlated with the children's phonological processing, reading readiness, and pre-literacy skills. Neural encoding of consonants in noise further predicted the children's performances on reading competence and a range of literacy tests when they grew older. When the data obtained from the group of children characterized by normal development were compared to the other children who had been diagnosed with a learning disability, the diagnostic group of children was found to have significantly poorer precision and stability of neural coding of consonants in noise when compared to the normal-developing children. Stability and precision of neural firing is a critical component for a child to receive and process the various features of speech stimuli consistently and efficiently.

2.5.2 *Effects of Linguistic Experience*

Neural structures of the auditory system are malleable such that a history of acoustic stimulation affects how neural circuits will respond to the subsequent incoming signals that may exist in a child's acoustic environment (Kral and Eggermont 2007; Shepard et al. 2013). Over time, changes of the neural structures in response to the acoustic signals of the listening environment will stay within the neural circuitry (Buonomano and Merzenich 1998; Kilgard and Merzenich 1998). This will enhance the auditory processing of the linguistic and paralinguistic factors that are important in the child's native language. Neural elements at subcortical levels, similar to those at the cortical level, have a tendency and preference to fine tune to the acoustic and linguistic parameters that occur often in the child's language. The brain's top-down processing and the frequent stimulation of the linguistic and paralinguistic parameters on the neuronal structures are thought to facilitate the functional reorganization of auditory neural circuits at the subcortical level (Bidelman et al. 2011; Song et al. 2008).

A child's linguistic environment plays an important role in the development of the neural networks at the subcortical level (Krizman et al. 2012, 2015). In modern societies, it is nearly impossible to find a child who has no exposure to any language at all. Even tribes deep in a jungle communicate with native languages of their own. As a result, the influence of a child's linguistic experience on how the brain works is very difficult to isolate and examine. This restriction gives researchers no choice but to focus primarily on the differences between two languages. Through behavioral measurements and cortical responses to the different features of two languages, researchers have found that a child's experience and exposure to a specific language enhances auditory circuitries to the linguistic features found within that language (Zhang et al. 2005). At the same time, the child's ability to process the acoustic features that are specific and unique to another language will deteriorate if they are not present in the native language. With increased experience, the child's brain becomes fine-tuned to the linguistic and paralinguistic features that are specific and important in the child's native language. Admittedly, human communications are not limited to spoken languages. For example, people with profound hearing loss often communicate through sign languages (Sacks 1989). Development of the neural circuitries inside the brain for individuals who use sign languages involves interactions among visual and other sensory inputs (Kral and Eggermont 2007; Kral et al. 2013) and is beyond the scope of this chapter. For clarity, the term "language" is used in the remainder of this chapter to refer to spoken language.

In many instances, a child's linguistic environment contains more than one language. If a child communicates with two different languages, this child is called a bilingual. From the viewpoint of the FFR in response to complex sounds, studies have shown that children who are born and raised in a bilingual community exhibit stronger neural encoding to the spectral and timing components of speech (Fig. 2.6) (Krizman et al. 2012, 2015). They also have better consistency of neural firing than children who are born and raised in a mono-linguistic environment. Although, in most cases, one language is predominantly conversed throughout the child's daily life, the existence of the non-predominant language and the additions of bilingual factors create an enriched linguistic environment. An environment as such not only facilitates the acquisition of a second language for the child, but also further enhances the structural and functional reorganization of the auditory neural circuitries at the subcortical level during childhood (Kraus and White-Schwoch 2015; Krishnan and Gandour, Chap. 3).

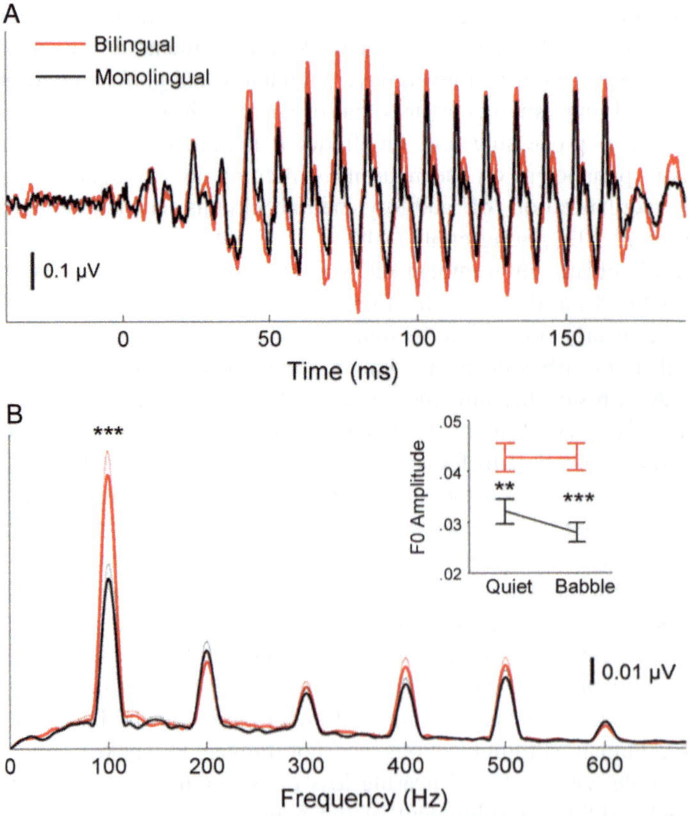

Fig. 2.6 Subcortical responses of bilinguals (*red*) and monolinguals (*black*) to the speech sound /da/ presented in multi-talker babble. (**A**) Bilinguals show a larger auditory brainstem response relative to monolinguals. (**B**) Amplitudes of the individual component frequencies in the steady-state (60–180 ms) region of the response to /da/ in multi-talker babble. *Thin lines* represent +1 standard error of the mean. *Inset in B* displays the mean amplitude (±1 standard error) of the F_0 in quiet and in multi-talker babble for bilinguals and monolinguals. For monolinguals, there is a decrease in the amplitude of the F_0 (100 Hz) when the stimulus is presented in multi-talker babble relative to when it is presented in quiet. In contrast, bilinguals show virtually no change in F_0 amplitude between the two conditions. *Asterisks* represent significance levels: $^{**}p < 0.005$, $^{***}p < 0.0001$. (Reproduced from Krizman et al. 2012 with permission. © Proceedings of the National Academy of Sciences of the USA)

Exposure to an additional language strengthens the auditory circuitry even further. Neural encoding and firing patterns are more resistant to the presence of background noise when a child has been exposed to an additional language (Krizman et al. 2012). For example, when speech sounds are presented in quiet, bilingual adolescents demonstrate larger FFR amplitudes and better consistency of neural firing. In the presence of background noise, such as multi-talker babble, the FFRs of bilinguals remain relatively stable. Monolinguals showed a decrement in the accuracy and consistency of tracking the temporal and spectral aspects of human speech.

Children who acquire two spoken languages simultaneously from birth exhibit stronger neural encoding patterns that are phase locked to the spectral components of speech than those who acquire two languages sequentially (Krizman et al. 2015). Simultaneous bilinguals also demonstrate a greater consistency of neural firing than those who acquire two languages sequentially (Fig. 2.7). Additionally, the greater number of years of bilingual experience, the stronger neural encoding and the better trial-by-trial consistency of neural firing will be. Children with more years of experience communicating in two languages demonstrate stronger FFRs and more consistent firing patterns than those who have fewer years of bilingual experience, signifying the idea that enhanced spectral encoding and neural consistency will emerge with increasing years of experience communicating in a bilingual environment during childhood.

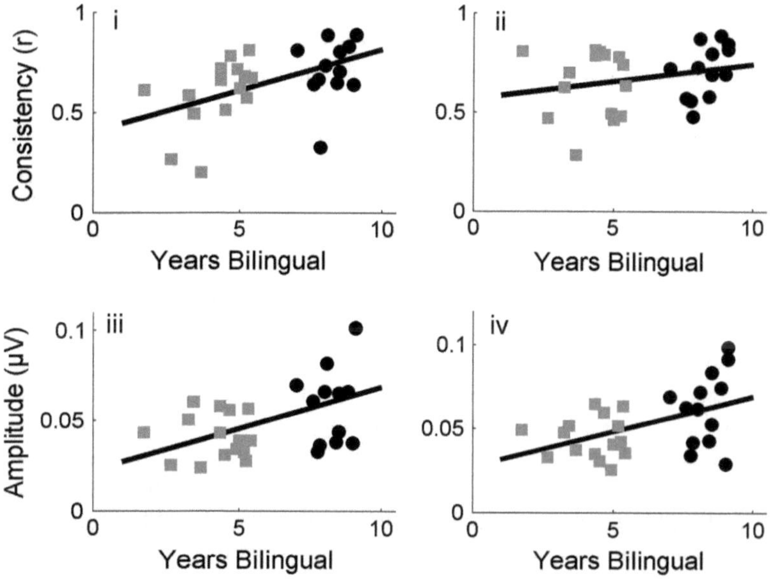

Fig. 2.7 Relationship between neural processing and years of bilingual experience. Response consistency for /ba/ (**i**) and /ga/ (**ii**) versus years of bilingual experience for the simultaneous (*black*) and sequential (*gray*) bilinguals plotted on the x-axis. The consistency to /ba/ relates with years of second language experience, while the consistency to /ga/ does not. F_0 encoding amplitude is shown for /ba/ (**iii**) and /ga/ (**iv**) versus years of bilingual experience for both groups. Both measures of F_0 encoding relate to the number of years of experience the child has speaking two languages. (Reproduced from Krizman et al. 2015 with permission of publisher. © Neuroscience Letters 2015)

2.6 Future Directions

Although the development of subcortical neural structures and the influence of linguistic experience during infancy and childhood have been extensively studied, a few issues remain.

The first issue is related to the listening experience of the fetus and its impact on the development of the auditory system. The human fetus becomes exposed to sounds of the linguistic environment while it is still in the mother's womb. It is hard to determine the influence of this linguistic exposure since there is no way to obtain an FFR in utero. Until recording FFRs in a living fetus becomes feasible, one can only derive conclusions indirectly. Currently, the earliest recording that can be made is during the immediate postnatal days. FFRs recorded in infants during their immediate postnatal days contain at least two components: one is the biological ability that comes with all newborns, and the other is the influence of the infant's listening experience when the infant is still in its mother's womb.

Although the FFR literature has provided partial and indirect evidence regarding the possible influence of prenatal listening experience on the development of auditory circuitry, a carefully designed study that includes the use of certain linguistic or musical training in pregnant women in a randomized-control study will be needed to properly address this issue. One possible approach is to develop an experimental protocol similar to that used in Partanen et al. (2013). A group of pregnant women could be recruited and asked to listen to a set of carefully designed sound material on a regular basis in order to familiarize the fetus with specific sound patterns (e.g., /tatata/ versus /tatota/). Electrophysiological responses could then be recorded in the neonates during their immediate postnatal days by using the specific sound patterns, and the responses could be compared to those elicited by other unrelated sound patterns.

The second issue is the lack of a normative database for each age group of normally developing infants and children. Completion and establishment of such a database for each age group is critical. Not only will this advance our understanding of the normal development of speech encoding at the subcortical level but also will allow the development of appropriate therapeutic and rehabilitative protocols for infants and children who are at risk of a specific disorder. Skoe and colleagues (2015) reported the development of subcortical auditory processing from 586 healthy participants across an extensive age range (ages 3 months to 72 years). This cross-sectional database is laudable and will be useful for future applications in the research and scientific realms. Additional databases supplementing the gaps in regards to the development of subcortical pitch processing during the immediate postnatal days and the first three months of life are warranted. Furthermore, a systematic large-scale multiple-site study, preferably prospective and longitudinal, will be needed to examine the characteristics and maturational trends of the FFR in normally developing infants and children across the various developmental stages of life. Infants and children who are born and raised in a non-tonal versus a tonal language environment and a mono-linguistic versus a multi-linguistic environment should all be considered. Upon the completion and establishment of the normative database for FFRs, infants and children at risk for a specific disorder can be examined and their test results can be compared with those published in the age-appropriate normative database. Researchers and clinicians can further design treatment protocols with successful outcome measurements targeted to the normative database.

The third issue is related to the shortage of useable computational models that are capable of capturing the characteristics and growth trends of the FFRs for infants and children. Computational models are beneficial because they not only can help to understand how speech sounds are processed at the subcortical level, but they also can help predict outcomes of specific measurements of the brain. Ideal computational models should have solid foundations based on auditory anatomy and physiology. To initiate the process of developing computational models for FFRs, researchers have started testing some algorithms. For example, a computational model that utilizes an exponential curve-fitting formula has been successfully applied to normal-hearing adults (Jeng et al. 2011a), and an automatic procedure has been developed for neonates (Jeng et al. 2013). When performing tests on neonates, infants, and difficult-to-test populations, the amount of time needed to complete a recording is of great importance. Preliminary results have shown that the exponential curve-fitting model provides a good fit to the FFR trends with an increasing number of sweeps. Thus, the testing time can be shortened by employing an appropriate exponential model and applying a pre-determined stopping criterion to complete an FFR recording. However, further testing and finding other specific models that will work for the various age groups of participants will be needed to corroborate our understanding of speech encoding at the subcortical level and to predict outcomes in a simulated environment. A cadre of experts in auditory electrophysiology, computer modeling and simulation, pediatric neuroscience, language development, and related fields will need to collaborate to resolve this issue.

The last issue is associated with the inherently small amplitude of the scalp-recorded FFR. This issue becomes particularly challenging when attempting to record an FFR in a newborn nursery where environmental interferences can be substantial or when trying to record an FFR in an infant or a child who is not in a state of rest. Technologies that are designed to reduce the influence of environmental and other unwanted physiological noises and to enhance the robustness of the elicited response are needed. Algorithms necessary to promote the detection of the presence or absence of an FFR, along with automation of the necessary signal-processing procedures, will be needed to facilitate the visibility of an FFR. A real-time assessment of the progression of the data recording is preferred. Once the FFR algorithms and methodology have been further improved for detecting the presence of an FFR and its interface has become user friendly for researchers and clinicians in the FFR community and related fields, puzzles related to the normal development of the auditory system and related pathologies can be researched and resolved in a timely manner.

2.7 Summary

Since subcortical neural structures were reported to be malleable with auditory experience in the early twentieth century, a tremendous amount of new information and discoveries have been added to our understanding of the development of the human auditory system (Krishnan et al. 2005; Wong et al. 2007). Preliminary results obtained during the past 10 years have demonstrated possible impacts of language exposure during the early stages of life on the development of speech representation at the subcortical level. Future work in this area will benefit from collaborations among related disciplines and will promote a deeper understanding of the underpinning mechanisms involved in typical and atypical development of the auditory system during infancy and childhood.

Compliance with Ethics Requirements Fuh-Cherng Jeng declared that he had no conflict of interest.

References

Abdala, C., & Folsom, R. C. (1995). The development of frequency resolution in humans as revealed by the auditory brain-stem response recorded with notched-noise masking. *The Journal of the Acoustical Society of America, 98*, 921–930.

Abrams, R. M., Gerhardt, K. J., Griffiths, S. K., Huang, X., & Antonelli, P. J. (1998a). Intrauterine sounds in sheep. *Journal of Sound and Vibration, 216*, 539–542.

Abrams, R. M., Griffiths, S. K., Huang, X., Sain, J., et al. (1998b). Fetal music perception: The role of sound transmission. *Music Perception, 15*, 307–317.

Aiken, S. J., & Picton, T. W. (2006). Envelope following responses to natural vowels. *Audiology and Neurotology, 11*, 213–232.

Aiken, S. J., & Picton, T. W. (2008). Envelope spectral frequency–following responses to vowel sounds. *Hearing Research, 245*, 35–47.

Anderson, S., Parbery-Clark, A., White-Schwoch, T., & Kraus, N. (2015). Development of subcortical speech representation in human infants. *The Journal of Acoustical Society of America, 137*, 3346–3355.

Bidelman, G. M., Gandour, J. T., & Krishnan, A. (2011). Musicians and tone-language speakers share enhanced brainstem encoding but not perceptual benefits for musical pitch. *Brain and Cognition, 77*, 1–10.

Brauer, J., Anwander, A., & Friederici, A. (2011). Neuroanatomical prerequisites for language functions in the maturing brain. *Cerebral Cortex, 21*, 459–466.

Buonomano, D. V., & Merzenich, M. M. (1998). Cortical plasticity: From synapses to maps. *Annual Review of Neuroscience, 21*, 149–186.

Butler, B. E., & Trainor, L. J. (2013). Brief pitch-priming facilitates infants' discrimination of pitch-evoking noise: Evidence from event-related potentials. *Brain and Cognition, 83*, 271–278.

Carral, V., Huotilainen, M., Ruusuvirta, T., Fellman, V., et al. (2005). A kind of auditory 'primitive intelligence' already present at birth. *European Journal of Neuroscience, 21*, 3201–3204.

Dinnsen, D. (1992). Variation in developing and fully developed phonetics inventories. In C. A. Ferguson, L. Menn, & C. Stoel-Gammon (Eds.), *Phonological development: Models, research, implications* (pp. 191–210). Timonium, MD: York Press.

Echteler, S. M., Arjmand, E., & Dallos, P. (1989). Developmental alterations in the frequency map of the mammalian cochlea. *Nature, 14*, 147–149.

Eggermont, J. J., & Moore, J. K. (2012). Morphological and functional development of the auditory nervous system. In L. Werner, R. R. Fay, & A. N. Popper (Eds.), *Human auditory development* (pp. 61–106). New York: Springer Science+Business Media.

Eimas, P. D., Siqueland, E. R., Jusczyk, P., & Vigorito, J. (1971). Speech perception in infants. *Science, 171*, 303–306.

Frenz, D. A., McPhee, J. R., & Van De Water, T. R. (2001). Structural and functional development of the ear. In A. F. Jahn & J. Santos-Sacchi (Eds.), *Physiology of the ear* (pp. 191–214). San Diego: Singular Thomson Learning.

Friederici, A. D., Friedrich, M., & Christophe, A. (2007). Brain responses in 4-month-old infants are already language specific. *Current Biology, 17*, 1208–1211.

Friedrich, M., & Friederici, A. D. (2010). Maturing brain mechanisms and developing behavioral language skills. *Brain and Language, 114*, 66–71.

Gardi, J., Salamy, A., & Mendelson, T. (1979). Scalp-recorded frequency-following responses in neonates. *Audiology: Journal of Auditory Communication, 18*(6), 494–506.

Gelman, S. R., Wood, S., Spellacy, W. N., & Abrams, R. M. (1982). Fetal movements in response to sound stimulation. *American Journal of Obstetrics and Gynecology, 143*, 484–485.

Gerhardt, K. J., & Abrams, R. M. (2000). Fetal exposures to sound and vibroacoustic stimulation. *Journal of Perinatology, 20*, S20–S29.

Griffiths, S. K., Brown, W. S., Jr., Gerhardt, K. J., Abrams, R. M., & Morris, R. J. (1994). The perception of speech sounds recorded within the uterus of a pregnant sheep. *The Journal of Acoustical Society of America, 96*, 2055–2063.

Grossmann, T., Striano, T., & Friederici, A. D. (2007). Developmental changes in infants' processing of happy and angry facial expressions: A neurobehavioral study. *Brain and Cognition, 64*, 30–41.

Hall, J. W., III. (2006). *New handbook of auditory evoked responses*. New York: Pearson.

He, C., Hotson, L., & Trainor, L. J. (2007). Mismatch responses to pitch changes in early infancy. *Journal of Cognitive Neuroscience, 19*, 878–892.

Hollich, G., Hirsh-Pasek, K., Golinkoff, R., Brand, R. J., et al. (2000). Breaking the language barrier: An emergentist coalition model for the origins of word learning. *Monographs of the Society for Research in Child Development* (Serial No. 262), *65*, 1–123.

Hornickel, J., & Kraus, N. (2013). Unstable representation of sound: A biological marker of dyslexia. *The Journal of Neuroscience, 33*, 3500–3504.

Huttenlocher, P. R., & Dabholkar, A. S. (1997). Regional differences in synaptogenesis in human cerebral cortex. *Journal of Comparative Neurology, 387*, 167–178.

Jakobson, R. (1968). *Child language, aphasia and phonological universals*. The Hague and Paris: Mouton.

Jeng, F.-C., Chung, H.-K., Lin, C.-D., Dickman, B. M., & Hu, J. (2011a). Exponential modeling of human frequency-following responses to voice pitch. *International Journal of Audiology, 50*, 582–593.

Jeng, F.-C., Hu, J., Dickman, B. M., Montgomery-Reagan, K., et al. (2011b). Cross-linguistic comparison of frequency-following responses to voice pitch in American and Chinese neonates and adults. *Ear and Hearing, 32*, 699–707.

Jeng, F.-C., Peris, K. S., Hu, J., & Lin, C.-D. (2013). Evaluation of an automated procedure for detecting frequency-following responses in American and Chinese neonates. *Perceptual and Motor Skills, 116*, 456–465.

Jeng, F.-C., Schnabel, E. A., Dickman, B. M., Hu, J., et al. (2010). Early maturation of frequency-following responses to voice pitch in infants with normal hearing. *Perceptual and Motor Skills, 111*, 765–784.

Jusczyk, P. (1997). *The discovery of spoken language*. Cambridge (MA): MIT Press.

Kilgard, M. P., & Merzenich, M. M. (1998). Plasticity of temporal information processing in the primary auditory cortex. *Nature Neuroscience, 1*, 727–731.

Kral, A., & Eggermont, J. J. (2007). What's to lose and what's to learn: Development under auditory deprivation, cochlear implants and limits of cortical plasticity. *Brain Research Reviews, 56*, 259–269.

Kral, A., Popper, A. N., & Fay, R. R. (2013). *Deafness*. New York: Springer Science+Business Media.

Kraus, N., & White-Schwoch, T. (2015). Unraveling the biology of auditory learning: A cognitive-sensorimotor-reward framework. *Trends in Cognitive Sciences, 19*, 642–654.

Krishnan, A., Xu, Y., Gandour, J., & Cariani, P. (2005). Encoding of pitch in the human brainstem is sensitive to language experience. *Cognitive Brain Research, 25*, 161–168.

Krizman, J., Marian, V., Shook, A., Skoe, E., & Kraus, N. (2012). Subcortical encoding of sound is enhanced in bilinguals and relates to executive function advantages. *Proceedings of the National Academy of Sciences of the USA, 109*, 7877–7881.

Krizman, J., Slater, J., Skoe, E., Marian, V., & Kraus, N. (2015). Neural processing of speech in children is influenced by extent of bilingual experience. *Neuroscience Letters, 585*, 48–53.

Kuhl, P. K. (1994). Learning and representation in speech and language. *Current Opinion in Neurobiology, 4*, 812–822.

Kuhl, P. K. (2004). Early language acquisition: cracking the speech code. *Nature Reviews Neuroscience, 5*, 831–843.

Kuhl, P. K. (2010). Brain mechanisms in early language acquisition. *Neuron, 67*, 713–727.

Kuhl, P., Conboy, B., Coffrey-Corina, S., Padden, D., et al. (2008). Phonetic learning as a pathway to language: New data and native language magnet theory expanded (NLM-e). *Philosophical Transactions of the Royal Society of London B: Biological Sciences, 363*, 979–1000.

Kuhl, P. K., Stevens, E., Hayashi, A., Deguchi, T., et al. (2006). Infants show a facilitation effect for native language phonetic perception between 6 and 12 months. *Developmental Science, 9*, F13–F21.

Kuhl, P. K., Williams, K. A., Lacerda, F., Stevens, K. N., & Lindblom, B. (1992). Linguistic experience alters phonetic perception in infants by 6 months of age. *Science, 255*, 606–608.

Locke, J. (1980). The prediction of child speech errors: Implications for a theory of acquisition. In G. H. Heni-komshian, J. F. Kavanagh, & C. A. Ferguson (Eds.), *Child phonology. I. Production*. New York: Academic Press.

Mattock, K., & Burnham, D. (2006). Chinese and English infants' tone perception: Evidence for perceptual reorganization. *Infancy, 10*, 241–265.

Mattock, K., Molnar, M., Polka, L., & Burnham, D. (2008). The developmental course of lexical tone perception in the first year of life. *Cognition, 106*, 1367–1381.

Moushegian, G., Rupert, A. L., & Stillman, R. D. (1973). Laboratory note. Scalp-recorded early responses in man to frequencies in the speech range. *Electroencephalography and Clinical Neurophysiology, 35*, 665–667.

Näätänen, R., Paavilainen, P., Rinne, T., & Alho, K. (2007). The mismatch negativity (MMN) in basic research of central auditory processing: A review. *Clinical Neurophysiology, 118*, 2544–2590.

Olsho, L. W., Koch, E. G., & Halpin, C. F. (1987). Level and age effects in infant frequency discrimination. *The Journal of the Acoustical Society of America, 82*, 454–464.

Panneton, R., & Newman, R. (2012). Development of speech perception. In L. Werner, R. R. Fay, & A. N. Popper (Eds.), *Human auditory development* (pp. 197–222). New York: Springer Science+Business Media.

Partanen, E., Kujala, T., Näätänen, R., Liitola, A., et al. (2013). Learning-induced neural plasticity of speech processing before birth. *Proceedings of the National Academy of Sciences of the USA, 110*, 15145–15150.

Querleu, D., Renard, X., Boutteville, C., & Crepin, G. (1989). Hearing by the human fetus? *Seminars in Perinatology, 13*, 409–420.

Romand, R., & Ehret, G. (1990). Development of tonotopy in the inferior colliculus. I. Electrophysiological mapping in house mice. *Developmental Brain Research, 54*, 221–234.

Rubel, E. W., & Ryals, B. M. (1983). Development of the place principle: Acoustic trauma. *Science, 4*, 512–514.

Russo, N., Nicol, T., Musacchia, G., & Kraus, N. (2004). Brainstem responses to speech syllables. *Clinical Neurophysiology, 115*, 2021–2030.

Sacks, O. (1989). *Seeing voices: A journal into the world of the deaf*. New York: Vintage Books.

Sano, M., Kaga, K., Kuan, C.-C., Ino, K., & Mima, K. (2007). Early myelination patterns in the brainstem auditory nuclei and pathway: MRI evaluation study. *International Journal of Pediatric Otorhinolaryngology, 71*, 1105–1115.

Shahidullah, S., & Hepper, P. G. (1994). Frequency discrimination by the fetus. *Early Human Development, 36*, 13–26.

Shepard, K. N., Kilgard, M. P., & Liu, R. (2013). Experience-dependent plasticity and auditory cortex. In Y. E. Cohen, A. N. Popper, & R. R. Fay (Eds.), *Neural correlates of auditory cognition* (pp. 293–327). New York: Springer Science+Business Media.

Skoe, E., & Kraus, N. (2010). Auditory brainstem response to complex sounds: A tutorial. *Ear and Hearing, 31*, 302–324.

Skoe, E., Krizman, J., Anderson, S., & Kraus, N. (2015). Stability and plasticity of auditory brainstem function across the lifespan. *Cerebral Cortex, 25*, 1415–1426.

Skoe, E., Krizman, J., & Kraus, N. (2013). The impoverished brain: Disparities in maternal education affect the neural response to sound. *The Journal of Neuroscience, 33*, 17221–17231.

Song, J. H., Skoe, E., Wong, P. C. M., & Kraus, N. (2008). Plasticity in the adult human auditory brainstem following short-term linguistic training. *Journal of Cognitive Neuroscience, 20*, 1892–1902.

Trainor, L. J., & Unrau, A. (2012). Development of pitch and music perception. In L. Werner, R. R. Fay, & A. N. Popper (Eds.), *Human auditory development* (pp. 223–254). New York: Springer Science+Business Media.

Tsao, F. M., Liu, H. M., & Kuhl, P. K. (2006). Perception of native and non-native affricate-fricative contrasts: Cross-language tests on adults and infants. *The Journal of the Acoustical Society of America, 120*, 2285–2294.

White-Schwoch, T., Davies, E. C., Thompson, E. C., Woodruff Carr, K., et al. (2015a). Auditory-neurophysiological responses to speech during early childhood: Effects of background noise. *Hearing Research, 328*, 34–47.

White-Schwoch, T., & Kraus, N. (2013). Physiologic discrimination of stop consonants relates to phonological skills in pre-readers: A biomarker for subsequent reading ability? *Frontiers in Human Neuroscience, 7*, 1–9.

White-Schwoch, T., Woodruff Carr, K., Thompson, E. C., Anderson, S., et al. (2015b). Auditory processing in noise: A preschool biomarker for literacy. *PLoS Biology, 13*(7), e1002196. Doi:10.1371/journal.pbio.1002196

Wong, P. C. M., Skoe, E., Russo, N. M., Dees, T., & Kraus, N. (2007). Musical experience shapes human brainstem encoding of linguistic pitch patterns. *Nature Neuroscience, 10*, 420–422.

Worden, F. G., & Marsh, J. T. (1968). Frequency-following (microphonic-like) neural responses evoked by sound. *Electroencephalography and Clinical Neurophysiology, 25*, 42–52.

Yip, M. (2002). *Tone*. Cambridge, UK: Cambridge University Press.

Zhang, Y., Kuhl, P. K., Imada, T., Kotani, M., & Tohkura, Y. (2005). Effects of language experience: Neural commitment to language-specific auditory patterns. *NeuroImage, 26*, 703–720.

Chapter 3
Shaping Brainstem Representation of Pitch-Relevant Information by Language Experience

Ananthanarayan Krishnan and Jackson T. Gandour

Abstract Pitch is a robust perceptual attribute that plays an important role in speech, language, and music. We present compelling evidence supporting the notion that long-term language experience enhances the neural representation of behaviorally relevant attributes of pitch in the brainstem. Pitch relevant neural activity in the brainstem is crucially dependent on specific dimensions or features of pitch contours. By focusing on specific properties of the auditory signal, irrespective of a speech or nonspeech context, it is argued that the neural representation of acoustic–phonetic features relevant to speech perception is already emerging in the brainstem and, importantly, can be shaped by experience. Such effects of language experience on sensory processing are compatible with a more integrated approach to language and the brain. Long-term language experience appears to shape an adaptive, integrated, distributed pitch-processing network. A theoretical framework for a neural network is proposed involving coordination between local, feedforward and feedback components that can account for experience-induced enhancement of pitch representations at multiple locations of the distributed pitch processing network. Feedback, feedforward connections provide selective gating of inputs to both cortical and subcortical structures to enhance neural representation of behaviorally relevant attributes of the stimulus and instantiate local mechanisms that exhibit enhanced sensitivity to behaviorally relevant pitch attributes. The focus on pitch processing in tonal languages notwithstanding, the findings presented here should be contextualized within the broader framework of language experience shaping subcortical processing.

A. Krishnan (✉) · J.T. Gandour
Department of Speech Language Hearing Sciences, Lyles-Porter Hall,
Purdue University, 715 Clinic Drive, West Lafayette, IN 47907-2122, USA
e-mail: rkrish@purdue.edu

J.T. Gandour
e-mail: gandour@purdue.edu

N. Kraus et al. (eds.), *The Frequency-Following Response*,
Springer Handbook of Auditory Research 61,
DOI 10.1007/978-3-319-47944-6_3

Keywords Auditory brainstem · Domain specificity · Experience-dependent plasticity · Feedback · Feedforward · Frequency-following response · Iterated rippled noise · Mandarin Chinese · Neural phase-locking · Pitch · Pitch acceleration · Pitch salience · Pitch strength

3.1 Introduction

Pitch is an important information-bearing perceptual component in the processing of speech, language, and music (Oxenham 2012). Functional brain imaging studies provide strong evidence for hierarchical processing of pitch (Kumar et al. 2007), starting in subcortical structures (Griffiths et al. 2001) and continuing up through Heschl's gyrus on to the planum polare and planum temporale (Gutschalk et al. 2007). Thus, pitch provides an excellent window for studying experience-dependent effects on both cortical and brainstem components of a well-coordinated, hierarchical processing network.

A complete understanding of the neural organization of language can only be achieved by viewing these processes as a set of hierarchical computations at different levels of biological abstraction—both cortical and subcortical. These computations are applied to representations that, in turn, are shaped by experience within a specific domain, such as language. Indeed, recent empirical data show that these neural representations of pitch, at both the brainstem and cortical level, are shaped by one's experience with language and music (Besson et al. 2011; Zatorre and Baum 2012).

While it is not known how language experience shapes subcortical and cortical stages of pitch processing, it is likely that the neural processes underlying such experience-dependent plasticity at each stage along the processing hierarchy are modulated by a coordinated interplay between ascending, descending, and local neural pathways that involve both sensory and cognitive components (Chandrasekaran and Kraus 2010; Kraus and White-Schwoch 2015). That is, feedback from language-dependent cortical processes shape early sensory-level processing to generate more precise outputs that facilitate stronger cortical representations of sound that, in turn, contribute to linguistic performance.

3.1.1 *Neural Bases of Pitch*

The neural bases of pitch perception are still a matter of debate. One view is that the auditory system extracts pitch from complex sounds by deriving a spectral profile from frequency-specific auditory input, followed by pattern-matching mechanisms (Cohen et al. 1995). A contrasting view is that the auditory system extracts pitch from the timing of auditory nerve fiber activity irrespective of frequency organization. These temporal models are based solely on the timing information available in the interspike intervals represented in simulated (Patterson et al. 1995; Meddis and O'Mard 1997) or actual (Cariani and Delgutte 1996a, b) auditory nerve activity. They derive a pitch estimate by pooling timing information across auditory nerve fibers without regard to the frequency-to-place mapping. More recently, a hybrid pitch encoding mechanism that uses both spectral and temporal information has been proposed (Cedolin and Delgutte 2005). Thus, neural phase locking related to voice fundamental frequency (F_0) plays a dominant role in the encoding of low pitch associated with complex sounds. Temporal encoding schemes provide a unified and parsimonious way of explaining a diverse range of pitch phenomena for at least pitch processing at or below the level of the inferior colliculus (IC) (Meddis and O'Mard 1997).

For any neural code of pitch to be useful it should be operational, at least in some form, at all levels of the processing hierarchy. There is evidence that neurons in primary auditory cortex exhibit temporal and spectral response properties that could enable these pitch-encoding schemes (Steinschneider et al. 1998; Lu et al. 2001), but whether they form a network with pitch-selective neurons to carry out this process is not known. In subcortical auditory structures, periodicity and pitch are often represented by regular temporal patterns of action potentials that are phase locked to the sound waveform. Whereas the most commonly observed code for periodicity and pitch within cortical neurons is a modulation of spike rates as a function of F_0, it is possible that the wider temporal integration window at the cortical level may render the auditory cortical neurons too sluggish to provide phase-locked representations of periodicity within the pitch range (Walker et al. 2011). Thus, it is not yet clear how cortical neurons transform the autocorrelation-like temporal analysis in the brainstem to a spike-rate code to extract pitch-relevant information. One possibility is that the temporal code is transformed into a response synchrony code whereby temporally coherent activity from the subcortical stages will produce greater spike rates, yielding larger response amplitudes at the cortical level. Analyses of the statistical properties of spike rates for virtual neural units have shown that frequency tuning and spike rate characteristics of their neural units are similar to those of auditory cortical neural units (Micheyl et al. 2013). Their findings indicate that sufficient statistical information is present in the population spike rate to account for small differences in frequency (pitch) and intensity (loudness).

3.1.2 Hierarchical Nature of Pitch Processing

Considerable progress has been made over the past decade in our understanding of the complex series of processing stages that are required to translate speech sounds into meaning at the level of the cerebral cortex. Functional imaging evidence points to multiple, parallel, hierarchically organized processing pathways that are related to speech processing in the cerebral cortex (Hickok and Poeppel 2007; Poeppel et al. 2008). Speech processing in the cortex also emerges from differential demands on distributed brain regions shared by both verbal and nonverbal auditory processing (Price et al. 2005).

In the case of pitch, functional imaging reveals hierarchical processing in subcortical regions along the auditory pathway. Encoding of temporal regularities of pitch begins as early as the cochlear nucleus but is not completed until the auditory cortex (Griffiths et al. 2001). The IC is reported to be more sensitive to changes in temporal regularity than the cochlear nucleus. Further evidence of a hierarchy of pitch processing is found in the cerebral cortex (Patterson et al. 2002). When the pitch is varied to produce a melody, activation moves beyond primary auditory cortex with relatively more activity in the right hemisphere.

Electrophysiological recordings are crucial for investigating questions about the hierarchy of pitch processing not only cortically but subcortically as well (Griffiths et al. 2004). In addition, the focus on language and pitch in the brainstem reflects the view that a complete understanding of the processing of linguistically relevant dimensions of the auditory signal can be achieved only within a framework involving a series of computations that apply to representations at different stages of processing. Hickok and Poeppel (2004) argue that early processing stages (e.g., within the brainstem) may perform transformations on the acoustic data that are relevant to linguistic as well as nonlinguistic auditory perception. Scott (2003) similarly argues for hierarchical processing at the cortical level, allowing for the possibility of differences in the degree of processing of speech and nonspeech stimuli.

3.1.3 Linguistic Functions of Pitch

Tone languages exploit phonologically contrastive pitch at the word or syllable level (Gandour 1994; Yip 2002). Such languages are common in the Far East and Southeast Asia. In Mandarin, for example, four words may comprise a minimal quadruplet, minimally distinguished by variations in pitch, but otherwise identical in terms of consonant and vowel segments. Mandarin has four lexical tones (Howie 1976): *yi*^{1} "clothing" high level [T1]; *yi*^{2} "aunt" high rising [T2]; *yi*^{3} "chair" low falling-rising [T3]; *yi*^{4} "easy" high falling [T4] (see Fig. 3.1, panel B). Such languages are to be distinguished from those in which pitch variations are usually not contrastive at the syllable or word level (e.g., English). In languages that are not tonal, however, variations in pitch may be used to signal stress and intonation patterns at post-lexical levels of representation. The crucial feature that differentiates between these two types of languages is whether or not pitch variations are contrastive in the lexicon. All languages use pitch variations for intonation, but fewer possibilities are available in tone languages because of co-occurring demands for pitch variation at the lexical level. Thus, tone languages not only provide a physiologic window to evaluate how neural representations of linguistically relevant pitch attributes emerge along the early stages of sensory processing in the hierarchy, but they may also shed light on the nature of interactions between early sensory levels and later, higher levels of cognitive processing in the human brain.

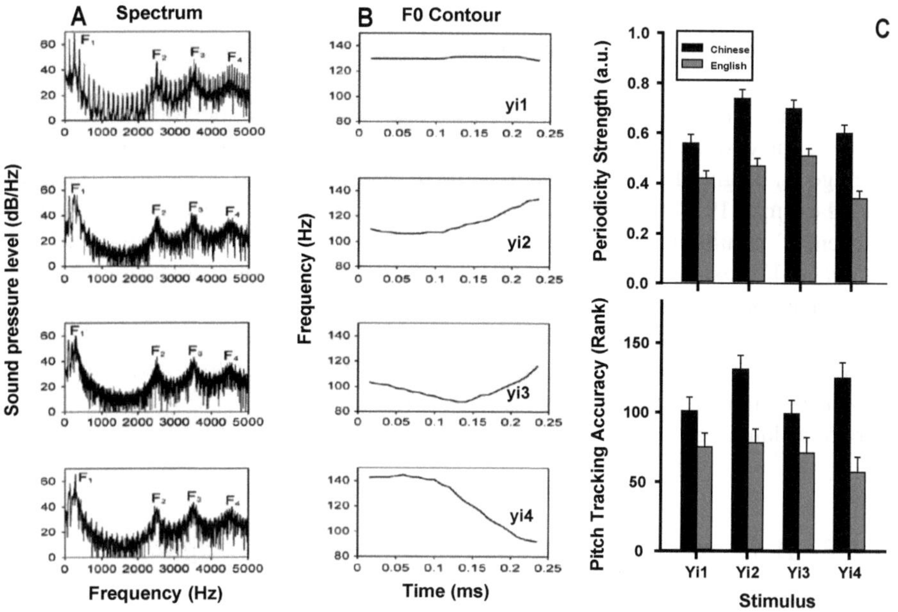

Fig. 3.1 (**A**) Stimulus spectrum for corresponding F_0 contours of the four Mandarin Chinese synthetic speech stimuli shown in **B** with invariant spectra across the four tones. (**B**) F_0 contours for Mandarin Chinese syllables identified in bottom right of each panel. (**C**) FFR periodicity strength (*top*) and FFR pitch tracking accuracy (*bottom*) of the four Mandarin lexical tones compared for Chinese and English listeners per lexical tone. Chinese listeners show greater periodicity strength and tracking accuracy across the four stimuli compared to English listeners. (Adapted from Krishnan et al. 2005, with permission of Elsevier B.V.)

3.1.4 Perceptual Dimensions of Pitch in Tone Languages

Voice F_0 is the most important acoustic correlate of tones, although there may be concomitant changes in duration, intensity, and phonation of tones as well. The primacy of F_0 as a cue to tonal identification of citation forms has been confirmed in perception tests using both natural and synthetic speech stimuli. In Mandarin, F_0 contours (shape of the temporal change in pitch over the duration of the stimulus) provide the dominant cue for tone recognition (Xu 1997). Both F_0 height and movement provide sufficient information for high intelligibility of tones in Thai (Abramson 1962), Mandarin (Howie 1976), and Cantonese (Khouw and Ciocca 2007). Rapid F_0 movements are required for high intelligibility of contour tones (Abramson 1978). Identification and discrimination tasks reveal classical patterns of categorical perception (CP) in Mandarin Chinese listeners for both speech and nonspeech stimuli varying along a linear rising or falling F_0 continuum (Xu et al. 2006a; Peng et al. 2010). CP for pitch direction depends on a listener's experience with a tone language, as shown by the lack of similar CP effects in nonnative listeners (Peng et al. 2010).

As a multidimensional perceptual attribute, pitch relies on several acoustic dimensions (e.g., height, direction). Psychophysical evidence comes primarily from cross-language multidimensional scaling (MDS) studies of dissimilarity ratings. Based on tone languages from the Far East and West Africa, including English as a control, three dimensions are reported to underlie a common perceptual space: average F_0 height, direction of F_0 movement, and magnitude of F_0 slope (Gandour and Harshman 1978; Gandour 1983). Their relative importance varies depending on a listeners' familiarity with specific types of pitch patterns that occur in their native language. For example, the perceptual saliency of the contour dimension is greater for native speakers of tone languages than for speakers of English, while English listeners give greater weight to the height dimension than do tone language speakers. Such differences in perceptual saliency suggest that long-term experience enhances listeners' attention to pitch dimensions that are phonetically relevant in a particular language.

3.1.5 A Physiologic Window into Speech and Language

The scalp-recorded human frequency-following response (FFR) provides a physiologic window into the early stages of subcortical processing of complex sounds. The FFR reflects sustained, phase-locked activity in a population of neural elements within the rostral brainstem, presumably the IC (Krishnan 2007). These responses can be recorded easily between scalp electrodes placed at high forehead and the seventh cervical vertebra (C7). The shorter latency of the FFR (around 6–9 ms) correlates well with activity from the IC region and is too early to reflect activity from cortical generators (Galbraith 2008). Furthermore, the nature of the auditory system makes it unlikely that the low-pass filtered phase-locked activity reflected in the FFR is of cortical origin (Akhoun et al. 2008). However, there is compelling evidence to suggest that this brainstem component is indeed subject to corticofugal modulation (Suga and Ma 2003; Banai et al. 2007).

The FFR is characterized by a waveform that follows the periodicities contained in both the envelope and the temporal fine structure of complex sounds (Fig. 3.2). The temporal and spectral characteristics of complex sounds that are preserved in the FFR can be extracted by frequency domain (spectral) and time domain autocorrelation analysis (Fig. 3.3), a measure of correlation between the original signal and temporally delayed versions of the response, yielding high correlations for periodicities harmonically related to the fundamental frequency (Krishnan et al. 2004, 2005). FFRs preserve spectrotemporal information relevant to the spectrum and pitch of steady-state (Krishnan and Plack 2011), dynamic speech (Krishnan et al. 2004, 2005), and nonspeech stimuli (Krishnan et al. 2009a, b). Importantly, the pitch-relevant information preserved in the FFR is strongly correlated with perceptual measures of pitch salience—a measure of the strength of the perceived pitch (Krishnan et al. 2010a; Bidelman and Krishnan 2011). These findings suggest

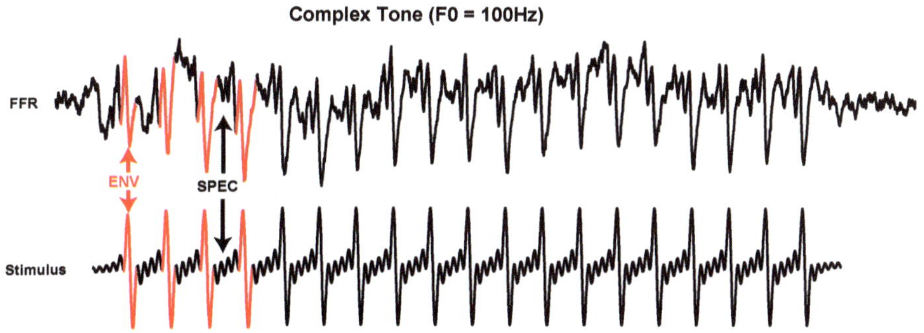

Fig. 3.2 Frequency-following responses (FFR, *top trace*) elicited by a complex sound (*bottom trace*). The response is characterized by neural phase-locking to the envelope (*ENV*, *red* for four cycles) periodicity and the temporal fine structure (*SPEC*, *double black arrow*). Stimulus onset is shifted to the right to achieve temporal match between the stimulus and the response. *ENV*, envelope; *SPEC*, spectral. (Reprinted from Krishnan and Gandour 2014, with permission from the Acoustical Society of Australia)

acoustic features relevant to pitch are preserved in the temporal pattern of phase-locked neural activity in the brainstem.

Gockel et al. (2011) observed that FFRs recorded using frequency-shifted complex tones presented monaurally did preserve pitch-relevant information, but that this information was similar to that measured in an auditory nerve model. Also, they failed to observe any pitch-relevant information in the FFRs to three-tone harmonic stimuli presented dichotically. They concluded that there was no additional pitch-relevant processing at the level of the brainstem. Several arguments may be presented to counter this inference. First, if the temporal code for pitch available at the brainstem level also utilizes autocorrelation-like analysis to determine the global distribution of interspike intervals from the temporal pattern of neural activity across a population of neurons, it would necessarily share certain fundamental attributes of the same temporal code operating at the level of the auditory nerve. Second, it is not clear that their dichotic stimuli produced the same pitch as when all harmonics are presented to the same ear. Notwithstanding, the salience of their stimuli would be quite weak. It is possible that FFR-related neural activity is not sufficiently robust to preserve the less salient pitch for their stimuli. Finally, the inferences of Gockel et al. (2011) cannot account adequately for the experience-dependent effects reflected in the FFR that are sensitive to specific attributes of dynamic pitch contours.

Within the past decade there has been increasing interest in the role of the auditory brainstem in speech processing. In terms of speech intelligibility, FFRs show increased amplitude in response to forward speech as compared to reversed speech, indicating that familiar phonetic and prosodic properties of forward speech selectively activate brainstem neurons (Galbraith et al. 2004). Using the /da/ syllable to elicit the brainstem response, Kraus and colleagues demonstrated how FFRs separately encode source and filter characteristics of the speech signal (Kraus and Nicol 2005;

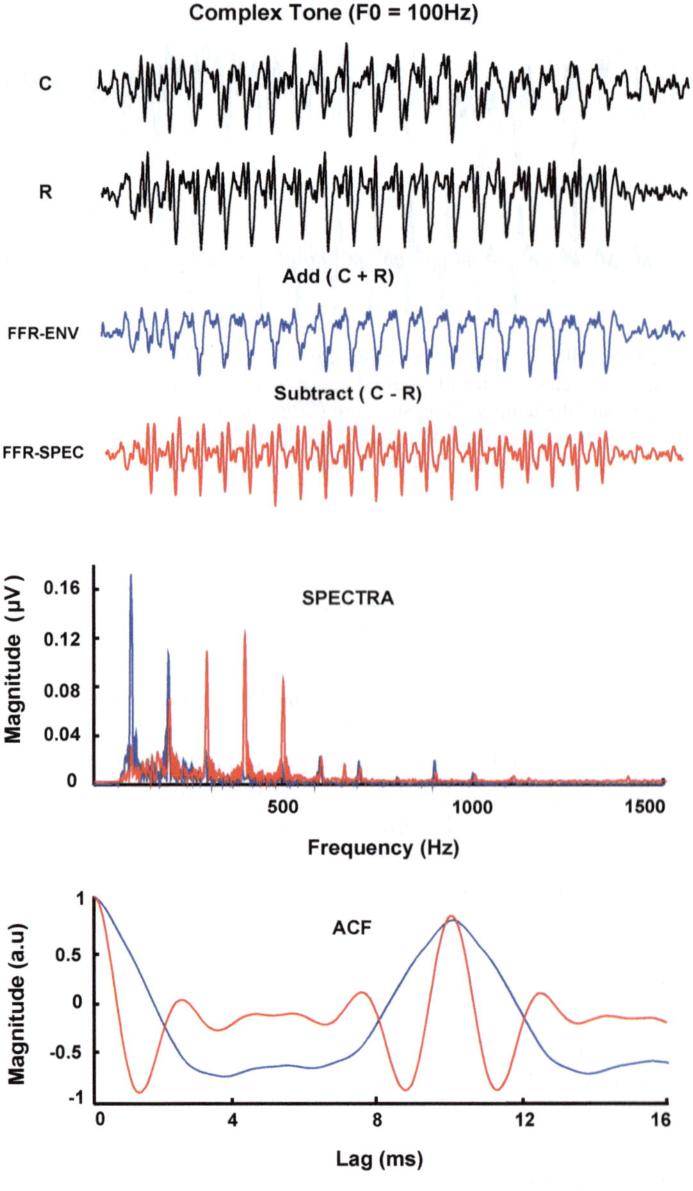

Fig. 3.3 FFR analyses: *Top two* FFR waveforms (*black*) are responses to condensation (***C***) and rarefaction (***R***) onset stimulus polarity. Addition of these waveforms results in the FFR-ENV waveform (*blue*). It is characterized by a prominent phase-locking to the envelope periodicity of the complex tone. Subtraction of the C and R responses (***C−R***) results in a FFR phase locked to the temporal fine structure (spectrum) of the complex tone (FFR-SPEC, *red*). Frequency domain analysis (SPECTRA, *middle panel*) shows that the envelope phase locking has a larger peak at F_0 (*blue*) as expected; whereas the temporal fine-structure phase locking shows multiple robust peaks at the harmonics of the complex tone. Using autocorrelation, the temporal analyses (ACF, *bottom panel*) show that there is a major peak at the fundamental periodicity of the complex tone for both responses. (Reprinted from Krishnan and Gandour 2014, with permission from the Acoustical Society of Australia)

Chandrasekaran and Kraus 2010) and how brainstem timing predicts cerebral asymmetry for speech (Abrams et al. 2006).

In terms of segmental features of speech, FFRs preserve spectral peaks corresponding to the first two formants (Fig. 3.4) of both steady-state vowels (Krishnan 1999, 2002) and time-variant consonants (Krishnan and Parkinson 2000; Plyler and Ananthanarayan 2001). Though FFRs are known to preserve pitch-relevant information about complex sounds that produce time-invariant pitch (Greenberg et al. 1987), the question arises as to how the brainstem handles suprasegmental features of speech that are characterized by time-variant pitch.

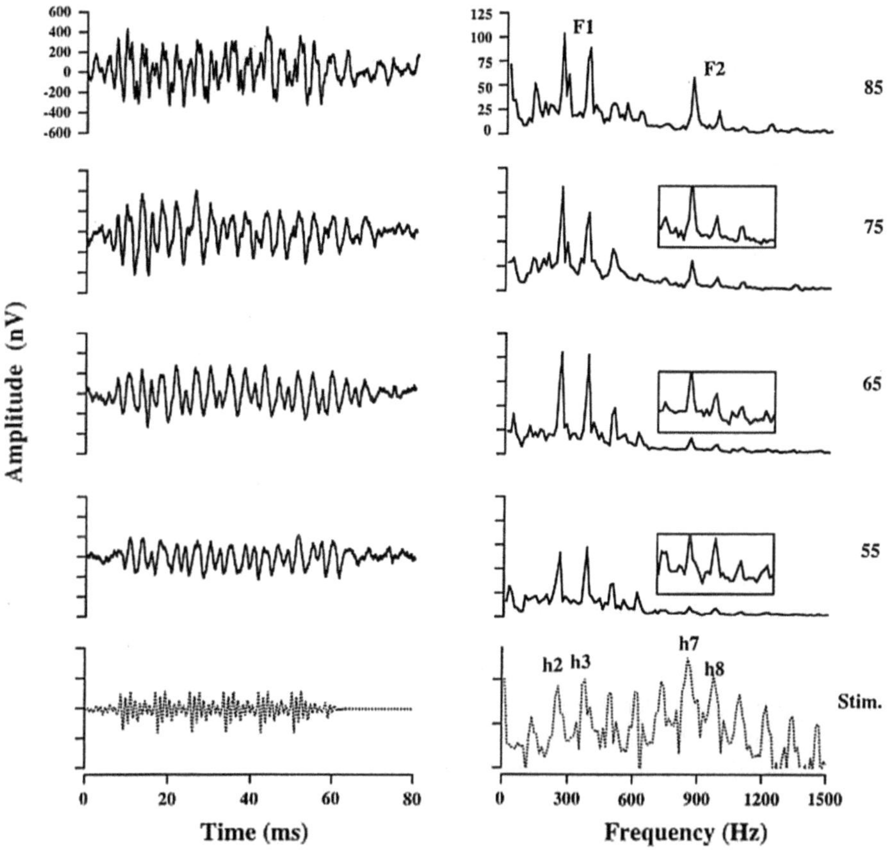

Fig. 3.4 Grand averaged FFR waveforms (*left*) and spectra (*right*) are plotted as a function of stimulus level (55–85 dB normal hearing level) for the English vowel /u/. The stimulus waveform (*left bottom*) and its spectrum, with F_1 (h2, h3) and F_2 (h7, h8) harmonics identified on the figure (*right bottom*). The amplified inset in the FFR spectral data clearly shows the F_2 harmonic peaks h7 and h8. Note the different amplitude scale for the stimulus spectrum. (Reprinted from Krishnan 2002, with permission of Elsevier B.V.)

Tone languages are especially advantageous for isolating the effects of encoding voice pitch at the level of the auditory brainstem (see Sect. 3.1.3). In Mandarin, for example, all four tones exhibit voice F_0 trajectories and harmonics that lie within the range of easily recordable FFRs (below 2 kHz). The relatively long duration of citation forms of lexical tones (200–350 ms) necessitates use of slower stimulus repetition rates. This, in turn, enables recording of robust FFRs with little or no neural adaptation.

3.1.6 Language-Dependent Processing in the Brainstem

Historically, the brainstem has not been considered to be a part of the brain worthy of interest when it comes to its contribution to speech/language processing. The conventional wisdom is that "processing operations conducted in the relay nuclei of the brainstem and thalamus are general to all sounds, and speech-specific operations probably do not begin until the signal reaches the cerebral cortex" (p. 100 in Scott and Johnsrude 2003). Though it is agreed that operations specific to speech perception are likely circumscribed to the cortex, experience-dependent modulation of pitch-relevant neural activity in the brainstem suggests that early sensory processing involves more than a simple transmission of pitch information from the ear to the cerebral cortex.

3.2 Language Experience Shapes Pitch-Relevant Information in the Brainstem

Long-term language experience enhances the neural representation of linguistically relevant pitch in the human brainstem, which is well before evoked neural activity relevant to pitch is detected in the auditory cortex. Indeed, neural representation of pitch-relevant attributes, as reflected in the FFR, may emerge as early as 6–10 ms after stimulus onset (Krishnan and Gandour 2009). In contrast, the pitch-related neural activity in the auditory cortex emerges at about 140–170 ms after stimulus onset (Krumbholz et al. 2003; Griffiths and Hall 2012).

3.2.1 Experience-Dependent Enhancement of Neural Representation of Pitch-Relevant Information

In the cerebral cortex, the neural substrates of pitch perception in the processing of lexical tones are shaped by language experience (Zatorre and Gandour 2008; Gandour and Krishnan 2014). Based on evidence from positron emission tomography (PET) and functional magnetic resonance imaging (fMRI) studies of pitch processing in Mandarin and Thai, it appears that pitch processing engages the left hemisphere only when the pitch patterns are of linguistic relevance (Wong 2002; Wong et al. 2004). These experiments all employed discrimination tasks, and thus likely reflect temporally aggregated neural events at relatively late attention-modulated stages of auditory processing. As reflected by the mismatch negativity (MMN: an early cortical event-related response associated with auditory discrimination), language experience similarly influences the early cortical automatic processing of linguistically relevant pitch contours (Chandrasekaran et al. 2007a, b); moreover, lexical tones are lateralized to the right hemisphere in contrast to the left lateralized consonants (Luo et al. 2008). The Luo et al. data suggest that hemispheric laterality effects are a result of specialized neural computations that apply to representations at different stages of auditory processing. The leftward asymmetry observed in a discrimination task likely reflects neural computations that occur downstream from preattentive auditory processing.

In animals, it is already well-established that experience-dependent neural plasticity is not limited to the cerebral cortex. Response properties and frequency maps in the IC of bats undergo change after auditory conditioning or focal electrical stimulation of the auditory cortex (Suga 1990, 1994). Auditory experience of altered interaural cues for localization in young owls leads to frequency-dependent changes in interaural time difference (ITD) tuning and frequency tuning of IC neurons (Gold and Knudsen 2000).

Other recent data also support experience-dependent neural plasticity at the level of the IC in humans. The latency of wave V in hearing-impaired listeners who use amplification is shorter than in those who do not (Philibert et al. 2005). Using the FFR, neural phase-locked activity improves after auditory training in children with learning impairments (Russo et al. 2005); pitch tracking accuracy of Mandarin tones is more accurate in nonnative musicians than nonmusicians (Wong et al. 2007); experience with sounds composed of acoustic elements relevant to speech leads to developmental changes in brainstem responses (Johnson et al. 2008); and pitch tracking accuracy improves in native English-speaking adults after undergoing short-term training on using Mandarin tones in word identification (Song et al. 2008b). Also relevant is the consequence of a disruption in the normal interaction between local processes and the corticofugal modulation of subcortical function, which contributes to plasticity. The deficits in brainstem encoding in children with a variety of language-based learning problems could very well reflect such a disruption in the ability of the corticofugal system to fine tune subcortical processes (Russo et al. 2008; Song et al. 2008a).

Preattentive stages of pitch processing in the brainstem can be influenced by language experience. A cross-language study was conducted to determine whether native speakers' long-term exposure and experience using pitch patterns in a tonal language has an influence on FFR response properties (Krishnan et al. 2005). FFRs were elicited by prototypical, curvilinear F_0 contours modeled after the four Mandarin lexical tones in a speech context (Fig. 3.1, panels A, B). They were presented to native speakers of Mandarin and nontone language speakers of

English. If driven by acoustic properties regardless of language experience, FFRs would be expected to be homogeneous across listeners. Results showed that both pitch strength (Fig. 3.1, panel C *top*) and pitch tracking accuracy (Fig. 3.1, panel C *bottom*), as measured by rank-transformed cross-correlation between stimulus and response pitch tracks, were greater for the Chinese group than for the English across all four Mandarin tones (Fig. 3.1, panel C *bottom*). Based on these findings, it appears that experience-driven adaptive neural mechanisms are involved subcortically that sharpen response properties of neurons tuned for processing linguistic pitch contours of a particular language. That is, language-dependent plasticity enhances or primes temporal intervals that carry linguistically relevant features of pitch contours.

From the perspective of auditory neuroethology, this adjustment in processing pitch contours of Mandarin tones is comparable to neural mechanisms that are developed for processing behaviorally relevant sounds in other nonprimate and nonhuman primate animals (Suga et al. 2003). Auditory processing is not limited to a simple representation of acoustic features of speech stimuli. Indeed, language-dependent operations may begin before the signal reaches the cerebral cortex.

While this chapter focuses primarily on the experience-dependent shaping of pitch-relevant information as it relates to tonal languages, it is important to view the information presented here in the broader context of how language experience shapes subcortical processing in general. For example, compelling evidence suggests that bilingual experience enhances subcortical representation of pitch-relevant information presented in speech sounds (Krizman et al. 2012) and/or neural response consistency (Krizman et al. 2014). Furthermore, these authors show a strong correlation between these improvements in subcortical representations and both attentional control and language proficiency, suggesting that these outcomes are related to experience-dependent strengthening of attentional control.

3.2.2 *Feature Specificity of Experience-Dependent Effects in the Brainstem*

If brainstem pitch processing is shaped by long-term language experience, what specific F_0 properties or features of the pitch stimuli, static or dynamic, are relevant? To what extent can a stimulus deviate from natural speech exemplars before exceeding the upper or the lower limit of linguistic sensitivity of brainstem neurons? For instance, linear ramps do not occur in natural speech because of physiological constraints of the speech production apparatus. Linear F_0 ramps (90–140 Hz, rising; 140–90 Hz, falling) were generated, similar to Mandarin T2 (rising) and T4 (falling) in direction but dissimilar in trajectories observed in natural speech (Fig. 3.5, panels A and B). By examining FFRs elicited by linear approximations of Mandarin T2 and T4 (Xu et al. 2006b), it was possible to assess the tolerance limits for priming

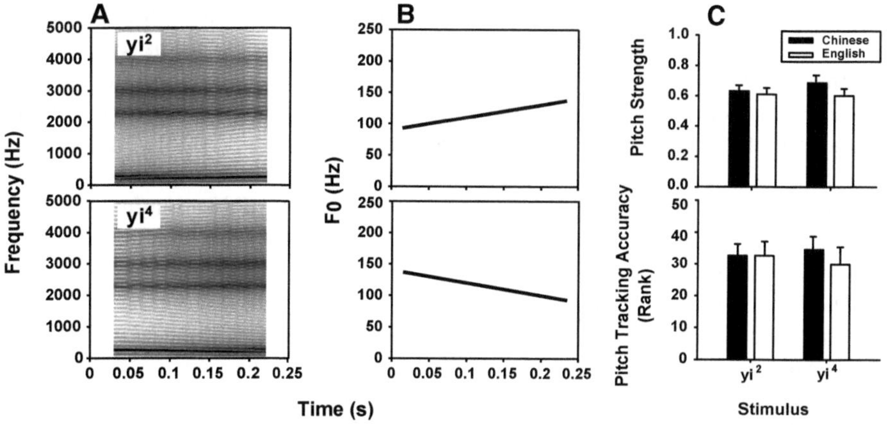

Fig. 3.5 (**A**) Stimulus spectrograms, (**B**) F_0 contours of Mandarin Chinese synthetic speech stimuli (*yi*^{2} "aunt", rising linear ramp; *yi*^{4} "easy", falling linear ramp), FFR pitch strength (**C** *top*), and FFR pitch tracking accuracy (**C** *bottom*). FFR pitch tracking accuracy revealed no significant main effects for either language group (*Chinese, English*) or pitch direction (*rising, falling*). (Adapted from Xu et al. 2006a, b, with permission of Lippincott Williams & Wilkins)

linguistically relevant features of the auditory signal involved in pitch extraction at the level of the brainstem. Results show no differences in FFR pitch strength or pitch tracking accuracy (Fig. 3.5, panel C top and bottom, respectively) between language groups (Chinese, English) or pitch direction (rising, falling).

It appears that no language-dependent effects are observed in response to linear rising or falling F_0 ramps because they are not part of native Chinese listeners' experience. Even though the F_0 ramps are dynamic, linear approximations of T2 and T4, they are constant in acceleration and deceleration, respectively. The fact that Mandarin and English FFRs are homogeneous in response to linear trajectories suggests that representations of pitch-relevant information in the brainstem are acutely sensitive to dynamic, curvilinear changes in trajectory throughout the duration of a pitch contour. In the auditory brainstem, neural mechanisms respond to specific dimensions of pitch contours to which native speakers have been exposed. Language dependent neuroplasticity occurs only when salient dimensions of pitch relevant to speech perception are present in the auditory signal.

Further support for feature specificity comes from FFRs recorded from Chinese and English participants in response to iterated rippled noise (IRN) homologs of pitch contours. The IRN stimuli preserve the perception of pitch, but do not have waveform periodicity or highly modulated stimulus envelopes that are characteristic of speech stimuli. An IRN stimulus is generated using a broadband noise, which is delayed and added to itself repeatedly and, therefore, does not have a prominent modulated envelope (Patterson et al. 1996; Yost 1996). The perceived pitch corresponds to the reciprocal of the delay, and the pitch salience increases with the number of iterations of the delay-and-add process. Increases in temporal regularity of steady-state IRN stimuli lead to better temporally locked neural activity in auditory structures from the cochlear nucleus to cortex (Griffiths et al. 1998; Shofner 1999). Importantly, a novel generalization of the IRN algorithm makes it possible to generate time-variant, dynamic curvilinear pitch contours representative of those that occur in natural speech (Swaminathan et al. 2008). The stimulus waveform (Fig. 3.6A), spectrogram (Fig. 3.6B), pitch contour (Fig. 3.6C), and the autocorrelation function (Fig. 3.6D) are illustrated for an IRN stimulus with a T2 pitch contour. Note the increase in temporal regularity with increase in the number of iteration steps.

The IRN homologs of a prototypical T2 were presented in contrast to three F_0 variants (two linear, one curvilinear) that do not occur in the Mandarin tonal space (Krishnan et al. 2009a). Of the two linear variants, one represented a linear ascending ramp and the other represented a tri-linear approximation of T2, preserving the major points of inflection in addition to onset and offset. The curvilinear variant was an inverted version of T2. No group differences in pitch strength were observed for any of these variants. The absence of language group effects in response to curvilinear and linear variants of T2 emphasizes that language-dependent neuroplasticity at the level of the brainstem extends only to those pitch patterns that actually occur in the Mandarin tonal space.

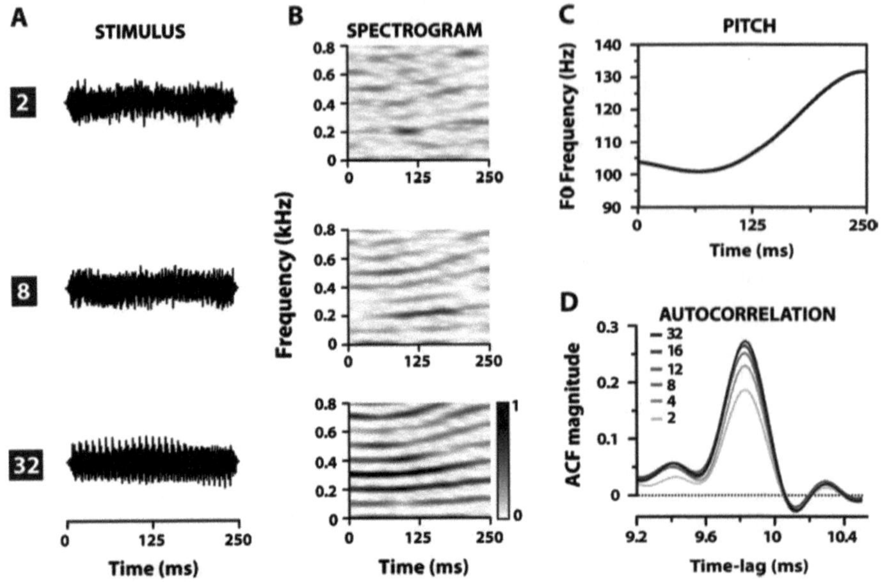

Fig. 3.6 (**A**) Iterated ripple noise (IRN) stimulus waveforms, (**B**) spectrograms, (**C**) F_0 contour and autocorrelation functions (**D**) plotted as a function of iteration steps (*2, 8, 32*). Note the increase in temporal regularity, clearer spectral bands, and increase in the peak magnitude of the autocorrelation function with increase in iteration steps. (Adapted from Krishnan et al. 2010a, with permission from Elsevier, B.V.)

3.2.3 Domain Specificity of the Experience-Dependent Effects in the Brainstem

3.2.3.1 Speech Versus Nonspeech

To address the question of domain specificity of experience-dependent effects on pitch processing in the brainstem, FFRs were recorded from native speakers of Mandarin and English speakers using IRN (nonspeech) homologs of the four Mandarin tones (Krishnan et al. 2009b). The Chinese group exhibited smoother pitch tracking than the English group in three (T2–4) out of the four tones (Fig. 3.7, left column). FFR pitch strength of 40-ms segments revealed that the Chinese group exhibited more robust pitch representation of those segments containing rapidly changing pitch movements across all four tones (Fig. 3.7, right column). These findings suggest that neural mechanisms underlying pitch representation are shaped by particular dimensions of the auditory stream rather than speech per se.

A discriminant analysis was used to determine the extent to which individual subjects can be classified into their respective language groups based on a weighted linear combination of their pitch strength of three 40-ms temporal intervals that were maximally differentiated in terms of slope (flat, rising, falling). About 83% of the subjects were correctly classified into their respective language groups. The average discriminant z scores of the Chinese group were larger than those of the English. Univariate tests of pitch strength confirmed that more dynamic changes in pitch (rising, falling) had a greater influence on the FFR responses of the Chinese group compared to the English, whereas less dynamic changes in pitch (flat) did not yield a language group effect. Pitch strength of the rising F_0 trajectory was the most important variable in discriminating listeners by language affiliation. Both psychoacoustic (Collins and Cullen 1978; Schouten 1985) and physiologic studies (Shore et al. 1987; Krishnan and Parkinson 2000) indicate better sensitivity for rising versus falling tones. Multidimensional scaling analyses showed that the perceptual dimension related to direction of pitch change is spatially distributed primarily in terms of rising versus non-rising F_0 movements (Gandour and Harshman 1978; Gandour 1983). This response asymmetry in FFRs presumably reflects greater neural synchrony (Shore and Nuttall 1985) and more coherent temporal response patterns to rising than to falling tones (Shore et al. 1987).

Such experimental findings support the view that at early stages of brain processing, particular features or dimensions of pitch patterns—regardless of the stimulus context in which they are embedded—shape neural mechanisms underlying speech perception. The role of the brainstem may be to facilitate cortical level processing of pitch-relevant information by optimally capturing those dimensions of the auditory signal that are of linguistic relevance. By focusing on tonal sections instead of the whole tone, it is possible to assess whether language-dependent effects are better conceptualized as applying to sections that exhibit certain acoustic features irrespective of tonal category.

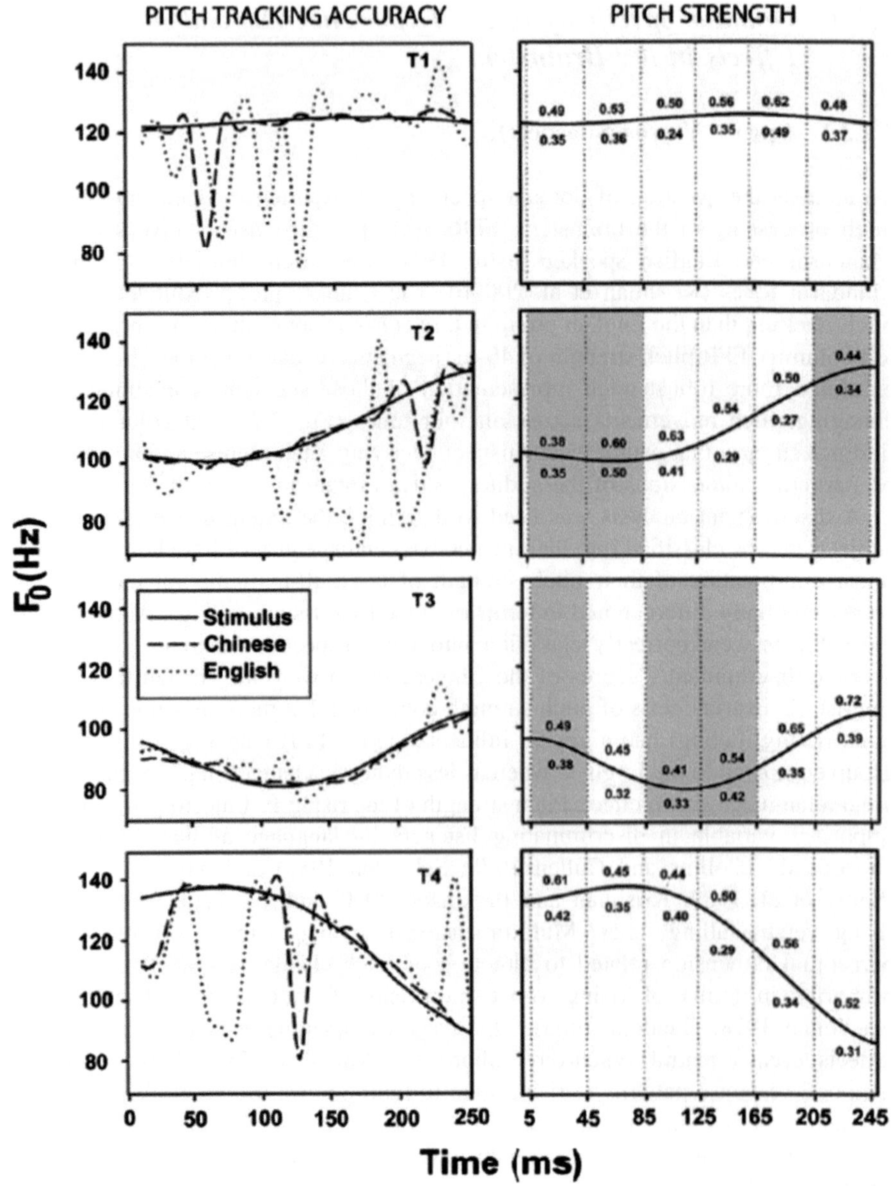

The question arises whether neural representation of pitch-relevant information in the brainstem is equally robust for speech and nonspeech stimuli. FFRs were recorded from Chinese and English listeners in response to four Mandarin tones presented in speech and nonspeech contexts (Swaminathan et al. 2008). Dynamic IRN stimuli preserve fine-grained measures of pitch representation at the level of the brainstem. However, given the relatively less robust temporal

◀ **Fig. 3.7** Pitch tracking accuracy of Iterated ripple noise (IRN) homologs of Mandarin tones (*left*) and pitch strength of tonal sections (*right*) derived from the grand averaged FFR waveforms of Chinese and English subjects. The four Mandarin tonal categories are represented as *T1*, *T2*, *T3*, and *T4*. Left panels show that the FFR-derived F_0 contours of the Chinese group (*dashed line*) more closely approximate those of the original IRN stimuli (*solid line*) when compared to the English group (*dotted line*). Right panels show that the pitch strength of the Chinese group (*value above the solid line*) is greater than that of the English group (*value below the solid line*). *Vertical dotted lines* demarcate six 40-ms sections within each F_0 contour: 5–45, 45–85, 85–125, 125–165, 165–205, and 205–245. Sections that yielded significantly larger pitch strength for the Chinese group relative to English are *unshaded*; those that did not are shaded in *gray*. (Reprinted from Krishnan et al. 2009b, with permission of MIT Press)

periodicity in the IRN waveform, pitch strength was observed to be greater for speech than nonspeech stimuli across language groups. Regardless of context, pitch strength of the Chinese group was greater than that of the English. Interestingly, group differences in pitch strength were not uniform throughout the duration of FFR responses. The Chinese group exhibited relatively more robust pitch representation of rapidly changing pitch segments. These findings support the view that at the level of the brainstem, neural mechanisms underlying pitch representation are shaped by particular features of the auditory stream rather than speech per se.

3.2.3.2 Speech Versus Music

Neural encoding of pitch in the auditory brainstem is shaped by long-term experience with language or music, implying that early sensory processing is subject to experience-dependent neural plasticity. The comparisons between the language and music domains reveal overall enhancement in brainstem FFRs elicited by either musical or linguistic pitch patterns in musicians and tone language speakers alike (Bidelman et al. 2011b, c). Thus, long-term pitch experience seems to improve the brain's ability to represent pitch-relevant information regardless of the domain of expertise. However, subtle differences in these sensory representations suggest a domain-specific sensitivity to acoustic features that are part of the experience in each domain. Musicians, for example, show enhanced responses when pitch patterns intersect discrete notes along the musical scale; tone language speakers, on the other hand, during rapidly changing portions of tonal contours (Bidelman et al. 2011a, c). Such cue weighting is consistent with the relative importance of these perceptual dimensions in their respective domains. These findings collectively suggest that both language and musical experience provide some mutual benefit to the neural representation of pitch-relevant information, but also they suggest that specific features of the acoustic signal are highlighted in subcortical responses depending on their perceptual salience and function within a listener's domain of expertise.

3.2.4 Experience-Dependent Effects are More Resilient to Signal Degradation

Most human communication occurs against a background of noise. It is important that the auditory system has a mechanism(s) in place to encode behaviorally relevant acoustic features of pitch that may be degraded in the presence of noise. By using IRN homologs of pitch contours associated with lexical tones, it is possible to systematically vary their degree of temporal regularity and, as a consequence, their pitch salience. The question then arises whether pitch representation in the brainstem is less vulnerable to systematic degradation in the temporal regularity of an IRN stimulus that represents a native pitch contour (Krishnan et al. 2010b). In response to IRN homologs of Mandarin T2 varying in pitch salience along a six-step continuum ranging from low to high, FFR pitch tracking accuracy is higher in the Chinese group relative to the English except for the three lowest steps along the continuum (Fig. 3.8, left panel). Also, FFR pitch strength is greater in the Chinese group, even in severely degraded stimuli, for sections of the response that exhibit rapid changes in pitch. Exponential time constants reveal that pitch strength growth emerges 2–3 times faster in Chinese than in English listeners as a function

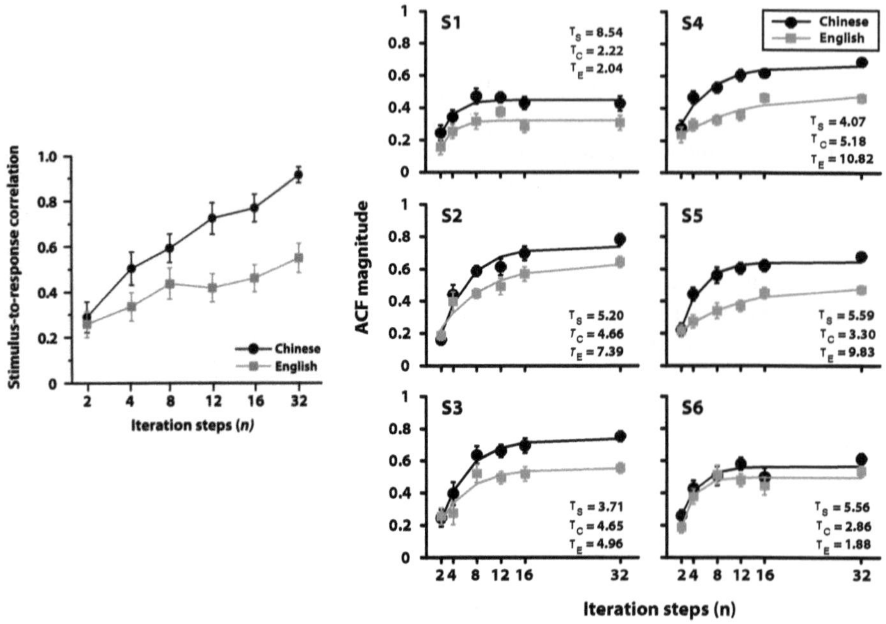

Fig. 3.8 Comparisons of FFR pitch tracking accuracy (*left panel*), and segmental periodicity strength (*right panel*) in response to iterated ripple noise (IRN) homologues of Mandarin Tone 2 (*T2*) as a function of iteration steps. Both pitch-tracking accuracy and periodicity strength (especially the rapidly changing segments 4 and 5 as reflected by the time constants) increase more rapidly with increasing pitch salience for the Chinese listeners. (Adapted from Krishnan et al. 2010b, with permission from Elsevier, B.V.)

of increasing temporal regularity of the stimulus (Fig. 3.8, right panels). These findings altogether suggest that experience-dependent brainstem mechanisms for pitch are especially sensitive to those dimensions of tonal contours that provide cues of high perceptual saliency in degraded as well as normal listening conditions.

Another way to degrade pitch-relevant information in the stimulus is to systematically increase the rate of pitch change in the dynamic portions of the stimulus. The question then arises whether language-related expertise in pitch encoding of linguistically relevant stimuli can transfer to pitch encoding of stimuli that are characterized by acceleration rates that do not occur in natural speech. Four click-train homologs of Mandarin T2 (Fig. 3.9, left panels) with maximum rates of

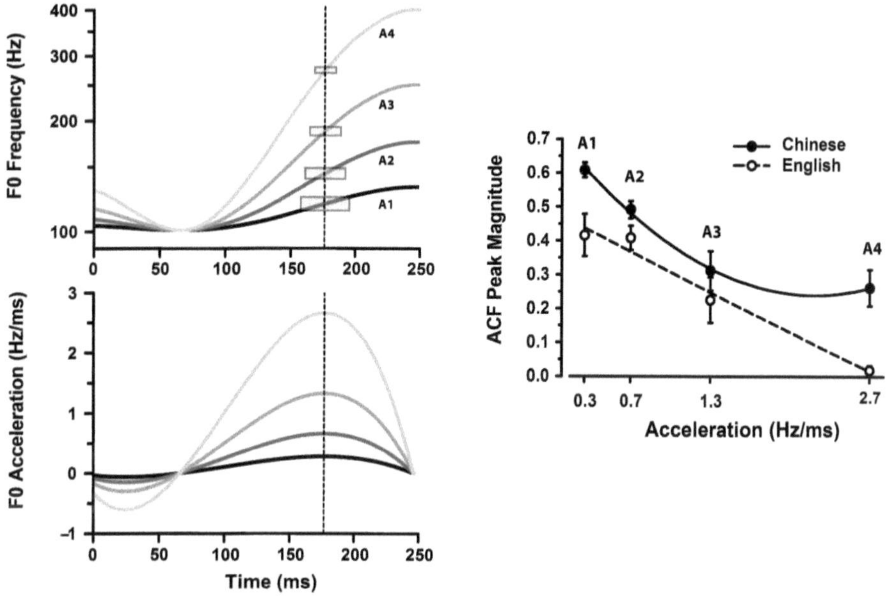

Fig. 3.9 Click train stimuli that are differentiated by varying degrees of rising acceleration. The F_0 contours of all four stimuli are displayed on a logarithmic scale spanning two octaves: 100.8 Hz, the minimum stimulus frequency, to 400 Hz (*left panel, top*). These stimuli represent a continuum of rates of acceleration from Mandarin Tone 2 (*A1*) rate in natural speech to an F_0 rate that falls well beyond the normal voice range (*A4*). The *vertical dotted line* at 177 ms defines the center of the analysis window for each stimulus (*left panel, bottom*). The location of this line is calculated from the maximum pitch acceleration. Group comparisons of pitch strength are shown within the region of maximum acceleration of the stimulus derived from FFR responses to click-train stimuli as a function of pitch acceleration (*right panel*). FFR pitch strength of the Chinese group—as measured by the magnitude of the normalized autocorrelation peak—is greater than that of the English in response to Mandarin Tone 2 (*A1*, 0.3 Hz/ms) as well as to a pitch pattern that does not occur in natural speech (*A4*, 2.7 Hz/ms). In the English group, pitch strength shows a steady, steep decline across the continuum, approaching zero at its opposite end (i.e., the absence of a phase-locked response). In the Chinese group, on the other hand, pitch strength exhibits a more gradual decline, but never approaches zero. Instead, pitch strength begins to level off once a pitch pattern moves clearly beyond the normal voice range (*A3*, 1.3 Hz/ms). (Adapted from Krishnan et al. 2010c, with permission of Elsevier B.V.)

pitch acceleration ranging from low (0.3 Hz/ms; Mandarin Tone 2) to high (2.7 Hz/ms; 2 octaves) were presented to Chinese and English listeners (Krishnan et al. 2010c). Regardless of language group, neural periodicity strength is greater in response to acceleration rates within or proximal to natural speech relative to those beyond its range (Fig. 3.9, right panel). Though both groups show decreasing pitch strength with increasing acceleration rates, pitch representations of the Chinese group are more resistant to degradation. These findings indicate that perceptually salient pitch cues associated with lexical tone influence brainstem pitch extraction not only in the speech domain but also in auditory signals that clearly fall outside the range of dynamic pitch to which a native listener is exposed.

While the focus of this review is on the influence of long-term language experience on the neural representation of pitch-relevant information in the auditory brainstem, there is growing empirical evidence suggesting that it also shapes pitch mechanisms at early sensory levels of processing in the auditory cortex, and that the hemispheric preference for processing pitch information may vary depending on the relative linguistic importance of specific temporal attributes of dynamic pitch (Krishnan et al. 2014a, b). As in the brainstem, cortical responses sharpen properties of neural elements to enable optimal representation of temporal attributes of native pitch contours.

3.3 Hierarchical Processing Underlies Experience-Dependent Pitch Processing

Human functional magnetic resonance imaging and lesion studies have suggested that an area in the vicinity of lateral Heschl's gyrus (HG) is specialized for pitch representation (Griffiths and Hall 2012). This region also appears to be important for computations relevant to the extraction of pitch in complex sounds (Zatorre and Belin 2001; Schönwiesner et al. 2005). However, growing evidence shows pitch-related neural activity in both primary auditory cortex as well as in the adjacent more lateral non-primary areas of HG. These studies suggest that pitch-relevant information is available in multiple areas of the auditory cortex: functional magnetic resonance imaging (Griffiths et al. 2010; Puschmann et al. 2010), direct cortical recordings (Patterson et al. 2002; Penagos et al. 2004), and magnetoencephalography (Krumbholz et al. 2003; Gutschalk et al. 2004).

How do these areas interact in a coordinated manner during pitch processing? To address this question, a predictive coding model of perception has been applied to depth-electrode recordings of pitch-relevant neural activity along HG (Rao and Ballard 1999; Kumar and Schönwiesner 2012). Essentially, the model of predictive coding of pitch posits that cortical areas contributing to pitch are organized hierarchically. The effective connection strengths between and within levels are continually adjusted in a recursive manner to optimize pitch representation at the higher level. Operationally, higher-level areas in the hierarchy contributing to pitch (lateral HG) use stored information of pitch to make a pitch prediction. This prediction is passed to the lower areas in the processing hierarchy (medial and middle HG) via top down connection(s) (Fig. 3.10, blue arrows, *#1*). The lower areas of the hierarchy then compute a prediction error (difference between the higher level prediction and the lower level representation), which is passed to the higher level via bottom-up connections (Fig. 3.10, red arrows, *#2*). The strength of these connections is continually adjusted in a recursive manner in order to minimize predictive error and to optimize representation at the higher level. The lateral connections (same level in the hierarchy) between middle and medial HG (Fig. 3.10, green bidirectional arrows, *#3*) are also subject to modulation and presumably play a role in reducing redundancy and making representations more efficient.

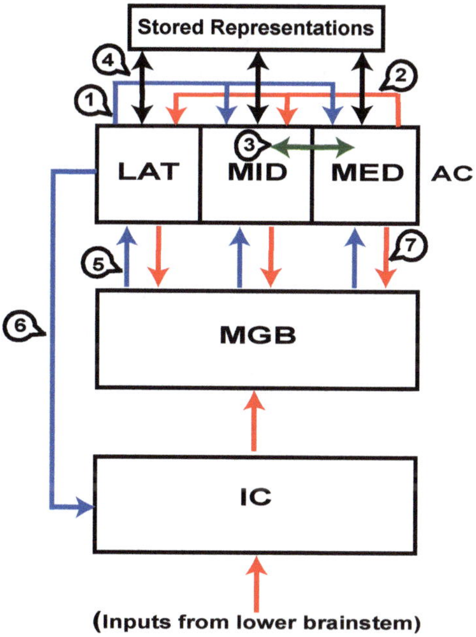

Fig. 3.10 Block diagram of the proposed predictive coding model for hierarchical processing of pitch at both subcortical and cortical levels. Operationally, higher-level areas in the hierarchy contributing to pitch (lateral HG) use stored representations of pitch to make a pitch prediction. This prediction is passed to the lower areas in the processing hierarchy (medial and middle HG) via top-down connection(s) (*blue arrows, #1*). The lower areas then compute a prediction error (difference between the higher level prediction and the lower level representation), which is passed to the higher level via bottom-up connections (*red arrows, #2*). The lateral connections (same level in the hierarchy) between middle and medial HG (*green bidirectional arrows, #3*) are also subject to modulation and presumably play a role in reducing redundancy and making representations more efficient. Inputs from subcortical (bottom-up) structures (*#5*) that are themselves subject to experience-dependent plasticity are presumably mediated by top-down connections (*#6, #7*). It is likely that these top-down connections in the hierarchy provide feedback to adjust the effective time scales of processing at each stage to optimally control the temporal dynamics of pitch processing. Language dependent changes at the early sensory level of processing in the auditory cortex may reflect interplay between sensory and cognitive processing (*black bidirectional arrows, #4*)

Dynamic causal modeling essentially tries to determine how the activity of one brain area changes the dynamics and/or responses of other areas. Using Bayesian model comparisons to determine the configuration(s) that best explain the data, Kumar and Schönwiesner (2012) showed that the lateral part of HG is at a higher level in the hierarchy compared to middle and medial HG, with the latter two at the same level. This is in agreement with evidence from depth-electrode recordings along HG. It confirms that middle and medial electrode contacts are indeed in the primary auditory cortex, whereas the lateral contacts are in non-primary auditory cortex (Brugge et al. 2009). Consistent with the predictions of the model, they also show that strength of connectivity varies with pitch salience such that the strength of the top-down connection from lateral HG to medial and middle HG increases with pitch salience, whereas the strength of the bottom-up connection from middle HG to lateral HG decreases. This distributed view of pitch processing, however, is not necessarily at odds with a single specialized pitch center. It is likely that lateral HG has more pitch-specific mechanisms and, therefore, plays a relatively greater role in pitch perception.

In this case, the predictive coding of the pitch model provides a framework to explain the language-dependent (cognitive) and language-universal (sensory) effects on pitch-related neural activity in the brainstem and auditory cortex. Changes at different stages of processing attributable wholly to acoustic properties of the stimulus implicate the recursive process (initial pitch prediction, error generation, error correction) in the representation of pitch. At this fundamental level of pitch processing, the hierarchical flow of processing and the connectivity strengths along the HG are essentially the same regardless of one's language background. For changes that are dependent on language experience, the initial pitch prediction at the level of the lateral HG is more precise for Chinese because of their access to stored information about lexical tones with a smaller error term. Consequently, the top-down connections from lateral HG to medial and middle HG and to the brainstem are stronger than the bottom-up connections from the medial and middle HG to the lateral HG and from the brainstem. The opposite would be true for English because of their less precise initial prediction. In addition, the recursive process itself would be expected to take a longer time for English relative to Chinese in determining pitch. Language experience, therefore, alters the nature of the interaction between functional components of the distributed network by modulating connection strengths.

It is clear that pitch processing in the auditory cortex is influenced by inputs from subcortical structures (Fig. 3.10, *#5*) that are themselves subject to experience-dependent plasticity, presumably mediated by top-down connections (Fig. 3.10, *#6*, *#7*). It is likely that these top-down connections in the hierarchy provide feedback to adjust the effective time scales of processing at each stage to optimally control the temporal dynamics of pitch processing (Balaguer-Ballester et al. 2009). Language-dependent changes at the sensory level of processing in the auditory cortex may reflect interplay between sensory and cognitive processing

(Fig. 3.10, black bidirectional arrows, *#4*). This model represents a unified, physiologically plausible, theoretical framework that includes both cortical and subcortical components in the hierarchical processing of pitch.

3.4 Summary and Directions for Future Research

The evidence presented herein suggests that neural activity relevant to pitch in the brainstem is crucially dependent on specific dimensions or features of pitch contours. By focusing on specific properties of the auditory signal, irrespective of a speech or nonspeech context, it is argued that the neural representation of acoustic-phonetic features relevant to speech perception is already emerging in the brainstem (≈8–10 ms after stimulus onset) and, importantly, can be shaped by experience. These sensory level auditory processes are tuned differentially to those features depending upon their linguistic relevance. Such effects of language experience on lower level sensory processing are compatible with a more integrated approach to language and the brain (Hickok and Poeppel 2004; Zatorre and Gandour 2008). The focus on pitch processing in tonal languages notwithstanding, these findings should be contextualized within the broader framework of language experience shaping subcortical processing.

While language experience shapes pitch processing at both subcortical and cortical levels, neural representations are transformed and fundamentally different at each biological level of abstraction. The representation of pitch-relevant information in the brainstem is more fine-grained spectrotemporally as it reflects sustained neural phase locking to pitch-relevant periodicities contained in the dynamic stimulus. In contrast, the cortical representation is coarser. That is, the cortical pitch-relevant neural activity reflects primarily a series of distinct transient temporal neural events marking only certain temporal attributes of the pitch contour. These differences notwithstanding, long-term language experience appears to shape adaptive, hierarchical pitch processing. Top-down connections provide selective gating of inputs to both cortical and subcortical structures to enhance neural representation of behaviorally relevant attributes of the stimulus and instantiate local mechanisms that exhibit enhanced representation of behaviorally relevant pitch attributes. The ability to record brainstem and cortical pitch-relevant responses concurrently may also provide a new window to evaluate the online interplay between feedforward and feedback components in the processing of pitch-relevant information at the level of the brainstem and the auditory cortex. While it is not known how language experience shapes subcortical and cortical stages of pitch processing, it is likely that the neural processes underlying such experience-dependent plasticity at each stage along the processing hierarchy are modulated by a coordinated interplay between ascending, descending, and local neural pathways that involve both sensory and cognitive components. The challenge is to develop experiments that systematically manipulate pitch attributes in order to optimally evaluate the relationship between representation of pitch-relevant information at the brainstem and cortical levels.

The results of these experiments are essential to further our understanding of the nature of interplay between cortical and subcortical functional components and the interactions between sensory and cognitive processes influencing pitch representation in what appears to be an integrated distributed processing network. Complementary studies using magnetoencephalography will be crucial to determine the anatomical sources of these components in an effort to shed more light on specific cortical generators contributing to pitch processing and how experience may shape these processes.

Acknowledgments The research was supported by NIH 5R01DC008549-06 (A.K.).

Compliance with Ethics Requirements Ananthanarayan Krishnan and Jackson T. Gandour declared that they had no conflicts of interest.

References

Abrams, D. A., Nicol, T., Zecker, S. G., & Kraus, N. (2006). Auditory brainstem timing predicts cerebral asymmetry for speech. *The Journal of Neuroscience, 26*(43), 11131–11137. Doi:10.1523/JNEUROSCI.2744-06.2006. (26/43/11131 [pii]).

Abramson, A. S. (1962). *The vowels and tones of standard Thai: Acoustical measurements and experiments*. Research Center in Anthropology, Folklore, and Linguistics, Pub. 20. Bloomington, IN: Indiana University.

Abramson, A. S. (1978). Static and dynamic acoustic cues in distinctive tones. *Language and Speech, 21*, 319–325.

Akhoun, I., Gallégo, S., Moulin, A., Ménard, M., et al. (2008). The temporal relationship between speech auditory brainstem responses and the acoustic pattern of the phoneme /ba/ in normal-hearing adults. *Clinical Neurophysiology, 119*, 922–933.

Balaguer-Ballester, E., Clark, N. R., Coath, M., Krumbholz, K., & Denham, S. L. (2009). Understanding pitch perception as a hierarchical process with top-down modulation. *PLoS Computational Biology, 5*(3), e1000301. Doi:10.1371/journal.pcbi.1000301

Banai, K., Abrams, D., & Kraus, N. (2007). Sensory-based learning disability: Insights from brainstem processing of speech sounds. *International Journal of Audiology, 46*(9), 524–532. Doi:10.1080/14992020701383035. (781872134 [pii]).

Besson, M., Chobert, J., & Marie, C. (2011). Language and music in the musician brain. *Language and Linguistics Compass, 5*(9), 617–634. Doi:10.1111/j.1749-818x.2011.00302

Bidelman, G. M., Gandour, J. T., & Krishnan, A. (2011a). Musicians demonstrate experience-dependent brainstem enhancement of musical scale features within continuously gliding pitch. *Neuroscience Letters, 503*(3), 203–207. Doi:10.1016/j.neulet.2011.08.036

Bidelman, G. M., Gandour, J. T., & Krishnan, A. (2011b). Musicians and tone-language speakers share enhanced brainstem encoding but not perceptual benefits for musical pitch. *Brain and Cognition, 77*(1), 1–10. Doi:10.1016/j.bandc.2011.07.006

Bidelman, G. M., Gandour, J. T., & Krishnan, A. (2011c). Cross-domain effects of music and language experience on the representation of pitch in the human auditory brainstem. *Journal of Cognitive Neuroscience, 23*(2), 425–434. Doi:10.1162/jocn.2009.21362

Bidelman, G. M., & Krishnan, A. (2011). Brainstem correlates of behavioral and compositional preferences of musical harmony. *NeuroReport, 22*(5), 212–216. Doi:10.1097/WNR.0b013e328344a689

Brugge, J. F., Nourski, K. V., Oya, H., Reale, R. A., et al. (2009). Coding of repetitive transients by auditory cortex on Heschl's gyrus. *Journal of Neurophysiology, 102*(4), 2358–2374. Doi:10.1152/jn.91346.2008

Cariani, P. A., & Delgutte, B. (1996a). Neural correlates of the pitch of complex tones. II. Pitch shift, pitch ambiguity, phase invariance, pitch circularity, rate pitch, and the dominance region for pitch. *Journal of Neurophysiology, 76*(3), 1717–1734.

Cariani, P. A., & Delgutte, B. (1996b). Neural correlates of the pitch of complex tones. I. Pitch and pitch salience. *Journal of Neurophysiology, 76*(3), 1698–1716.

Cedolin, L., & Delgutte, B. (2005). Pitch of complex tones: Rate-place and interspike interval representations in the auditory nerve. *Journal of Neurophysiology, 94*(1), 347–362. Doi:10.1152/jn.01114.2004. (01114.2004 [pii]).

Chandrasekaran, B., Gandour, J. T., & Krishnan, A. (2007a). Neuroplasticity in the processing of pitch dimensions: A multidimensional scaling analysis of the mismatch negativity. *Restorative Neurology and Neuroscience, 25*(3–4), 195–210.

Chandrasekaran, B., & Kraus, N. (2010). The scalp-recorded brainstem response to speech: Neural origins and plasticity. *Psychophysiology, 47*(2), 236–246. Doi:10.1111/j.1469-8986.2009.00928.x. (PSYP928 [pii]).

Chandrasekaran, B., Krishnan, A., & Gandour, J. T. (2007b). Mismatch negativity to pitch contours is influenced by language experience. *Brain Research, 1128*(1), 148–156. Doi:10.1016/j.brainres.2006.10.064. (S0006-8993(06)03179-9 [pii]).

Cohen, M. A., Grossberg, S., & Wyse, L. L. (1995). A spectral network model of pitch perception. *The Journal of the Acoustical Society of America, 98*(2), 862–879.

Collins, M. J., & Cullen, J. K., Jr. (1978). Temporal integration of tone glides. *The Journal of the Acoustical Society of America, 63*(2), 469–473.

Galbraith, G. C. (2008). Deficient brainstem encoding in autism. *Clinical Neurophysiology, 119*(8), 1697–1700. Doi:10.1016/j.clinph.2008.04.012

Galbraith, G. C., Amaya, E. M., de Rivera, J. M., Donan, N. M., et al. (2004). Brain stem evoked response to forward and reversed speech in humans. *NeuroReport, 15*(13), 2057–2060.

Gandour, J. T. (1983). Tone perception in Far Eastern languages. *Journal of Phonetics, 11*, 149–175.

Gandour, J. T. (1994). Phonetics of tone. In R. Asher & J. Simpson (Eds.), *The encyclopedia of language & linguistics* (Vol. 6, pp. 3116–3123). New York: Pergamon Press.

Gandour, J. T., & Harshman, R. A. (1978). Cross language differences in tone perception: A multidimensional scaling investigation. *Language and Speech, 21*(1), 1–33.

Gandour, J. T., & Krishnan, A. (2014). Neural bases of lexical tone. In H. Winskel & P. Padakannaya (Eds.), *Handbook of South and Southeast Asian psycholinguistics* (pp. 339–349). Cambridge, UK: Cambridge University Press.

Gockel, H. E., Carlyon, R. P., Mehta, A., & Plack, C. J. (2011). The frequency following response (FFR) may reflect pitch-bearing information but is not a direct representation of pitch. *Journal of the Association for Research in Otolaryngology, 12*(6), 767–782. Doi:10.1007/s10162-011-0284-1

Gold, J. I., & Knudsen, E. I. (2000). A site of auditory experience-dependent plasticity in the neural representation of auditory space in the barn owl's inferior colliculus. *Journal of Neuroscience, 20*(9), 3469–3486.

Greenberg, S., Marsh, J. T., Brown, W. S., & Smith, J. C. (1987). Neural temporal coding of low pitch. I. Human frequency-following responses to complex tones. *Hearing Research, 25*(2–3), 91–114.

Griffiths, T. D., Buchel, C., Frackowiak, R. S., & Patterson, R. D. (1998). Analysis of temporal structure in sound by the human brain. *Nature Neuroscience, 1*(5), 422–427.

Griffiths, T. D., & Hall, D. A. (2012). Mapping pitch representation in neural ensembles with fMRI. *The Journal of Neuroscience, 32*(39), 13343–13347. Doi:10.1523/JNEUROSCI.3813-12.2012

Griffiths, T. D., Kumar, S., Sedley, W., Nourski, K. V., et al. (2010). Direct recordings of pitch responses from human auditory cortex. *Current Biology, 20*(12), 1128–1132. Doi:10.1016/j.cub.2010.04.044

Griffiths, T. D., Uppenkamp, S., Johnsrude, I., Josephs, O., & Patterson, R. D. (2001). Encoding of the temporal regularity of sound in the human brainstem. *Nature Neuroscience, 4*(6), 633–637.

Griffiths, T. D., Warren, J. D., Scott, S. K., Nelken, I., & King, A. J. (2004). Cortical processing of complex sound: A way forward? *Trends in Neurosciences, 27*(4), 181–185. Doi:10.1016/j.tins.2004.02.005. (S0166223604000566 [pii]).

Gutschalk, A., Patterson, R. D., Scherg, M., Uppenkamp, S., & Rupp, A. (2004). Temporal dynamics of pitch in human auditory cortex. *Neuroimage, 22*(2), 755–766. Doi:10.1016/j.neuroimage.2004.01.025. (S1053811904000680 [pii]).

Gutschalk, A., Patterson, R. D., Scherg, M., Uppenkamp, S., & Rupp, A. (2007). The effect of temporal context on the sustained pitch response in human auditory cortex. *Cerebral Cortex, 17*(3), 552–561. Doi:10.1093/cercor/bhj180

Hickok, G., & Poeppel, D. (2004). Dorsal and ventral streams: A framework for understanding aspects of the functional anatomy of language. *Cognition, 92*(1–2), 67–99.

Hickok, G., & Poeppel, D. (2007). The cortical organization of speech processing. *Nature Reviews Neuroscience, 8*(5), 393–402. Doi:10.1038/nrn2113. (nrn2113 [pii]).

Howie, J. M. (1976). *Acoustical studies of Mandarin vowels and tones*. New York: Cambridge University Press.

Johnson, K. L., Nicol, T., Zecker, S. G., & Kraus, N. (2008). Developmental plasticity in the human auditory brainstem. *The Journal of Neuroscience, 28*(15), 4000–4007. Doi:10.1523/JNEUROSCI.0012-08.2008

Khouw, E., & Ciocca, V. (2007). Perceptual correlates of Cantonese tones. *Journal of Phonetics, 35*(1), 104–117. Doi:10.1016/j.wocn.2005.10.003

Kraus, N., & Nicol, T. (2005). Brainstem origins for cortical 'what' and 'where' pathways in the auditory system. *Trends in Neurosciences, 28*(4), 176–181.

Kraus, N., & White-Schwoch, T. (2015). Unraveling the biology of auditory learning: A cognitive-sensorimotor-reward framework. *Trends in Cognitive Sciences, 19*(11), 642–654. Doi:10.1016/j.tics.2015.08.017

Krishnan, A. (1999). Human frequency-following responses to two-tone approximations of steady-state vowels. *Audiology and Neuro-Otology, 4*(2), 95–103.

Krishnan, A. (2002). Human frequency-following responses: Representation of steady-state synthetic vowels. *Hearing Research, 166*(1–2), 192–201.

Krishnan, A. (2007). Human frequency following response. In R. F. Burkard, M. Don, & J. J. Eggermont (Eds.), *Auditory evoked potentials: Basic principles and clinical application* (pp. 313–335). Baltimore: Lippincott Williams & Wilkins.

Krishnan, A., Bidelman, G. M., & Gandour, J. T. (2010a). Neural representation of pitch salience in the human brainstem revealed by psychophysical and electrophysiological indices. *Hearing Research, 268*(1–2), 60–66. Doi:10.1016/j.heares.2010.04.016

Krishnan, A., & Gandour, J. T. (2009). The role of the auditory brainstem in processing linguistically relevant pitch patterns. *Brain and Language, 110*(3), 135–148. Doi:10.1016/j.bandl.2009.03.005. (S0093-934X(09)00042-X [pii]).

Krishnan, A., & Gandour, J. T. (2014). Language experience shapes processing of pitch relevant information in the human brainstem and auditory cortex: Electrophysiological evidence. *Acoustics Australia, 42*(3), 187–199

Krishnan, A., Gandour, J. T., Ananthakrishnan, S., & Vijayaraghavan, V. (2014a). Cortical pitch response components index stimulus onset/offset and dynamic features of pitch contours. *Neuropsychologia, 59*, 1–12. Doi:10.1016/j.neuropsychologia.2014.04.006

Krishnan, A., Gandour, J. T., & Bidelman, G. M. (2010b). Brainstem pitch representation in native speakers of Mandarin is less susceptible to degradation of stimulus temporal regularity. *Brain Research, 1313*, 124–133. Doi:10.1016/j.brainres.2009.11.061

Krishnan, A., Gandour, J. T., Bidelman, G. M., & Swaminathan, J. (2009a). Experience-dependent neural representation of dynamic pitch in the brainstem. *NeuroReport, 20*(4), 408–413. Doi:10.1097/WNR.0b013e3283263000

Krishnan, A., Gandour, J. T., Smalt, C. J., & Bidelman, G. M. (2010c). Language-dependent pitch encoding advantage in the brainstem is not limited to acceleration rates that occur in natural speech. *Brain and Language, 114*(3), 193–198. Doi:10.1016/j.bandl.2010.05.004

Krishnan, A., Gandour, J. T., & Suresh, C. H. (2014b). Cortical pitch response components show differential sensitivity to native and nonnative pitch contours. *Brain and Language, 138*, 51–60. Doi:10.1016/j.bandl.2014.09.005

Krishnan, A., & Parkinson, J. (2000). Human frequency-following response: Representation of tonal sweeps. *Audiology and Neuro-Otology, 5*(6), 312–321.

Krishnan, A., & Plack, C. J. (2011). Neural encoding in the human brainstem relevant to the pitch of complex tones. *Hearing Research, 275*(1–2), 110–119. Doi:10.1016/j.heares.2010.12.008

Krishnan, A., Swaminathan, J., & Gandour, J. T. (2009b). Experience-dependent enhancement of linguistic pitch representation in the brainstem is not specific to a speech context. *Journal of Cognitive Neuroscience, 21*(6), 1092–1105. Doi:10.1162/jocn.2009.21077

Krishnan, A., Xu, Y., Gandour, J. T., & Cariani, P. (2005). Encoding of pitch in the human brainstem is sensitive to language experience. *Brain Research. Cognitive Brain Research, 25*(1), 161–168. Doi:10.1016/j.cogbrainres.2005.05.004. (S0926-6410(05)00123-0 [pii]).

Krishnan, A., Xu, Y., Gandour, J. T., & Cariani, P. A. (2004). Human frequency-following response: representation of pitch contours in Chinese tones. *Hearing Research, 189*(1–2), 1–12. Doi:10.1016/S0378-595(03)00402-7. (S037859503004027 [pii]).

Krizman, J., Marian, V., Shook, A., Skoe, E., & Kraus, N. (2012). Subcortical encoding of sound is enhanced in bilinguals and relates to executive function advantages. *Proceedings of the National Academy of Sciences of the USA, 109*(20), 7877–7881. Doi:10.1073/pnas.1201575109

Krizman, J., Skoe, E., Marian, V., & Kraus, N. (2014). Bilingualism increases neural response consistency and attentional control: Evidence for sensory and cognitive coupling. *Brain and Language, 128*(1), 34–40. Doi:10.1016/j.bandl.2013.11.006

Krumbholz, K., Patterson, R. D., Seither-Preisler, A., Lammertmann, C., & Lutkenhoner, B. (2003). Neuromagnetic evidence for a pitch processing center in Heschl's gyrus. *Cerebral Cortex, 13*(7), 765–772.

Kumar, S., & Schönwiesner, M. (2012). Mapping human pitch representation in a distributed system using depth-electrode recordings and modeling. *The Journal of Neuroscience, 32*(39), 13348–13351. Doi:10.1523/JNEUROSCI.3812-12.2012

Kumar, S., Stephan, K. E., Warren, J. D., Friston, K. J., & Griffiths, T. D. (2007). Hierarchical processing of auditory objects in humans. *PLoS Computational Biology, 3*(6), e100. Doi:10.1371/journal.pcbi.0030100

Lu, T., Liang, L., & Wang, X. (2001). Temporal and rate representations of time-varying signals in the auditory cortex of awake primates. *Nature Neuroscience, 4*(11), 1131–1138. Doi:10.1038/nn737

Luo, F., Wang, Q., Kashani, A., & Yan, J. (2008). Corticofugal modulation of initial sound processing in the brain. *The Journal of Neuroscience, 28*(45), 11615–11621. Doi:10.1523/JNEUROSCI.3972-08.2008. (28/45/11615[pii]).

Meddis, R., & O'Mard, L. (1997). A unitary model of pitch perception. *Journal of the Acoustical Society of America, 102*(3), 1811–1820.

Micheyl, C., Schrater, P. R., & Oxenham, A. J. (2013). Auditory frequency and intensity discrimination explained using a cortical population rate code. *PLoS Computational Biology, 9*(11), e1003336. Doi:10.1371/journal.pcbi.1003336

Oxenham, A. J. (2012). Pitch perception. *The Journal of Neuroscience, 32*(39), 13335–13338. Doi:10.1523/JNEUROSCI.3815-12.2012

Patterson, R. D., Allerhand, M. H., & Giguere, C. (1995). Time-domain modeling of peripheral auditory processing: A modular architecture and a software platform. *Journal of the Acoustical Society of America, 98*(4), 1890–1894.

Patterson, R. D., Handel, S., Yost, W. A., & Datta, A. J. (1996). The relative strength of the tone and noise components in iterated ripple noise. *The Journal of the Acoustical Society of America, 100*(5), 3286–3294.

Patterson, R. D., Uppenkamp, S., Johnsrude, I. S., & Griffiths, T. D. (2002). The processing of temporal pitch and melody information in auditory cortex. *Neuron, 36*(4), 767–776.

Penagos, H., Melcher, J. R., & Oxenham, A. J. (2004). A neural representation of pitch salience in nonprimary human auditory cortex revealed with functional magnetic resonance imaging. *The Journal of Neuroscience, 24*(30), 6810–6815. Doi:10.1523/JNEUROSCI.0383-04.2004. (24/30/6810 [pii]).

Peng, G., Zheng, H.-Y., Gong, T., Yang, R.-X., et al. (2010). The influence of language experience on categorical perception of pitch contours. *Journal of Phonetics, 38*, 616–624.

Philibert, B., Collet, L., Vesson, J. F., & Veuillet, E. (2005). The auditory acclimatization effect in sensorineural hearing-impaired listeners: Evidence for functional plasticity. *Hearing Research, 205*(1–2), 131–142.

Plyler, P. N., & Ananthanarayan, A. K. (2001). Human frequency-following responses: Representation of second formant transitions in normal-hearing and hearing-impaired listeners. *Journal of the American Academy of Audiology, 12*(10), 523–533.

Poeppel, D., Idsardi, W. J., & van Wassenhove, V. (2008). Speech perception at the interface of neurobiology and linguistics. *Philosophical Transactions of the Royal Society of London B: Biological Sciences, 363*(1493), 1071–1086. Doi:10.1098/rstb.2007.2160. (TM42571U1117682 [pii]).

Price, C., Thierry, G., & Griffiths, T. (2005). Speech-specific auditory processing: Where is it? *Trends in Cognitive Sciences, 9*(6), 271–276.

Puschmann, S., Uppenkamp, S., Kollmeier, B., & Thiel, C. M. (2010). Dichotic pitch activates pitch processing centre in Heschl's gyrus. *Neuroimage, 49*(2), 1641–1649. Doi:10.1016/j.neuroimage.2009.09.045

Rao, R. P., & Ballard, D. H. (1999). Predictive coding in the visual cortex: A functional interpretation of some extra-classical receptive-field effects. *Nature Neuroscience, 2*(1), 79–87. Doi:10.1038/4580

Russo, N. M., Nicol, T. G., Zecker, S. G., Hayes, E. A., & Kraus, N. (2005). Auditory training improves neural timing in the human brainstem. *Behavioural Brain Research, 156*(1), 95–103.

Russo, N. M., Skoe, E., Trommer, B., Nicol, T., et al. (2008). Deficient brainstem encoding of pitch in children with autism spectrum disorders. *Clinical Neurophysiology, 119*(8), 1720–1731. Doi:10.1016/j.clinph.2008.01.108

Schönwiesner, M., Rubsamen, R., & von Cramon, D. Y. (2005). Hemispheric asymmetry for spectral and temporal processing in the human antero-lateral auditory belt cortex. *European Journal of Neuroscience, 22*(6), 1521–1528. Doi:10.1111/j.1460-9568.2005.04315.x

Schouten, M. E. (1985). Identification and discrimination of sweep tones. *Perception and Psychophysics, 37*(4), 369–376.

Scott, S. K. (2003). How might we conceptualize speech perception? The view from neurobiology. *Journal of Phonetics, 31*, 417–422.

Scott, S. K., & Johnsrude, I. S. (2003). The neuroanatomical and functional organization of speech perception. *Trends in Neurosciences, 26*(2), 100–107.

Shofner, W. P. (1999). Responses of cochlear nucleus units in the chinchilla to iterated rippled noises: Analysis of neural autocorrelograms. *Journal of Neurophysiology, 81*(6), 2662–2674.

Shore, S. E., Clopton, B. M., & Au, Y. N. (1987). Unit responses in ventral cochlear nucleus reflect cochlear coding of rapid frequency sweeps. *Journal of the Acoustical Society of America, 82*(2), 471–478.

Shore, S. E., & Nuttall, A. L. (1985). High-synchrony cochlear compound action potentials evoked by rising frequency-swept tone bursts. *The Journal of the Acoustical Society of America, 78*(4), 1286–1295.

Song, J. H., Banai, K., & Kraus, N. (2008a). Brainstem timing deficits in children with learning impairment may result from corticofugal origins. *Audiology and Neuro-Otology, 13*(5), 335–344. Doi:10.1159/000132689

Song, J. H., Skoe, E., Wong, P. C. M., & Kraus, N. (2008b). Plasticity in the adult human auditory brainstem following short-term linguistic training. *Journal of Cognitive Neuroscience, 20*(10), 1892–1902.

Steinschneider, M., Reser, D. H., Fishman, Y. I., Schroeder, C. E., & Arezzo, J. C. (1998). Click train encoding in primary auditory cortex of the awake monkey: Evidence for two mechanisms subserving pitch perception. *The Journal of the Acoustical Society of America, 104*(5), 2935–2955.

Suga, N. (1990). Biosonar and neural computation in bats. *Scientific American, 262*(6), 60–68.

Suga, N. (1994). Processing of auditory information carried by complex species specific sounds. In M. S. Gazzaniga & E. Bizzi (Eds.), *The cognitive neurosciences* (pp. 295–318). Cambridge, MA: MIT Press.

Suga, N., & Ma, X. (2003). Multiparametric corticofugal modulation and plasticity in the auditory system. *Nature Reviews Neuroscience, 4*(10), 783–794.

Suga, N., Ma, X., Gao, E., Sakai, M., & Chowdhury, S. A. (2003). Descending system and plasticity for auditory signal processing: Neuroethological data for speech scientists. *Speech Communication, 41*(1), 189–200.

Swaminathan, J., Krishnan, A., & Gandour, J. T. (2008). Pitch encoding in speech and nonspeech contexts in the human auditory brainstem. *NeuroReport, 19*(11), 1163–1167. Doi:10.1097/WNR.0b013e3283088d31. (00001756-200807160-00017 [pii]).

Walker, K. M., Bizley, J. K., King, A. J., & Schnupp, J. W. (2011). Cortical encoding of pitch: Recent results and open questions. *Hearing Research, 271*(1–2), 74–87. Doi:10.1016/j.heares.2010.04.015

Wong, P. C. (2002). Hemispheric specialization of linguistic pitch patterns. *Brain Research Bulletin, 59*(2), 83–95.

Wong, P. C., Parsons, L. M., Martinez, M., & Diehl, R. L. (2004). The role of the insular cortex in pitch pattern perception: The effect of linguistic contexts. *The Journal of Neuroscience, 24*(41), 9153–9160.

Wong, P. C., Skoe, E., Russo, N. M., Dees, T., & Kraus, N. (2007). Musical experience shapes human brainstem encoding of linguistic pitch patterns. *Nature Neuroscience, 10*(4), 420–422. Doi:10.1038/nn1872. (nn1872 [pii]).

Xu, Y. (1997). Contextual tonal variations in Mandarin. *Journal of Phonetics, 25*, 61–83.

Xu, Y., Gandour, J. T., & Francis, A. L. (2006a). Effects of language experience and stimulus complexity on the categorical perception of pitch direction. *Journal of the Acoustical Society of America, 120*(2), 1063–1074.

Xu, Y., Krishnan, A., & Gandour, J. T. (2006b). Specificity of experience-dependent pitch representation in the brainstem. *NeuroReport, 17*(15), 1601–1605. Doi:10.1097/01.wnr.0000236865.31705.3a. (00001756-200610230-00008 [pii]).

Yip, M. (2002). *Tone*. New York: Cambridge University Press.

Yost, W. A. (1996). Pitch of iterated rippled noise. *The Journal of the Acoustical Society of America, 100*(1), 511–518.

Zatorre, R. J., & Baum, S. R. (2012). Musical melody and speech intonation: Singing a different tune. *PLoS Biology, 10*(7), e1001372. Doi:10.1371/journal.pbio.100137

Zatorre, R. J., & Belin, P. (2001). Spectral and temporal processing in human auditory cortex. *Cerebral Cortex, 11*(10), 946–953.

Zatorre, R. J., & Gandour, J. T. (2008). Neural specializations for speech and pitch: Moving beyond the dichotomies. *Philosophical Transactions of the Royal Society of London B: Biological Sciences, 363*(1493), 1087–1104. Doi:10.1098/rstb.2007.2161. (J412P80575385013 [pii]).

Chapter 4
Short-Term Learning and Memory: Training and Perceptual Learning

Samuele Carcagno and Christopher J. Plack

Abstract The frequency-following response (FFR) is a sustained auditory-evoked potential that reflects the phase locking of neurons in the auditory brainstem to periodicities in the waveform of a sound. Studies have shown that short-term auditory training can improve the robustness and/or accuracy of this phase locking. FFR plasticity has been investigated using training tasks that are thought to involve some form of auditory temporal coding, including fundamental-frequency discrimination training, training to identify Mandarin lexical tones, and training to identify speech in noise. The results of these studies have shown that improvements in the trained task are often accompanied by FFR plasticity. This suggests that subcortical auditory processing is not hardwired but can be modified by training even in adulthood. The FFR has also been shown to change following auditory-cognitive training protocols in special populations of listeners who may have subcortical auditory processing deficits, such as children with language-based learning disabilities, elderly listeners, and listeners with sensorineural hearing loss. The results of these studies provide promising evidence that subcortical auditory plasticity could be harnessed to ameliorate auditory processing deficits. It has been hypothesized that this learning-induced subcortical plasticity may be guided by efferent cortical feedback; however, the mechanisms of FFR plasticity remain largely unclear.

Keywords Aging · Brainstem · Efferent system · Evoked potentials · F_0 discrimination · Learning disabilities · Lexical tones · Pitch · Plasticity · Speech-in-noise · Temporal coding

S. Carcagno (✉) · C.J. Plack
Department of Psychology, Lancaster University, Lancaster LA1 4YF, UK
e-mail: s.carcagno@lancaster.ac.uk

C.J. Plack
e-mail: chris.plack@manchester.ac.uk

C.J. Plack
Manchester Centre for Audiology and Deafness, University of Manchester, Manchester Academic Health Science Centre, Manchester M13 9PL, UK

N. Kraus et al. (eds.), *The Frequency-Following Response*,
Springer Handbook of Auditory Research 61,
DOI 10.1007/978-3-319-47944-6_4

4.1 Introduction

Sensory systems can change the way they process incoming external stimuli to extract from the environment information that is relevant for the organism. While some of these changes are fast and transient (e.g., adaptation of neural responses), perceptual learning reflects changes that generally occur over longer time scales and are long lasting (Goldstone et al. 2012). Perceptual learning is likely to play a crucial role in the acquisition of several skills, such as learning a foreign language (Lively et al. 1994), learning to play a musical instrument (Herholz and Zatorre 2012), or learning to detect tumors in X-rays (Sowden et al. 2000). Although the acquisition of some perceptual skills is thought to be limited to specific developmental time windows (Hensch 2004), mature perceptual systems retain a high degree of plasticity. This plasticity may be important not only to acquire new skills in adulthood but also to make effective use of novel or restored sensory stimulation following the introduction of sensory prostheses, such as cochlear implants (Fu and Galvin 2007; Moore and Shannon 2009), or to offset some of the deleterious effects of aging on perceptual systems (Anderson et al. 2013c; Bower et al. 2013).

The frequency-following response (FFR) has proved to be an invaluable tool to study perceptual learning and neural plasticity at the early stages of auditory processing. The initial interest in the FFR as a tool to study auditory plasticity at the level of the brainstem came from the finding that the FFRs of Mandarin speakers track the time-varying pitch of some Mandarin tones more accurately than the FFRs of English speakers (Krishnan et al. 2005; Krishnan and Gandour, Chap. 3). This finding could be interpreted as evidence of subcortical plasticity as a result of the extensive practice that Mandarin speakers have in the identification of Mandarin tones. However, the interpretation of such comparisons between experts and non-experts is equivocal because two nonrandomly selected groups of individuals may differ not only because of the amount of practice they have in a certain domain, but also for other reasons (Monaghan et al. 1998). For example, differences in FFR pitch-tracking ability could be potentially caused by genetic factors rather than by experience-related plasticity. This hypothesis is not a priori less plausible than the neuroplasticity hypothesis. Certain pitch processing abilities, such as the ability to detect a mistuned tone in a familiar melody, are highly heritable (Drayna et al. 2001). Also, the adoption of tone languages is associated with the frequency in the population of specific alleles of two genes related to brain growth, and this association is hard to explain by geographical or historical factors (Dediu and Ladd 2007). Genetic differences could explain both the behavioral and the neurophysiological advantage of Mandarin speakers in processing the dynamic pitch trajectory of Mandarin tones. Later studies of FFR differences between Mandarin and English speakers provided some evidence that these differences were due to experience, rather than to other factors, by showing that they were specific for tones with the same fundamental frequency (F_0) contour as Mandarin tones (Xu et al. 2006). However, it is difficult to rule out genetic influences.

Musicians also have been found to have better FFR pitch tracking of tones than nonmusicians (Wong et al. 2007). This finding is consistent with the idea that extensive practice in discriminating and identifying tones leads to neuroplastic changes resulting in more accurate FFR pitch tracking. However, as for the comparison between Mandarin and English speakers, it is difficult to rule out the possibility that FFR differences between musicians and nonmusicians are due to factors other than neural plasticity.

Evidence for experience-dependent plasticity also has been provided by the results of short-term training studies in which FFR responses are recorded before and after a period of training in a tone discrimination or identification task. Unlike experiments comparing experts to nonexperts, short-term training studies can provide direct evidence of experience-related neuroplastic changes. These training experiments have additional advantages compared to experiments comparing experts to nonexperts, but they also have certain disadvantages. One advantage of training experiments is that they can determine whether or not FFR plasticity occurs in adulthood. Both Mandarin speakers and musicians usually start their experience in the discrimination and identification of Mandarin or musical tones during childhood. The superior FFR pitch tracking accuracy that these listeners show compared to English speakers without musical practice could be due to plastic changes that can happen only during a time-critical developmental window in childhood. Training experiments instead can test the hypothesis of adult neuroplasticity by directly training adults and testing for changes in the FFR. The question of whether the mature auditory system is malleable to experience-related changes is not only of theoretical interest. The potential for neuroplasticity in adulthood could pave the way for training interventions that are aimed at enhancing auditory skills in normal adults or remedy auditory deficits in special populations, such as elderly listeners with difficulties in hearing speech in noise or people with learning difficulties.

The disadvantages of training studies are that: (1) they are time consuming because usually participants have to attend multiple training sessions in addition to the sessions in which FFR responses are recorded; and (2) the amount of training that participants can get in a few sessions cannot match the often lifelong experience that Mandarin speakers and musicians have in processing tonal stimuli. As a consequence, the training effects elicited by short-term training studies may not be as large as the performance differences observed between expert and nonexperts.

The FFR is well suited for the study of auditory plasticity for several reasons. First of all, in the absence of auditory training, the FFR has good test-retest reliability (Song et al. 2011a, b). As a consequence, it is relatively easy to detect potential training-induced FFR changes. Another desirable feature of the FFR is that it can be recorded while participants are not paying attention to the stimuli (e.g., while watching a movie or even while sleeping). Under these conditions it is possible to exclude general changes in attention or higher-level cognitive processes as causes of FFR changes. Therefore, it is more likely that potential training-induced FFR changes reflect changes in sensory processing. Also, given the close correspondence between the FFR and certain stimulus features (Kraus et al. 2009), the interpretation of FFR changes following training is generally easier than the interpretation of changes in the amplitude and/or latency of cortical event-related potential (ERP) components. In particular, given that the FFR reflects periodicities present in the stimulus, it is relatively straightforward to interpret FFR changes following training as enhancements in the fidelity with which these periodicities are represented in the brainstem. The FFR also offers practical advantages. FFR recordings with a good signal-to-noise ratio (SNR) can be obtained relatively quickly (e.g., for a single stimulus a good FFR response can be obtained with about 15 min or less of electroencephalogram recordings). The time needed to set up the electrodes with a single channel configuration (three electrodes: positive, negative, and ground), which has been commonly used in the literature, is also short. Recently, multichannel configurations have also been used for FFR recordings (Bharadwaj and Shinn-Cunningham 2014). While these multichannel setups require more time for electrode placement, with the use of multichannel denoising techniques they can achieve the same SNR as single channel configurations but with a reduced recording time. Therefore, these multichannel setups could be used to measure the time course of FFR changes within a recording session with a better time resolution than is possible with single channel setups.

The FFR also has certain limitations. Like all ERPs recorded at the scalp, it reflects the summed activity of many neurons and, therefore, can only give a relatively coarse picture of neural changes following training. On a positive note, however, the generators of the FFR have been localized to a relatively restricted region of the brainstem (Bidelman 2015). Like all ERPs, the FFR can be degraded by myogenic noise. Therefore, in studies of neuroplasticity care must be taken to avoid potential biases stemming from systematic differences in myogenic noise levels between sessions and/or between groups, depending on the study design. In particular, appropriate control groups should be used in between-subjects designs, or control stimuli in within-subjects designs.

Another limitation of the FFR is that due to frequency dependent delays of the traveling wave in the basilar membrane, the FFR to a low-frequency stimulus may reflect the activation of neurons with characteristic frequencies higher by an octave or more (Ananthanarayan and Durrant 1992; Dau 2003), while perceptual responses may result mainly from the activation of neurons with center frequencies closer to the stimulus frequency (Gockel et al. 2015). As a result, training-induced behavioral changes mediated by neurons with characteristic frequencies close to the stimulus frequency may not be reflected in the FFR. Finally, two practical limitations of the FFR in the study of neuroplasticity arise from the fact that the FFR can be readily recorded only in response to sounds of moderate or high intensity (sound levels greater than about 40–45 dB above perceptual threshold) and by tones with frequencies lower than about 2 kHz (Krishnan 2007). These limitations constrain the range of stimuli that can be used in studies of FFR plasticity.

The FFR is a sustained potential that reflects the phase locking of neural populations in the rostral brainstem to the envelope and/or fine structure of a sound waveform (Krishnan 2007). Therefore, the FFR reflects temporal information about a sound available at the level of the brainstem. Several auditory processes, such as sound localization (Grothe et al. 2010) and frequency coding (Moore 2008), are thought to rely on precise temporal coding at the level of the brainstem. Thus, the FFR can be used as an index of temporal coding fidelity at the level of the brainstem to investigate whether and how temporal coding changes as a result of training-induced neural plasticity. According to temporal theories, the pitch of a sound is coded temporally in the peripheral auditory system (de Cheveigné 2005). Studies of FFR plasticity have commonly used pitch-discrimination or pitch-identification tasks for training. Accurate coding of temporal information is thought to play an important role also in more complex auditory skills, such as speech understanding in noise (Drennan et al. 2007; Moore 2008). Tasks involving speech understanding in noise have also been used in studies of FFR plasticity. Section 4.2 will summarize the main findings on the effects of short-term auditory training on behavioral and cortical electrophysiological responses, with a focus on pitch-discrimination training.

4.2 Effects of Short-Term Training on Behavioral and Cortical Responses

Many basic auditory skills, such as frequency and intensity discrimination (Wright and Zhang 2009) and complex auditory skills, such as speech understanding in noise (Engen 2012; Song et al. 2012), improve with practice. These improvements can result from changes at different processing levels. At the lowest level the response properties of early sensory neurons may change to sharpen the encoding of basic stimulus features (Gilbert et al. 2001). At a midlevel, the connections between sensory neurons and central decision areas may be re-weighted to enhance the internal representation of stimulus features that are crucial for the performance of the training task (Dosher and Lu 1999; Petrov et al. 2005). At the highest level, nonperceptual changes in decision processes, response biases, memory, and attentional resources can also influence the outcome of perceptual training (Amitay et al. 2014; Jones et al. 2015). Neurophysiological studies can help to pinpoint the source of training-induced changes in behavioral measures of performance by measuring neural responses at different processing levels before and after training. Caution needs to be exercised when drawing inferences from these studies. Changes of neural responses at higher processing stages may not reflect plasticity at those stages but rather reflect plasticity occurring downstream at lower processing stages. This argument also applies in reverse. The information flow does not simply proceed in a feedforward fashion from low-level sensory areas toward more central areas of the brain. Efferent feedback from central areas can modulate the responses of neurons in low-level sensory areas (Tsodyks and Gilbert 2004; Kraus and White-Schwoch 2015). Therefore, changes of neural responses in low-level sensory areas may not necessarily reflect plasticity at this level of processing (Watanabe and Sasaki 2015).

An important aspect of perceptual learning is its degree of specificity versus generalization. From a theoretical point of view, specificity of learning with respect to a given stimulus attribute or a given task can give important insights on the mechanisms underlying learning. If perceptual learning in a given task reflects a sharpening of tuning of early sensory neurons this learning should transfer to a different task in which performance is limited by the same noise in the sensory encoding phase. Task-specific learning, on the other hand, would indicate that learning affected later processing stages (Petrov et al. 2005). Specificity of learning with respect to a given characteristic of the trained stimulus has sometimes been used to infer the locus of training-induced plasticity. For example, in the visual system, specificity of learning to certain stimulus attributes, such as its retinal location or stimulus orientation, has been interpreted as a sign that learning is mediated by plasticity of primary sensory areas in which neural responses are best tuned to these basic perceptual attributes. However, this interpretation is equivocal (Mollon and Danilova 1996). Specificity of learning with respect to a low-level stimulus attribute such as retinal location, for example, may occur also as a result of a re-weighting of the connections between primary sensory areas and central decision areas (Dosher and Lu 1999; Petrov et al. 2005).

The degree of specificity of perceptual learning also has important implications from an applied point of view. Perceptual learning that is highly stimulus specific or task specific has little value in real-world scenarios in which organisms are confronted with stimuli that vary continuously along the trained perceptual dimension and appear in different contexts. For therapeutic or rehabilitative purposes, perceptual learning should generalize to a wide range of levels of the trained perceptual attribute and to different contexts. Some recent studies have started to investigate factors that promote the generalization of perceptual learning across different levels of the trained perceptual dimension and across presentation contexts. Deveau and Seitz (2014) have proposed that the use of a wide stimulus set, multisensory stimuli, reinforcement, and the engagement of attention are all factors that promote generalization of perceptual learning. The duration of training may also play an important role in the generalization of learning to untrained stimuli. Wright et al. (2010) have proposed that generalization of perceptual learning to untrained stimuli may lag behind stimulus-specific learning, and that increasing the duration of training may be an effective means to achieve greater generalization of learning to untrained stimuli.

One of the perceptual attributes that has been most widely studied in the context of auditory plasticity is pitch. Naturally occurring periodic sounds consist of several harmonically related frequencies, each an integer multiple of an F_0. The pitch of these sounds (i.e., the perceived tonal height) is commonly determined by their F_0, which usually corresponds to their envelope repetition rate. Pitch plays a crucial role in the perception of both speech and nonspeech sounds. In most languages the F_0 conveys prosodic information as well as cues to the speaker's gender and emotional state (Rosen 1992; Skuk and Schweinberger 2014). In tone languages, the F_0 is also used to convey lexical information; for example, in Mandarin Chinese two otherwise similar syllables take on a different meaning depending on the pitch contour. In music the F_0 plays a fundamental role in the composition of melodies and harmonies. The F_0 also provides important cues for the perceptual grouping and segregation of auditory streams (Carlyon 2004). The ability to discriminate between sounds with a different F_0 or a different F_0 contour improves dramatically with practice. For this reason, as well as for the importance that pitch plays in hearing, tasks that rely on accurate pitch perception often have been used to study training-induced neural plasticity in the auditory system.

Several studies have found that training for multiple hours on an F_0-discrimination task leads to large improvements in performance on the task (see Wright and Zhang 2009 for a review). Micheyl et al. (2006) measured F_0 discrimination thresholds in a group of musicians and a group of naive listeners. They found that musicians' thresholds were on average six times lower than the thresholds of naive listeners. However, the thresholds of nonmusicians decreased progressively as they trained on the task and became as low as those of musicians after 4–8 hours of training. These results indicate that short-term training effects on F_0 discrimination are large and can match in magnitude the long-term effects of musical training after only a few hours of practice.

The behavioral improvements in pitch discrimination after training are often paralleled by changes of transient auditory-evoked potentials (AEPs) and auditory-evoked fields (AEFs), respectively derived from electroencephalographic and magnetoencephalographic recordings that reflect the activation of cortical areas in response to pitch-evoking sounds. The amplitude of the *N1m*, a sensory component of the AEFs peaking around 100 ms after the stimulus onset, has been found to increase after frequency-discrimination training (Menning et al. 2000), although the *N1*, the electric counterpart of the N1m, has not been found to change (Atienza et al. 2002; Bosnyak et al. 2004). A component of the AEPs whose amplitude has been found to increase after frequency-discrimination training is the *N1c*, a negative deflection occurring around 150 ms after stimulus onset at temporal electrode sites (Bosnyak et al. 2004). Several studies have also found increases in the amplitude of the *P2* component of the AEPs, a positive deflection peaking around 180 ms after stimulus onset (Bosnyak et al. 2004; Carcagno and Plack 2011a). The functional significance of these P2 changes has been a matter of debate because increases in the amplitude of the P2 have also been found following passive exposure to sounds rather than active learning (Sheehan et al. 2005). In a study by Carcagno and Plack (2011a), however, these P2 amplitude changes extended also to untrained stimuli, suggesting that stimulus exposure per se could not explain increases of the P2. Recently, it has been suggested that P2 changes are associated with some aspect of the training procedure itself rather than with perceptual learning (Tremblay et al. 2014). The mismatch negativity (MMN), a difference wave derived by subtracting the AEPs to a rarely presented "deviant" tone from the AEPs to a frequently presented "standard" tone, has also been found to increase in amplitude after training on pitch-related tasks (Atienza et al. 2002; Tong et al. 2009). Carcagno and Plack (2011a), however, did not find MMN changes after training participants on an F_0-discrimination task despite the fact that the authors demonstrated that the measured MMNs had an adequate sensitivity to detect equivalent changes in performance produced by changes to the stimuli rather than by training. The reasons for this discrepancy remain unclear but may be related to differences in the task (deviance detection versus F_0 discrimination) used for the training, suggesting that MMN-amplitude increases may reflect changes of higher-level melodic processing functions rather than low-level sensory discrimination processes.

Overall, AEP and AEF studies indicate that short-term training affects the cortical processing of pitch-evoking sounds at several levels of processing. The functional significance of these changes, however, is not yet well understood. Given that AEPs and AEFs in these studies were recorded while participants were passively listening to the stimuli, rather than being actively engaged in a task, it is likely that these AEP/AEF changes reflect changes of low-level sensory processes. One possibility is that they reflect reorganization of cortical tonotopic maps. Studies of tonotopic map plasticity following perceptual discrimination training in non-human animals are few, and the evidence for such tonotopic reorganization is mixed (Recanzone et al. 1993; Brown et al. 2004). Recent results suggest that an expanded cortical representation of the trained stimulus accelerates learning in the initial stages of training. However, this map expansion is subsequently renormalized and is not necessary for maintaining good performance in a learned discrimination task (Reed et al. 2011). It is also possible that training-induced changes in cortical AEP/AEF responses reflect, at least in part, training-induced changes at more peripheral levels of processing.

4.3 Effects of Short-Term Training on FFR Responses

Until about a decade ago, experience-related auditory plasticity was studied mainly using cortical responses. Following the findings that Mandarin speakers (Krishnan et al. 2005) as well as musicians (Wong et al. 2007) show enhanced subcortical encoding of Mandarin tones, suggesting that experience-related plasticity may also occur in brainstem structures, several studies have investigated the effects of training on the subcortical encoding of sounds using the FFR. Several training tasks have been used to investigate FFR plasticity, including pitch and pitch-contour discrimination, identification of lexical tones, and speech-in-noise comprehension. This section will summarize the FFR changes that have been found after training on these tasks in normal-hearing listeners as well as in special populations of listeners with hearing difficulties.

4.3.1 Effects of Pitch-Discrimination Training

As mentioned in Sect. 4.2, short-term training on a pitch discrimination task can lead to large improvements in pitch discrimination after only a few hours of training. Carcagno and Plack (2011b) trained three groups of listeners on an F_0

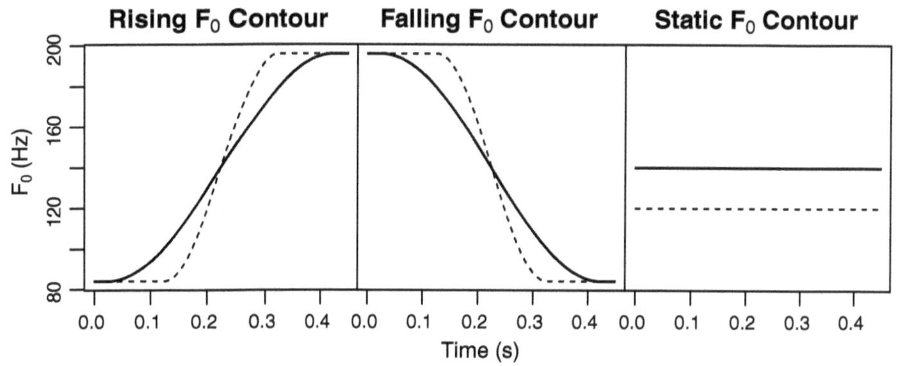

Fig. 4.1 F_0 contours of the stimuli used in the study by Carcagno and Plack (2011b). The *solid lines* represent the F_0 contours of the standard stimuli. The *dashed lines* show examples of the comparison stimuli that had to be discriminated from the standard stimuli. For the stimuli with a dynamic F_0 contour (first two panels) the comparison stimuli had a higher FM rate. For the stimulus with a static F_0 (right panel) the comparison stimuli had a lower F_0

discrimination task for 10 hours and compared the difference between their pre-training and post-training FFR responses to those of an untrained control group. Each group was trained with one of three different stimuli (depicted in Fig. 4.1).

All three stimuli consisted of harmonic complex tones band-pass filtered between 2 and 3 kHz, so as to include only unresolved harmonics (see Plack and Oxenham 2005 for a definition of resolvability). The FFR reflects temporal information, thus, improvements on a temporal pitch-encoding mechanism are likely to be reflected in the FFR (improvements on a place pitch-encoding mechanism, based on cochlear frequency selectivity, would not necessarily be reflected in the FFR). The use of complex tones consisting exclusively of unresolved harmonics forced listeners to use a temporal pitch-encoding mechanism to perform the task because these tones do not contain useful spectral information to extract their pitch. Two of the stimuli (S-Up and S-Down) had a dynamic F_0 contour, while the third stimulus (S-Static) had a static F_0 contour. For the dynamic stimuli, the task was to detect a difference in the rate of change of F_0. For the static stimulus, the task was to detect an F_0 difference.

Before and after training the behavioral discrimination thresholds as well as the FFR responses to each stimulus were measured for all listeners (Figs. 4.2A, B). For each trained group, the F_0-discrimination performance for the trained stimulus improved more than for the control group, indicating that training was effective in eliciting F_0 discrimination learning. For listeners trained with a dynamic F_0 contour, these performance improvements were partly specific to the trajectory (rising versus falling) of the F_0 contour with which they were trained. Collapsing the data of the stimuli with a dynamic F_0 contour (S-Up and S-Down) together and comparing them to the data from the stimulus with an unmodulated F_0 contour (S-Static), it was also apparent that threshold improvements were partly specific to the modulation (dynamic versus static) of the F_0 contour used during training.

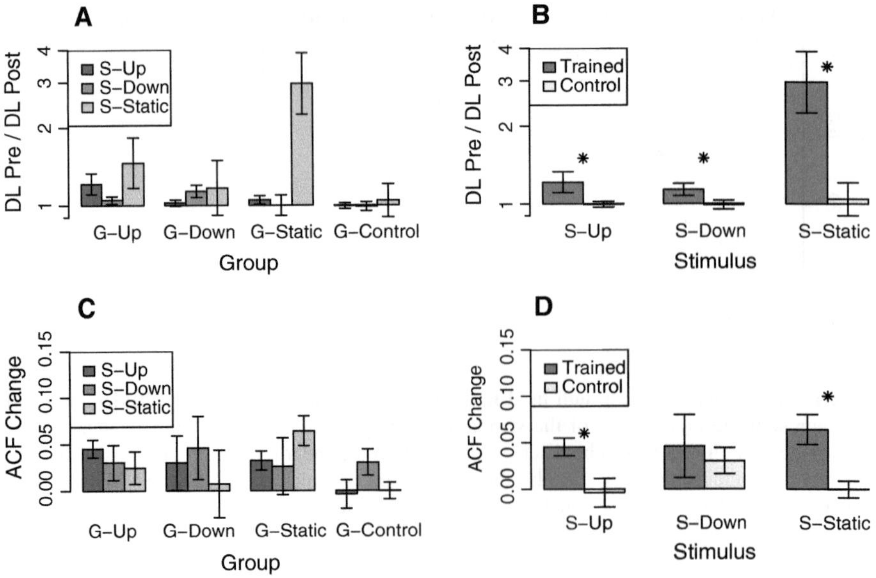

Fig. 4.2 Changes in the behavioral and FFR measures in the study of Carcagno and Plack (2011b). (**A**) Ratio of the behavioral discrimination thresholds before and after training. *G-Up*, *G-Down*, *G-Static*, and *G-Control* labels denote respective training groups (e.g., G-Up trained on the S-Up stimulus). Values >1 indicate improvements. (B) Plot as in **A**, but for a set of planned contrasts comparing changes for each trained group on the trained stimulus to changes in the control group for the same stimulus. (**C**) Difference of the autocorrelation function (ACF) value at the time lags corresponding to the inverse of the F_0 of the stimuli between post-training and pre-training FFR assessment sessions. (**D**) Plot as in **C**, but for a set of planned contrasts comparing changes for each trained group on the trained stimulus to changes in the control group for the same stimulus. The *asterisks* in **B** and **D** denote statistically significant differences ($p < 0.05$). (Reproduced with permission from Carcagno and Plack 2011b)

The robustness of FFR phase locking to the sound envelope of each stimulus before and after training was assessed by averaging the value of the autocorrelation function (ACF) at the time lag corresponding to the inverse of the stimulus F_0 in seven time windows of 64-ms length. Figure 4.2 (C, D) shows the difference between the post-training and pre-training average value of the ACF for each stimulus and group. The ACF value for the trained stimulus increased significantly more for participants trained with the S-Up and S-Static stimuli than for participants in the control group. For participants trained with the S-Down stimulus, the ACF change for the trained stimulus was not significantly larger than for control participants. As for threshold changes, ACF changes were partly specific to the modulation (dynamic versus static) of the F_0 contour of the trained stimulus, while the specificity of ACF changes to the F_0 trajectory (rising versus falling) was not statistically significant. The behavioral threshold improvements were found to correlate with the ACF changes for participants trained with the S-Up and S-Static stimuli, while the correlation for participants trained with the S-Down stimulus was not significant.

Overall, the results of this study indicate that the robustness of subcortical temporal encoding can improve as a result of short-term F_0 discrimination training. The correlation between FFR and threshold changes indicates that perceptual improvements are associated with FFR changes. However, it should be kept in mind that these correlations were relatively small, accounting for about 14–20% of the variance in threshold improvements for participants trained with the S-Up and S-Static stimuli. These data suggest that although the subcortical changes in temporal encoding measured by the FFR make a significant contribution to improvements in pitch discrimination after training, these behavioral improvements may depend in large part either on subcortical changes that are not captured by the FFR or by changes occurring at higher levels of processing.

4.3.2 Effects of Training on the Identification of Lexical Tones

Several studies have used tasks in which participants are presented with pseudo-words consisting of syllables onto which the F_0 contour of a Mandarin lexical tone is superimposed, and they are trained to associate these pseudo-words with drawings of English nouns. In order to successfully perform this task, participants must be able to follow the dynamic F_0 contour of each Mandarin tone used and discriminate it from the others. Song et al. (2008) trained 23 English speakers in a lexical tone identification task for eight sessions each lasting for about 30 min. The FFR responses to lexical tones were recorded before and after training. Mandarin uses four lexical tones: *T1* (high-level), *T2* (rising), *T3* (falling-rising), and *T4* (falling), with different F_0 contours (depicted in Fig. 4.3).

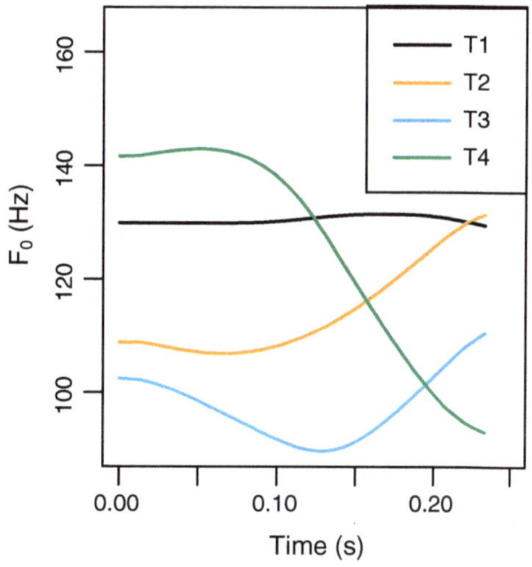

Fig. 4.3 Prototypical F_0 contours of the four Mandarin lexical tones. The F_0 contours were derived from the synthesized speech monosyllable /yi/ following the procedures used by Krishnan et al. (2005). (Data courtesy of Ananthanarayan Krishnan)

Three of these tones (T1, T2, and T3) were used during training. The percentage of correct identifications across all three tones improved from 22% after the first training session to 89% after the last training session. FFR pitch-tracking accuracy, assessed by measuring the distance between the F_0 trajectories extracted from spectrograms of the FFR and stimuli waveforms, improved after training for the T3 tone. This improvement was paralleled by an increase in SNR measured by comparing the power at the F_0 during the presentation of the stimulus to the power at the same frequency in the pre-stimulus baseline window. No significant post-training changes were found for the T1 and T2 tones. Although English is not a tonal language, F_0 modulations similar to T1 and T2 occur frequently in English at the syllable level as intonational markers, while F_0 modulations following the T3 pattern do not occur in English at the syllable level. Song et al. (2008) argued that the familiarity of English speakers with the T1 and T2 pitch contours may explain why FFR pitch tracking did not appear to improve for these stimuli and improved only for the T3 tone, which was the least familiar to the participants. Other studies, however, have subsequently found training effects also for these tones.

Chandrasekaran et al. (2012) investigated the impact of individual differences in inferior colliculus (IC) activity on lexical tone learning assessed behaviorally and electrophysiologically via the FFR. The authors trained 18 English speakers on a lexical tone identification task for nine sessions. Before training fMRI scans of the IC were collected in response to repeated lexical tones (e.g., T1-T1-T1-T1, tone repeat condition) or alternating lexical tones (e.g., T1-T3-T4-T2, tone change condition). These scans were used to divide the participants into a repetition suppression (*RS*) group, showing less activation in the tone repeat condition than in the tone change condition, and a repetition enhancement (*RE*) group, showing the reverse pattern. RS is generally associated with improved neural encoding of the repeated signal and, according to neural sharpening models (Grill-Spector et al. 2006), reflects a sparser and more efficient representation of the signal (fewer neural units are needed to encode it). Before and after training, FFR amplitude and FFR pitch tracking accuracy (based on stimulus-to-response correlations of F_0 trajectories derived by autocorrelograms) were measured for tone T2. FFR pitch tracking was measured separately for the initial portion of the stimulus (in which the F_0 was nearly static) and for the second part of the stimulus (in which the F_0 was rising).

Activation of the IC (RS versus RE) was predictive of pre-training FFR amplitude, with participants in the RS group showing smaller FFR amplitudes (possibly reflecting greater neural efficiency) than participants in the RE group. The RS participants also had better FFR pitch-tracking accuracy than the RE participants for the dynamic portion of the stimulus. This is consistent with the idea that an efficient representation of the stimulus is associated with better encoding of the stimulus.

Performance in the lexical tone-identification task was initially similar for the two groups but, from the third session onward, participants of the RS group showed greater improvements in the task than participants of the RE group. After training, FFR pitch-tracking accuracy improved significantly for both groups on the dynamic-pitch portion of the stimulus, while no training-related changes were seen for the static-F_0 portion of the stimulus. The FFR amplitude, on the other hand, decreased significantly for the RS group, but not for the RE group. In addition to providing further evidence of greater post-training accuracy of FFR pitch tracking, these results indicate that pre-training FFR amplitude may be an index of stimulus encoding efficiency at the level of the brainstem and may be predictive of learning success in a lexical tone-identification task.

A number of studies have shown that the FFR is sensitive to the probability of presentation of a stimulus (see Escera, Chap. 5). Skoe et al. (2014) investigated how these probability effects interact with the effects of short-term training on the accuracy of F_0 encoding. The authors trained 12 participants for nine sessions using a lexical tone identification task. FFR pitch-tracking accuracy was assessed by taking the distance between the F_0 contours of the FFR and stimulus waveforms, which were derived from autocorrelograms. FFR pitch-tracking accuracy was measured before and after training and compared to that of a control group consisting of 13 participants who did not receive training. Stimuli (tones T1 and T4) were presented in an oddball paradigm (80% standards: 20% deviants) to investigate the effects of stimulus probability. Before training, FFR pitch tracking was more accurate for frequently presented standard sounds than for deviants. After training, collapsing responses across standard and deviant tones, FFR pitch-tracking accuracy improved for the trained group but not for the control group. Additionally, the effect of stimulus probability for tone T1 reversed after training: post-training FFR pitch-tracking accuracy was greater when T1 tones were presented as deviants compared to when they were presented as standard tones. In addition to providing further evidence that FFR pitch-tracking accuracy can be improved by training, these results suggest that the effects of training interact in a complex way with contextual effects of stimulus probability.

4.3.3 Effects of Training to Understand Speech in Noise

FFR plasticity has also been investigated using general speech-in-noise training protocols. Song et al. (2012) tested 28 participants that trained at home for 20 sessions (approximately 30 min per session) using the Listening and Communication Enhancement (LACE®; Neurotone, Inc., Redwood City, CA) program and 32 participants who did not receive training. The LACE® program included tasks related to comprehension of degraded speech, cognitive skills, and communication strategies. Before and after training, FFR responses to the syllable /da/ were recorded in quiet and in a six-talker babble noise. Participants were also tested for their ability to hear speech in noise using the Quick Speech-in-Noise Test (QuickSIN™; Etymotic Research, Elk Grove Village, IL) and the Hearing-in-Noise Test (HINT; Biologic Systems Corp., Mundelein, IL). A subset of the participants was re-tested approximately six months after the end of the program to check for retention of training effects. The FFR was analyzed separately for the formant transition region (20–60 ms) and the steady-state vowel region (60–180 ms). The average of the FFR response measured from FFT spectra at frequencies corresponding to the first two harmonics (F_0 and H2) of the stimulus was also analyzed separately from the average of the FFR response at frequencies corresponding to harmonics 3 to 10 of the stimulus.

After training, trained participants showed larger improvements than control participants on the QuickSIN^{TM} and HINT tests. These behavioral improvements were paralleled by a larger increase of the FFR responses in noise at frequencies corresponding to the first two harmonics of the stimulus in the trained group compared to the control group. This effect was present for both the steady state and the formant transition region, although it was significantly greater for the latter. No significant differences between the trained and control group were found for the FFR responses to the syllable in quiet. Both the behavioral and the neurophysiological improvements were in large part retained six months after the training. Another interesting finding was that the strength of F_0 and H2 encoding measured by the FFR at pretest predicted the improvement in some of the speech-in-noise measures; participants who had better FFR encoding of F_0 and H2 at pretest showed the largest amount of learning. The results of this study are important because, in addition to showing that a general speech-in-noise training protocol can improve subcortical encoding of temporal information, they also provide evidence of long-term retention of both behavioral and FFR training effects.

4.3.4 Studies on Special Populations

The finding that subcortical auditory processing is not static but can be manipulated by training suggests that sensory deficits caused by degraded subcortical temporal processing may be partly remedied by training. A number of studies have tested this hypothesis by measuring subcortical responses before and after a period of auditory training in special populations of listeners with hearing difficulties.

Children with language-based learning problems (e.g., dyslexia) often show concomitant auditory perception problems. Russo et al. (2005) trained nine children with learning disabilities for 35 to 40 one-hour sessions with *Earobics*® (Cognitive Concepts, Inc., Evanston, IL), a commercial auditory training program. The training consisted of interactive computer games that tapped into phonological awareness, auditory skills, and language processing skills. Auditory stimuli were presented both in quiet and with background noise. Brainstem and cortical responses to the syllable /da/, as well as measures of auditory and cognitive processing, were evaluated before and after training. A control group, consisting of five children with learning disabilities and five normal children, was also evaluated on these measures across a time period equivalent to the training period but did not undergo training. After training, the correlation between the FFR component of the response recorded in quiet and the FFR component of the response recorded in noise increased significantly more for the experimental group compared to the control group. This increased inter-response correlation reflected mainly an improved phase locking of the response in noise. Children in the training group also showed improvements in several behavioral tests, including the Incomplete Words, Auditory Processing, and Sentences-in-Noise tests. It is unclear how many of the children tested by Russo et al. (2005) specifically had a language-based learning disability. However, language-based learning disabilities constitute a sizable portion of learning disabilities, and given the extensive literature linking these disabilities to subcortical timing deficits (for reviews see Chandrasekaran and Kraus 2012; Reetzke, Xie, and Chandrasekaran, Chap. 10), the results of Russo et al. (2005) are a promising line of work for the development of training interventions for children with these disabilities.

Additional evidence for the plasticity of subcortical auditory processing in children with language-based learning disabilities comes from a study of Hornickel et al. (2012). In this study, the FFR to the syllables /ba/, /da/, and /ga/ was recorded at the beginning and at the end of one academic year in 38 children (8–14 years old) diagnosed with dyslexia. Reading and phonological awareness tests were also administered at each time point. During the academic year, 19 children assigned to the treatment group wore *FM systems* listening devices in the classroom, while the remaining 19 children in the control group did not. FM systems are radio systems that broadcast the voice of a speaker wearing a microphone directly to the ears of a listener wearing a receiver, thus bypassing background noise and enhancing the acoustic clarity of the speaker's voice. At the end of the academic year, children in the treatment group had improved scores on the reading and phonological awareness tests, while children in the control group did not show significant improvements in these tests. At the end of the year, the consistency of the FFR, as measured by the correlation between the average response obtained during the first and the second half of the session, improved significantly in the formant transition region of the syllables (7–60 ms) for the children in the treatment group. No significant changes in FFR response consistency were found for children in the control group. Interestingly, the greatest improvements in FFR consistency were observed in a subset of children who showed the greatest improvements in the phonological awareness tests. Overall, these results suggest that the use of FM systems listening devices may improve both reading abilities and subcortical responses to sounds in children with dyslexia. The authors of the study hypothesize that these changes are mediated by the improved clarity of the acoustic signal provided by the FM system. The increased acoustic clarity of the acoustic signal may strengthen sound-to-meaning relationships and, through the action of the efferent system, may lead to a fine-tuning of the temporal encoding of sounds at a subcortical level.

Elderly listeners represent another population of listeners that might benefit from auditory perceptual training. Recent studies indicate that age-related visual processing deficits can be partially offset by perceptual training (Andersen et al. 2010; Polat et al. 2012). Similarly, reversals of age-related deficits have been observed in the cortical responses of rats after auditory perceptual training (de Villers-Sidani et al. 2010). Subcortical auditory temporal processing degrades as a result of aging. FFR phase locking to pure tones (Clinard et al. 2010; Marmel et al. 2013) and musical dyads (Bones and Plack 2015) has been found to decline as a function of age. Also, the timing of peaks in the subcortical response to syllables (Anderson et al. 2012; Clinard and Tremblay 2013) has been found to be delayed as a result of aging (see Anderson, Chap. 11).

Anderson et al. (2013c) investigated whether a home-based cognitive training program could offset some of these age-related deficits. The authors trained 35 elderly participants (ages 55–70) for 8 weeks (40 hours of training at home) with *Brain Fitness*^{TM} (Posit Science Corporation, San Francisco, CA), an auditory-based cognitive training program, while 32 control participants in the same age range watched a series of educational DVDs (about art, science, history and other topics) and completed multiple-choice tests on the topics of the DVDs. The *Brain Fitness*^{TM} program contains exercises in which the formant transition region of speech syllables is adaptively stretched or compressed. The exercises are presented at different levels (isolated syllables, syllables within words, sentences, and stories). Before and after training, brainstem responses to the syllable /da/ were recorded, as well as measures of speech perception in noise (QuickSIN^{TM} test), auditory short-term memory, and processing speed. After an initial stop burst of 5 ms, the syllable had a constant F_0 of 100 Hz with peaks occurring every 10 ms. After training, the timing of the FFR peaks in the formant transition region of the syllable in noise was significantly earlier for the trained group than for the control group. Across the overall length of the stimulus presented in noise, the FFR inter-peak variability, measured as the standard deviation of latency differences between adjacent peaks, was also significantly reduced after training in the experimental group compared to the control group. These neurophysiological changes were paralleled by significant improvements in all behavioral measures (QuickSIN^{TM} test, auditory short-term memory, and speed processing) in the trained group compared to the control group.

Anderson et al. (2014) re-tested 30 of the 35 participants in the auditory training group and all 32 participants of the active control group in a follow-up study six months after the training to determine whether or not behavioral and neurophysiological changes would be retained in the long term. The results of this follow-up study revealed that subcortical improvements in response peak timing were largely maintained after six months. Behaviorally, improvements in speed processing were also maintained, but gains in auditory short-term memory and speech reception thresholds were not. The dissociation between retention of subcortical peak timing changes and changes in speech reception thresholds suggests that the two were not directly related in this study. Improvements in speech reception thresholds immediately after the training may have instead resulted from cognitive components of the training. In sum, the results of the studies of Anderson et al. (2013c, 2014) indicate that perceptual learning can lead to subcortical plasticity also in elderly listeners and can ameliorate some age-related perceptual auditory deficits. However, more work needs to be done to understand the relationship between neurophysiological and behavioral changes following training and how the benefits of training can be retained over time.

Additional evidence of subcortical plasticity following training in elderly listeners comes from a study by Anderson et al. (2013b) in which two groups of elderly participants, one with normal hearing and one with mild hearing loss, were trained for eight weeks with *Brain Fitness*^{TM}, and their performance changes in a series of behavioral and electrophysiological tests were compared to those of age and audiometrically matched active control groups as in the study by Anderson et al. (2013c). Because listeners with hearing loss have abnormally enlarged responses to sound envelopes (Anderson et al. 2013a), this study focused on the relative representation of envelope and fine structure in the FFR. The pre-training results confirmed the presence of enlarged FFR envelope responses in the group of listeners with hearing loss. These enlarged FFR envelope responses were reduced in the group of hearing-impaired participants after training compared to the age and audiometrically matched control group. No significant changes occurred for the normal-hearing participants. This reduction of the FFR envelope response was accompanied by small improvements in speech-reception thresholds. The effect of training on the FFR in this study, a reduction of the FFR envelope response, is opposite to the enhancement of the FFR envelope response found in two previous studies of short-term FFR plasticity (Carcagno and Plack 2011b; Song et al. 2012). The cause of this difference is unclear. It may be related to the different populations of listeners employed in these studies (young normal-hearing versus elderly hearing-impaired listeners) or to differences in the training tasks and stimuli employed.

4.3.5 Summary of Empirical Findings on FFR Plasticity

Table 4.1 presents an overview of the short-term training studies on FFR plasticity that have been conducted to date. For each study the table lists the stimuli and training task used and specifies whether for each stimulus a significant training effect on the FFR was found on at least one of the measures used to quantify the FFR changes. It should be kept in mind that this is only a qualitative and crude way to summarize FFR training effects, but a full-blown meta-analytic treatment is beyond the scope of this chapter. The table also includes the population of listeners trained in each study and whether their results were compared to those of a control group or not.

Overall, the studies reviewed above indicate that short-term training on pitch discrimination or pitch identification leads to increased FFR pitch-tracking accuracy. The term "FFR pitch tracking" is used in this chapter to indicate the fact that FFR periodicities follow periodicities present in the stimulus. However, it should be noted that the FFR is not a direct measure of the subcortical representation of pitch; rather, it reflects subcortical timing information that may be used by the auditory system to encode pitch (Gockel et al. 2011). It is currently unclear how specific

Table 4.1 Summary of short-term training FFR studies

Study	Stimulus	Task	Treatment group	Control group	Follow-up
Carcagno and Plack (2011b)	Rising F_0–Sig. Falling F_0–NS Static F_0–Sig.	F_0 discrimination	Young adults (N = 27; 9 per training group)	Young adults (N = 12) Passive	No
Song et al. (2008)	T1–NS T2–NS T3–Sig.	Lexical tone identification	Young adults (N = 23)	No	No
Chandrasekaran et al. (2012)	T2 (rising segment)–Sig. T2 (static segment)–NS	Lexical tone identification	Young adults (N = 18)	No	No
Hornickel et al. (2012)	/ba/–/da/–/ga/–Sig.	Wearing FM assisting listening devices	Children with dyslexia (N = 19)	Children with dyslexia (N = 19)	No
Skoe et al. (2014)	T1–Sig. T4–Sig.	Lexical tone identification	Young adults (N = 12)	Young adults (N = 13) Passive	No
Song et al. (2012)	/da/–Sig.	Speech-in-noise comprehension	Young adults (N = 28)	Young adults (N = 32) Passive	6 months Maintained
Russo et al. (2005)	/da/–Sig.	Phonological, auditory, and language processing: *Earobics*®	Children with learning disabilities (N = 9)	Children with (N = 5) and without (N = 5) learning disabilities	No
Anderson et al. (2013c)	/da/–Sig.	Auditory cognitive training: *Brain Fitness*™	Elderly adults (N = 35)	Elderly adults (N = 32) Active	6 months Largely maintained
Anderson et al. (2013b)	/da/–Sig.	Auditory cognitive training: *Brain Fitness*™	Elderly adults with (N = 15) or without (N = 14) hearing loss	Elderly adults with (N = 15) or without (N = 14) hearing loss Active	No

Stimulus column: stimuli used and whether training had a significant (*Sig.*) or not significant (*NS*) effect for at least one of the measures used to quantify FFR changes. *Treatment group*: specifies the population tested and the number of participants. *Control group*: specifies whether a control group was used or not and, if applicable, the number of participants included in the control group and whether the participants in the control group were engaged in an alternative training task (*Active*) or not (*Passive*). *Follow-up*: indicates the time at which any follow-up test was performed and whether FFR changes measured immediately after training were maintained at the follow-up test these improvements in FFR pitch-tracking accuracy are with respect to the trained F_0 contour. The only study that investigated this issue (Carcagno and Plack 2011b) found that improvements were specific to the presence/absence of a modulation in the F_0 contour, but they did not find evidence of specificity with respect to the trajectory (rising or falling) of a modulated F_0 contour. It also remains unclear whether certain F_0 contours are more likely to elicit FFR training effects compared to others. Stimuli with dynamically modulated F_0 contours seem more likely to elicit FFR training effects compared to stimuli with a static F_0 contour, possibly as a result of the fact that tracking a dynamically changing F_0 is more challenging and susceptible to tracking errors than encoding a static F_0. There is no clear evidence of larger FFR training effects for a particular F_0 contour among stimuli with dynamically modulated F_0 contours. For some Mandarin tones training effects were found in some studies (T2: Chandrasekaran et al. 2012; T1: Skoe et al. 2014) but not all (Song et al. 2008). This apparent discrepancy may reflect methodological differences (e.g., differences in the measures used to summarize FFR training effects), but it should also be kept in mind that in single studies with relatively modest sample sizes effect sizes may be either overestimated or underestimated. Only a meta analysis combining the results of several studies could clarify whether FFR training effects for these tones are smaller than for other Mandarin tones.

Speech perception training tasks seem to have positive effects on subcortical temporal encoding, resulting in more robust encoding of stimulus-related periodicities. These training benefits appear to be greater for stimuli presented in noise. Again this may reflect the fact that the encoding of stimulus features in noise is a more challenging task for the auditory system and more susceptible to encoding errors that may be remedied by training.

With the exception of the studies of Anderson et al. (2013b, c), none of the studies reviewed above used an active control group. This makes it difficult to determine whether some of the measured changes reflect practice in the specific training task or general effects of auditory training. Some of the studies did not have a control group at all (Song et al. 2008; Chandrasekaran et al. 2012), leaving open the possibility that the measured changes were not a direct consequence of training. Furthermore, in the study of Hornickel et al. (2012), even though a control group was included, FFR changes in the treatment group were not directly compared to FFR changes in the control group. Instead, a statistically significant FFR change in the treatment group, and a lack of statistically significant FFR change in the control group were taken as evidence of FFR plasticity. Unfortunately, this kind of comparison, although suggestive of such an effect, does not warrant such a conclusion (Nieuwenhuis et al. 2011). Given that the measured changes (e.g., accuracy of FFR pitch tracking) are closely linked to the training task (e.g., F_0 discrimination) it may seem unlikely that they would be caused by factors other than the training itself. However, FFR pitch tracking may improve not only as a result of changes in specific temporal encoding mechanisms but also as a result of a generic reduction in physiological noise, for example, due to the participant being more relaxed in later test sessions. Therefore, it is important that future studies of FFR training-induced changes use appropriate controls to identify the specific causes of FFR changes.

4.4 Mechanisms of Auditory Brainstem Plasticity in Short-Term Learning

There is currently no comprehensive mechanistic model of auditory brainstem plasticity induced by perceptual learning. This is perhaps not surprising because, besides the fact that interest in subcortical plasticity is relatively new, such models are likely to involve complex interactions between brainstem nuclei and cortical structures via the efferent system.

In vitro studies in brain slices indicate that cells from auditory brainstem nuclei display synaptic plasticity in the form of long-term potentiation and long-term depression (reviewed by Tzounopoulos and Kraus 2009). While to the best knowledge of the authors there have been no neurophysiological studies of subcortical plasticity induced by perceptual-discrimination training in nonhuman animals, several studies have investigated subcortical auditory plasticity induced by associative learning (for a discussion of differences between perceptual and associative learning see Weinberger 2008). These studies have shown that after auditory fear conditioning with a tone, the best frequency of IC neurons in the big brown bat (*Eptesicus fuscus*) shifted toward the frequency of the conditioned tone. These best-frequency shifts are short term, lasting only about 3–3.5 hours after the association phase (Suga and Ma 2003). Inactivation of the auditory cortex during the association phase blocks these best-frequency shifts, indicating that plasticity is mediated by corticofugal modulation via the efferent system. However, IC tuning shifts persist if the auditory cortex is inactivated after the conditioning phase, indicating that they do not depend on online corticofugal modulation (Gao and Suga 1998; Suga and Ma 2003). The importance of the auditory efferent system in mediating auditory plasticity has also been highlighted by studies showing that destruction of cortico-collicular neurons in ferrets selectively disrupts the ability to learn an auditory spatial localization task (Bajo et al. 2010; Bajo and King 2012).

Changes in the FFR after short-term training may occur either as a result of plastic changes that are local to brainstem circuits and/or as a result of online modulation by the auditory cortex via the efferent system. In the second case, FFR changes could not be considered the results of subcortical plasticity, as subcortical responses would passively reflect plasticity occurring at the cortical level. Krishnan and Gandour (2009) and Krishnan et al. (2010) have argued against this case, citing the fact that efferent activation is sluggish with respect to FFR responses that instead have short onset latencies (6–9 ms). For example, one well-studied efferent effect, the medial olivo-cochlear reflex, has an onset latency of about 25 ms (Backus and Guinan 2006). If the stimuli used for probing training effects had a static F_0, online efferent effects in this latency range could be useful to track the F_0 of later portions of the stimuli. However, it is unlikely that efferent effects in this latency range would be of much help in tracking the time-varying F_0 contours of the stimuli used in many FFR training studies unless the top-down efferent modulation used a memory trace of the stimulus to predict its time-varying F_0 and facilitate responses phase locked to the predicted F_0. Chandrasekaran et al. (2014) have emphasized the idea that efferent feedback may modulate brainstem activity on the basis of predictive coding schemes that attempt to anticipate the ongoing stimulation on the basis of preceding regularities in the stimulus stream. While there is disagreement on the relevance of efferent modulation for FFR changes after the training phase, most authors think that the efferent system is likely to be involved during the training phase. During the training phase efferent feedback could modulate brainstem activity online, as mentioned before, but the feedback could also be delayed and contingent on the outcome (correct or incorrect response) of a training trial.

The increased synchrony of FFR responses to stimulus-related periodicities induced by training and the apparent increase in SNR-based or autocorrelation-based measures may reflect either a greater accuracy of phase locking of single fibers to the periodicities in the stimulus or a greater proportion of fibers phase locking to stimulus-related periodicities. An increase in the proportion of neurons phase locking to stimulus-related periodicities may be the result of either the recruitment of additional fibers phase locking to these periodicities or the inhibition of fibers firing at different periods. Because the electroencephalogram reflects the summed response of many neurons, scalp-recorded FFRs cannot distinguish between these options. Nonetheless, it would be interesting to determine if increases in FFR SNR at the signal frequencies (or increases in ACF values at time lags related to the signal frequencies) after training are due to increased responses at the signal frequencies or decreased responses at the noise frequencies. It should be noted that a decreased response at the noise frequencies could reflect neuroplastic changes caused by training (e.g., inhibition of fibers not phase locked to the stimulus periodicities) but may also reflect a generic reduction of background physiological noise unrelated to neuroplastic changes, which could be caused instead by the participants being in a more relaxed state in the post-training FFR session. The use of a control group protects against this confound because participants in the control group should also be more relaxed at the post-training FFR session. However, this protection is not full because the greater familiarity that participants in the training group have with the stimuli used during the FFR recording may lead them to a more relaxed state compared to participants in the control group. Future studies should address this issue to ensure that measured FFR training effects truly reflect neuroplastic changes.

4.5 Summary

The research summarized in this chapter suggests that subcortical temporal processing changes as a result of training in pitch discrimination/identification tasks and in speech perception tasks. Training in these tasks leads to better tracking of stimulus-related periodicities and/or better timing of neural responses to sounds. This research, along with research on contextual effects on brainstem auditory processing (see Escera, Chap. 5) changes the view of auditory brainstem nuclei as static processing modules. Brainstem auditory processing appears instead to be dynamic because it shows short-term adaptation to the stimulus context, as well as long-term training induced plasticity. In a way this is not surprising given the richness and complexity of subcortical auditory processing compared to the more circumscribed role that subcortical processing plays for other senses such as vision (King and Nelken 2009).

The idea that subcortical auditory processing can be modified by short-term training has important practical implications. Subcortical auditory processing is disrupted in certain populations of listeners such as children with language-based learning impairments and the elderly. Some of the studies reviewed in this chapter provide promising evidence that some of these deficits can be ameliorated by perceptual training. Further research is necessary to explore the full potential of auditory training to address nonperipheral hearing deficits and to find optimal training paradigms whose benefits translate to improvements in everyday communication settings that are maintained over time.

Although large inter-individual differences in the amount of perceptual learning are often noted in the literature, few studies have been designed to understand the causes of these differences. The results of Chandrasekaran et al. (2012) and Song et al. (2012), showing relationships between pre-training FFR metrics and perceptual learning, represent nice examples of how electrophysiological studies can shed light on the nature of these inter-individual differences. More studies are necessary to better understand the nature of inter-individual differences in perceptual learning. This would be important not only to improve theoretical understanding of perceptual learning but also for the development of individually tailored training programs in applied settings.

Much work remains to be done in order to understand how FFR plasticity occurs, both at the level of single neurons and at the level of neural circuits. This work will likely need to unravel complex interactions between subcortical nuclei and the efferent feedback from cortical areas. Other aspects of FFR plasticity that have not been fully explored are its degree of specificity with respect to particular stimulus features and its degree of generalization to untrained stimuli. Short-term training studies, although expensive in terms of time and resources, will be fundamental to answering these questions.

Compliance with Ethics Requirements Samuele Carcagno and Christopher J. Plack declared that they had no conflict of interest.

References

Amitay, S., Zhang, Y.-X., Jones, P. R., & Moore, D. R. (2014). Perceptual learning: Top to bottom. *Vision Research, 99*, 69–77.

Ananthanarayan, A. K., & Durrant, J. D. (1992). The frequency-following response and the onset response: Evaluation of frequency specificity using a forward-masking paradigm. *Ear and Hearing, 13*(4), 228–232.

Andersen, G. J., Ni, R., Bower, J. D., & Watanabe, T. (2010). Perceptual learning, aging, and improved visual performance in early stages of visual processing. *Journal of Vision, 10*(13), article 4, 1–13.

Anderson, S., Parbery-Clark, A., White-Schwoch, T., & Kraus, N. (2012). Aging affects neural precision of speech encoding. *The Journal of Neuroscience, 32*(41), 14156–14164.

Anderson, S., Parbery-Clark, A., White-Schwoch, T., Drehobl, S., & Kraus, N. (2013a). Effects of hearing loss on the subcortical representation of speech cues. *The Journal of the Acoustical Society of America, 133*(5), 3030–3038.

Anderson, S., White-Schwoch, T., Choi, H. J., & Kraus, N. (2013b). Training changes processing of speech cues in older adults with hearing loss. *Frontiers in Systems Neuroscience, 7*. Doi:10.3389/fnsys.2013.00097

Anderson, S., White-Schwoch, T., Parbery-Clark, A., & Kraus, N. (2013c). Reversal of age-related neural timing delays with training. *Proceedings of the National Academy of Sciences of the U S A, 110*(11), 4357–4362.

Anderson, S., White-Schwoch, T., Choi, H. J., & Kraus, N. (2014). Partial maintenance of auditory-based cognitive training benefits in older adults. *Neuropsychologia, 62*, 286–296.

Atienza, M., Cantero, J. L., & Dominguez-Marin, E. (2002). The time course of neural changes underlying auditory perceptual learning. *Learning & Memory, 9*(3), 138–150.

Backus, B. C., & Guinan, J. J. (2006). Time-course of the human medial olivocochlear reflex. *The Journal of the Acoustical Society of America, 119*(5), 2889–2904.

Bajo, V. M., & King, A. J. (2012). Cortical modulation of auditory processing in the midbrain. *Frontiers in Neural Circuits, 6*. Doi:10.3389/fncir.2012.00114

Bajo, V. M., Nodal, F. R., Moore, D. R., & King, A. J. (2010). The descending corticocollicular pathway mediates learning-induced auditory plasticity. *Nature Neuroscience, 13*(2), 253–260.

Bharadwaj, H. M., & Shinn-Cunningham, B. G. (2014). Rapid acquisition of auditory subcortical steady state responses using multichannel recordings. *Clinical Neurophysiology, 125*(9), 1878–1888.

Bidelman, G. M. (2015). Multichannel recordings of the human brainstem frequency-following response: Scalp topography, source generators, and distinctions from the transient ABR. *Hearing Research, 323*, 68–80.

Bones, O., & Plack, C. J. (2015). Losing the music: Aging affects the perception and subcortical neural representation of musical harmony. *The Journal of Neuroscience, 35*(9), 4071–4080.

Bosnyak, D. J., Eaton, R. A., & Roberts, L. E. (2004). Distributed auditory cortical representations are modified when non-musicians are trained at pitch discrimination with 40 Hz amplitude modulated tones. *Cerebral Cortex, 14*(10), 1088–1099.

Bower, J. D., Watanabe, T., & Andersen, G. J. (2013). Perceptual learning and aging: Improved performance for low-contrast motion discrimination. *Frontiers in Psychology, 4*. Doi:10.3389/fpsyg.2013.00066

Brown, M., Irvine, D. R. F., & Park, V. N. (2004). Perceptual learning on an auditory frequency discrimination task by cats: Association with changes in primary auditory cortex. *Cerebral Cortex, 14*(9), 952–965.

Carcagno, S., & Plack, C. J. (2011a). Pitch discrimination learning: Specificity for pitch and harmonic resolvability, and electrophysiological correlates. *Journal of the Association for Research in Otolaryngology, 12*(4), 503–517.

Carcagno, S., & Plack, C. J. (2011b). Subcortical plasticity following perceptual learning in a pitch discrimination task. *Journal of the Association for Research in Otolaryngology, 12*, 89–100.

Carlyon, R. P. (2004). How the brain separates sounds. *Trends in Cognitive Sciences, 8*(10), 465–471.

Chandrasekaran, B., & Kraus, N. (2012). Biological factors contributing to reading ability: Subcortical auditory function. In A. A. Benasich & R. H. Fitch (Eds.), *Developmental dyslexia: Early precursors, neurobehavioral markers and biological substrates* (pp. 83–98). Baltimore: Paul H. Brookes Publishing.

Chandrasekaran, B., Kraus, N., & Wong, P. C. M. (2012). Human inferior colliculus activity relates to individual differences in spoken language learning. *Journal of Neurophysiology, 107*(5), 1325–1336.

Chandrasekaran, B., Skoe, E., & Kraus, N. (2014). An integrative model of subcortical auditory plasticity. *Brain Topography, 27*(4), 539–552.

Clinard, C. G., & Tremblay, K. L. (2013). Aging degrades the neural encoding of simple and complex sounds in the human brainstem. *Journal of the American Academy of Audiology, 24* (7), 590–599.

Clinard, C. G., Tremblay, K. L., & Krishnan, A. R. (2010). Aging alters the perception and physiological representation of frequency: Evidence from human frequency-following response recordings. *Hearing Research, 264*(1–2), 48–55.

Dau, T. (2003). The importance of cochlear processing for the formation of auditory brainstem and frequency following responses. *The Journal of the Acoustical Society of America, 113*(2), 936–950.

de Cheveigné, A. (2005). Pitch perception models. In R. Fay, A. N. Popper, C. J. Plack, & A. J. Oxenham (Eds.), *Pitch: Neural coding and perception* (pp. 169–233). New York: Springer.

Dediu, D., & Ladd, D. R. (2007). Linguistic tone is related to the population frequency of the adaptive haplogroups of two brain size genes, ASPM and microcephalin. *Proceedings of the National Academy of Sciences of the U S A, 104*(26), 10944–10949.

Deveau, J., & Seitz, A. R. (2014). Applying perceptual learning to achieve practical changes in vision. *Frontiers in Psychology, 5*(1166), 1–6.

de Villers-Sidani, E., Alzghoul, L., Zhou, X., Simpson, K. L., et al. (2010). Recovery of functional and structural age-related changes in the rat primary auditory cortex with operant training. *Proceedings of the National Academy of Sciences of the U S A, 107*(31), 13900–13905.

Dosher, B. A., & Lu, Z. L. (1999). Mechanisms of perceptual learning. *Vision Research, 39*(19), 3197–3221.

Drayna, D., Manichaikul, A., de Lange, M., Snieder, H., & Spector, T. (2001). Genetic correlates of musical pitch recognition in humans. *Science, 291*(5510), 1969–1972.

Drennan, W. R., Won, J. H., Dasika, V. K., & Rubinstein, J. T. (2007). Effects of temporal fine structure on the lateralization of speech and on speech understanding in noise. *Journal of the Association for Research in Otolaryngology, 8*(3), 373–383.

Engen, K. J. V. (2012). Speech-in-speech recognition: A training study. *Language and Cognitive Processes, 27*(7–8), 1089–1107.

Fu, Q.-J., & Galvin, J. J., 3rd. (2007). Perceptual learning and auditory training in cochlear implant recipients. *Trends in Amplification, 11*(3), 193–205.

Gao, E., & Suga, N. (1998). Experience-dependent corticofugal adjustment of midbrain frequency map in bat auditory system. *Proceedings of the National Academy of Sciences of the U S A, 95* (21), 12663–12670.

Gilbert, C. D., Sigman, M., & Crist, R. E. (2001). The neural basis of perceptual learning. *Neuron, 31*(5), 681–697.

Gockel, H. E., Carlyon, R. P., Mehta, A., & Plack, C. J. (2011). The frequency following response (FFR) may reflect pitch-bearing information but is not a direct representation of pitch. *Journal of the Association for Research in Otolaryngology, 12*(6), 767–782.

Gockel, H. E., Krugliak, A., Plack, C. J., & Carlyon, R. P. (2015). Specificity of the human frequency following response for carrier and modulation frequency assessed using adaptation. *Journal of the Association for Research in Otolaryngology, 16*(6), 747–762.

Goldstone, D. R. L., Braithwaite, D. W., & Byrge, L. A. (2012). Perceptual learning. In P. D. N. M. Seel (Ed.), *Encyclopedia of the sciences of learning* (pp. 2580–2583). New York: Springer.

Grill-Spector, K., Henson, R., & Martin, A. (2006). Repetition and the brain: Neural models of stimulus-specific effects. *Trends in Cognitive Sciences, 10*(1), 14–23.

Grothe, B., Pecka, M., & McAlpine, D. (2010). Mechanisms of sound localization in mammals. *Physiological Reviews, 90*(3), 983–1012.

Hensch, T. K. (2004). Critical period regulation. *Annual Review of Neuroscience, 27*, 549–579.

Herholz, S. C., & Zatorre, R. J. (2012). Musical training as a framework for brain plasticity: Behavior, function, and structure. *Neuron, 76*(3), 486–502.

Hornickel, J., Zecker, S. G., Bradlow, A. R., & Kraus, N. (2012). Assistive listening devices drive neuroplasticity in children with dyslexia. *Proceedings of the National Academy of Sciences of the U S A, 109*(41), 16731–16736.

Jones, P. R., Moore, D. R., Shub, D. E., & Amitay, S. (2015). The role of response bias in perceptual learning. *Journal of Experimental Psychology: Learning, Memory, and Cognition, 41*(5), 1456–1470.

King, A. J., & Nelken, I. (2009). Unraveling the principles of auditory cortical processing: Can we learn from the visual system? *Nature Neuroscience, 12*(6), 698–701.

Kraus, N., Skoe, E., Parbery-Clark, A., & Ashley, R. (2009). Experience-induced malleability in neural encoding of pitch, timbre, and timing. *Annals of the New York Academy of Sciences, 1169*, 543–557.

Kraus, N., & White-Schwoch, T. (2015). Unraveling the biology of auditory learning: A cognitive-sensorimotor-reward framework. *Trends in Cognitive Sciences, 19*(11), 642–654.

Krishnan, A. (2007). Frequency-following response. In R. F. Burkard, J. J. Eggermont, & M. Don (Eds.), *Auditory evoked potentials: Basic principles and clinical applications* (pp. 313–333). Philadelphia: Lippincott Williams.

Krishnan, A., & Gandour, J. T. (2009). The role of the auditory brainstem in processing linguistically-relevant pitch patterns. *Brain and Language, 110*(3), 135–148.

Krishnan, A., Xu, Y., Gandour, J., & Cariani, P. (2005). Encoding of pitch in the human brainstem is sensitive to language experience. *Cognitive Brain Research, 25*(1), 161–168.

Krishnan, A., Gandour, J. T., & Bidelman, G. M. (2010). The effects of tone language experience on pitch processing in the brainstem. *Journal of Neurolinguistics, 23*(1), 81–95.

Lively, S. E., Pisoni, D. B., Yamada, R. A., Tohkura, Y., & Yamada, T. (1994). Training Japanese listeners to identify English /r/ and /l/. III. Long-term retention of new phonetic categories. *The Journal of the Acoustical Society of America, 96*(4), 2076–2087.

Marmel, F., Linley, D., Carlyon, R. P., Gockel, H. E., et al. (2013). Subcortical neural synchrony and absolute thresholds predict frequency discrimination independently. *Journal of the Association for Research in Otolaryngology, 14*(5), 757–766.

Menning, H., Roberts, L. E., & Pantev, C. (2000). Plastic changes in the auditory cortex induced by intensive frequency discrimination training. *NeuroReport, 11*(4), 817–822.

Micheyl, C., Delhommeau, K., Perrot, X., & Oxenham, A. J. (2006). Influence of musical and psychoacoustical training on pitch discrimination. *Hearing Research, 219*(1–2), 36–47.

Mollon, J. D., & Danilova, M. V. (1996). Three remarks on perceptual learning. *Spatial Vision, 10* (1), 51–58.

Monaghan, P., Metcalfe, N. B., & Ruxton, G. D. (1998). Does practice shape the brain? *Nature, 394*(6692), 434.

Moore, B. C. J. (2008). The role of temporal fine structure processing in pitch perception, masking, and speech perception for normal-hearing and hearing-impaired people. *Journal of the Association for Research in Otolaryngology, 9*(4), 399–406.

Moore, D. R., & Shannon, R. V. (2009). Beyond cochlear implants: Awakening the deafened brain. *Nature Neuroscience, 12*(6), 686–691.

Nieuwenhuis, S., Forstmann, B. U., & Wagenmakers, E.-J. (2011). Erroneous analyses of interactions in neuroscience: A problem of significance. *Nature Neuroscience, 14*(9), 1105–1107.

Petrov, A. A., Dosher, B. A., & Lu, Z.-L. (2005). The dynamics of perceptual learning: An incremental reweighting model. *Psychological Review, 112*(4), 715–743.

Plack, C. J., & Oxenham, A. J. (2005). The Psychophysics of Pitch. In R. Fay, A. N. Popper, C. J. Plack, & A. J. Oxenham (Eds.), *Pitch: Neural coding and perception*. New York: Springer.

Polat, U., Schor, C., Tong, J.-L., Zomet, A., et al. (2012). Training the brain to overcome the effect of aging on the human eye. *Scientific Reports, 2*, 278. Doi:10.1038/srep00278

Recanzone, G. H., Schreiner, C. E., & Merzenich, M. M. (1993). Plasticity in the frequency representation of primary auditory cortex following discrimination training in adult owl monkeys. *The Journal of Neuroscience, 13*(1), 87–103.

Reed, A., Riley, J., Carraway, R., Carrasco, A., et al. (2011). Cortical map plasticity improves learning but is not necessary for improved performance. *Neuron, 70*(1), 121–131.

Rosen, S. (1992). Temporal information in speech: Acoustic, auditory and linguistic aspects. *Philosophical Transactions of the Royal Society of London B: Biological Sciences, 336*(1278), 367–373.

Russo, N. M., Nicol, T. G., Zecker, S. G., Hayes, E. A., & Kraus, N. (2005). Auditory training improves neural timing in the human brainstem. *Behavioural Brain Research, 156*(1), 95–103.

Sheehan, K. A., McArthur, G. M., & Bishop, D. V. M. (2005). Is discrimination training necessary to cause changes in the P2 auditory event-related brain potential to speech sounds? *Cognitive Brain Research, 25*(2), 547–553.

Skoe, E., Chandrasekaran, B., Spitzer, E. R., Wong, P. C. M., & Kraus, N. (2014). Human brainstem plasticity: The interaction of stimulus probability and auditory learning. *Neurobiology of Learning and Memory, 109*, 82–93.

Skuk, V. G., & Schweinberger, S. R. (2014). Influences of fundamental frequency, formant frequencies, aperiodicity, and spectrum level on the perception of voice gender. *Journal of Speech, Language, and Hearing Research, 57*(1), 285–296.

Song, J. H., Nicol, T., & Kraus, N. (2011a). Test-retest reliability of the speech-evoked auditory brainstem response. *Clinical Neurophysiology, 122*(2), 346–355.

Song, J. H., Nicol, T., & Kraus, N. (2011b). Reply to Test–retest reliability of the speech-evoked ABR is supported by tests of covariance. *Clinical Neurophysiology, 122*(9), 1893–1895.

Song, J. H., Skoe, E., Banai, K., & Kraus, N. (2012). Training to improve hearing speech in noise: Biological mechanisms. *Cerebral Cortex, 22*(5), 1180–1190.

Song, J. H., Skoe, E., Wong, P. C. M., & Kraus, N. (2008). Plasticity in the adult human auditory brainstem following short-term linguistic training. *Journal of Cognitive Neuroscience, 20*(10), 1892–1902.

Sowden, P. T., Davies, I. R., & Roling, P. (2000). Perceptual learning of the detection of features in X-ray images: A functional role for improvements in adults' visual sensitivity? *Journal of Experimental Psychology: Human Perception and Performance, 26*(1), 379–390.

Suga, N., & Ma, X. (2003). Multiparametric corticofugal modulation and plasticity in the auditory system. *Nature Reviews Neuroscience, 4*(10), 783–794.

Tong, Y., Melara, R. D., & Rao, A. (2009). P2 enhancement from auditory discrimination training is associated with improved reaction times. *Brain Research, 1297*, 80–88.

Tremblay, K. L., Ross, B., Inoue, K., McClannahan, K., & Collet, G. (2014). Is the auditory evoked P2 response a biomarker of learning? *Frontiers in Systems Neuroscience, 8*(28), 1–13.

Tsodyks, M., & Gilbert, C. (2004). Neural networks and perceptual learning. *Nature, 431*(7010), 775–781.

Tzounopoulos, T., & Kraus, N. (2009). Learning to encode timing: Mechanisms of plasticity in the auditory brainstem. *Neuron, 62*(4), 463–469.

Watanabe, T., & Sasaki, Y. (2015). Perceptual learning: Toward a comprehensive theory. *Annual Review of Psychology, 66*, 197–221.

Weinberger, N. M. (2008). Cortical plasticity in associative learning and memory. In J. H. Byrne (Ed.), *Learning and memory: A comprehensive reference. Vol. 3. Memory systems* (pp. 187–218). Oxford: Academic Press.

Wong, P. C. M., Skoe, E., Russo, N. M., Dees, T., & Kraus, N. (2007). Musical experience shapes human brainstem encoding of linguistic pitch patterns. *Nature Neuroscience, 10*(4), 420–422.

Wright, B. A., & Zhang, Y. (2009). A review of the generalization of auditory learning. *Philosophical Transactions of the Royal Society of London B: Biological Sciences, 364*(1515), 301–311.

Wright, B. A., Wilson, R. M., & Sabin, A. T. (2010). Generalization lags behind learning on an auditory perceptual task. *The Journal of Neuroscience, 30*(35), 11635–11639.

Xu, Y., Krishnan, A., & Gandour, J. T. (2006). Specificity of experience-dependent pitch representation in the brainstem. *NeuroReport, 17*(15), 1601–1605.

Chapter 5
The Role of the Auditory Brainstem in Regularity Encoding and Deviance Detection

Carles Escera

Abstract How does a listener perceive the auditory world and make sense from the myriad of concurrent sounds in the noisy and complex soundscape impinging our ears as a continuous flow? A major emerging view in cognitive auditory neuroscience is that the auditory system implements a pervasive mechanism by which dynamic auditory input is modeled into neural traces of regularities that allow the system to derive perceptual auditory objects. A large number of studies that used auditory sequences of various statistical complexities and that were performed with a range of neuroscience methods (e.g., neuro-imaging, electroencephalography and auditory evoked potentials, single neuron recordings) together have shown that regularity encoding and deviance detection is a key property of the auditory cortex. Furthermore, recordings in the inferior colliculus (IC) and the auditory thalamus of experimental animals have disclosed stimulus-specific adaptation (SSA) at these levels of the auditory pathway, challenging the corticocentric view of regularity encoding and deviance detection. Together with recent experiments using oddball sequences to measure early auditory evoked potentials, such as the middle latency response (MLR) and particularly the frequency-following response (FFR), those studies support the emerging view that regularity encoding and deviance detection are a key functional properties of the entire auditory system from at least the IC to high-order auditory cortical regions, and that the subcortical auditory pathway can implement certain forms of "primitive intelligence", thereby contributing to auditory cognition.

Keywords Auditory cognition · Change detection · cABR · FFR · Frequency-following response · MLR · Mismatch negativity · MMN · Novelty detection · Oddball paradigm · Repetition suppression · Roving-standard paradigm · SSA · Stimulus statistics

---

C. Escera (✉)
Brainlab-Cognitive Neuroscience Research Group, Department of Clinical Psychology and Psychobiology, Faculty of Psychology, University of Barcelona, P. Vall d'Hebron 171, ES-08035-Barcelona Catalonia, Spain
e-mail: cescera@ub.edu

N. Kraus et al. (eds.), *The Frequency-Following Response*,
Springer Handbook of Auditory Research 61,
DOI 10.1007/978-3-319-47944-6_5

5.1 Introduction

How does a listener perceive the auditory world and make sense from the continuous flow of the myriad concurrent sounds in the noisy and complex soundscape impinging on our ears? A major emerging view in cognitive auditory neuroscience is that dynamic auditory input is modeled as neural traces of regularities that allow the formation of perceptual auditory objects. Indeed, sounds do not occur in isolation but are generally integrated into more complex sound patterns, as in speech, music, animal vocalizations, or common sounds such as a cell-phone ringtone. In such cases, temporal integration of ongoing sensory input plays an important role in organizing the acoustic background and thus guiding perception (Bregman 1990; Winkler 2007).

Modeling the auditory scene in search of regularities is essential not only to organize the acoustic background into meaningful percepts but also to predict future sensory events (Friston 2005; Winkler et al. 2009) and to guide attention involuntarily to potential relevant events outside the focus of attention (Escera et al. 1998; Escera and Corral 2007). This major view in cognitive neuroscience has emerged from the successful combination of the empirical, cognitive psychophysiological research that made use of mismatch negativity (MMN) (Näätänen et al. 2007), the auditory evoked potential (AEP) derived from the electroencephalogram (EEG), and several other methods. The theoretical neuroscience approach has led to the formulation of predictive coding as a general theory of perceptual inference (Friston et al. 2006). While all this research was conducted on auditory cortical responses (Deouell 2007), and all the theoretical formulation refers to the cerebral cortex (Friston 2005), neurophysiological investigations in humans and animal models show that the subcortical auditory pathway contributes to these predictive processes and, by extension, auditory cognition.

This chapter begins with a review of studies on a particular form of auditory plasticity that can be regarded as *on-line*, which focuses on how the auditory system captures the ongoing stimulation "on the fly", hence adapting to the moment on a temporal scale of only a few seconds (Sects. 5.2–5.5). Section 5.6 briefly considers how the auditory system's plasticity incorporates temporal scales that range from minutes to the entire life and covers how these two forms of auditory plasticity interact with each other. The integration of these two areas of research supports the emerging view of *regularity encoding* as a key property of the whole auditory system.

5.2 Regularity Encoding in Auditory Cortex

A broadly used approach to examine whether the acoustic environment has been internalized into neural traces is by means of *oddball* auditory sequences. In these sequences, a repetitive ("standard" or "common") stimulus is presented with a high probability of occurrence, whereas a different stimulus, referred to as "deviant" or "rare," occurs only occasionally (Fig. 5.1A). These latter stimuli elicit a typical human AEP, the MMN, peaking at 100–150 ms from deviant sound onset. The brain's neurophysiologic response to such rare stimuli is taken as evidence that the auditory system has built up a neural representation of the preceding sound regularity (Näätänen and Winkler 1999; Winkler et al. 2009).

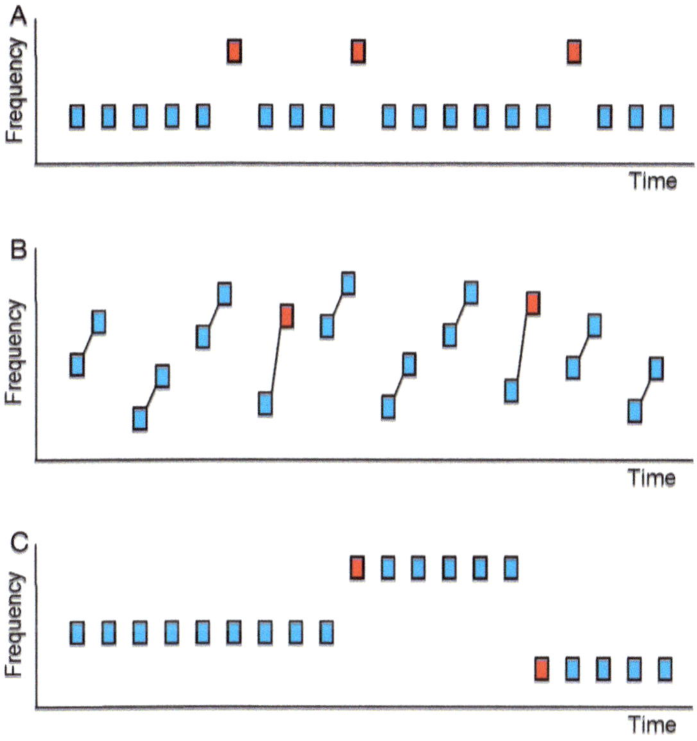

Fig. 5.1 Schematic illustration of different approaches to studying regularity encoding in the auditory system. (**A**) In *simple oddball* sequences, a "standard" stimulus is repeated with high probability (*blue*) at regular or random inter-stimulus intervals, whereas a "rare" or "deviant" stimulus differing in any physical feature (frequency in the plot) occurs with low probability and unexpectedly (*red*). Notice that complex sounds, such as phonemes, syllables, or musical notes or chords can also be used in simple oddball sequences. (**B**) In *complex oddball* sequences, the regularity is not defined by mere stimulus repetition but by the contingency between successive discrete sounds. In this example using tone pairs, standard events (*blue*) are defined by a frequency relationship of the second tone, for example, a semi-tone higher than its pair, whereas the deviant event (*red*) features a relationship that is double in pitch. Notice that frequency varies across the entire spectrum for the different pairs precluding the encoding of any particular pitch as the regularity. (**C**) In the *roving-standard* sequence, a particular stimulus is repeated for a number of times and then it changes, for example, in its frequency. The first stimulus after the feature change is a deviant (*red*), whereas after it is repeated at least twice it becomes a standard (*blue*). With this approach it is possible to study not only deviant responses but also how regularity encoding evolves as a function of the number of standard stimulus repetitions

A large body of EEG evidence suggests the MMN has cortical generators in supratemporal and prefrontal cortices (Deouell 2007) in agreement with functional neuroimaging data. Indeed, using oddball sequences in blocked or event-related functional magnetic resonance imaging studies has confirmed the involvement of the both the temporal and prefrontal cortices in the neural representation of sound regularities (Opitz et al. 1999; Sabri et al. 2004).

Encoding auditory regularities does not occur only for simple acoustic feature repetitions (e.g., frequency, intensity, or duration) and complex discrete stimuli (e.g., speech sounds) but also for complex contingencies between single auditory events, such as the frequency relationship between two tones within a pair or the combination of two sound features (e.g., pitch and duration, Fig. 5.1B) (Paavilainen 2013). These more complex kinds of regularity, and particularly those defined by the relationship between stimulus features that evolve in time but vary along that feature dimension, have supported the view that the auditory cortex implements pre-attentive cognitive operations to make predictions about the near future, a kind of "primitive intelligence" in audition (Näätänen et al. 2001, 2010). Given the cortical nature of the evidence considered above, the function of regularity encoding and deviance detection has been suggested to pertain to high level cognition. However, recordings in the midbrain and the auditory thalamus of experimental animals have disclosed stimulus repetition effects at these stations of the auditory pathway, challenging the corticocentric view of regularity encoding and deviance detection. As discussed in Sect. 5.6, even the auditory brainstem can show such forms of primitive intelligence.

Another, more direct approach to investigate regularity encoding in audition is by using the so-called *roving-standard* sequences (Fig. 5.1C), in which trains consisting each of a different number of repetitive stimuli are isochronously presented and a particular stimulus feature is changed in every train. This particular type of sequence allows not only studying deviance-related responses (by comparing the brain response across the first tone within the trains to that across the last trains' tones), but also how regularity encoding evolves as a function of stimulus repetition. The use of this approach reveals *repetition suppression* (Desimone 1996) as the mechanism of regularity encoding, viewed as the reduction of prediction error (Grill-Spector et al. 2006) in the predictive coding framework introduced in Sect. 5.1. Studies of the human AEP correlate of repetition suppression have revealed that the number of repetitions (Haenschel et al. 2005) and temporal predictability (Costa-Faidella et al. 2011a) are key factors in cortical auditory repetition suppression (Recasens et al. 2015).

5.3 Neurophysiologic Mechanisms in Regularity Encoding

A step forward in understanding regularity encoding in the auditory system was provided by single-unit and multi-unit recordings in animals, which have revealed the existence of *stimulus-specific adaptation* (SSA) (Ulanovsky et al. 2003) at different levels of the auditory system. SSA neurons rapidly reduce their firing rates after a few repetitions of a sound, but robust responses are restored to a rare or deviant stimulus. Interestingly, SSA shares a number of properties with the MMN cortical potential introduced in Sect. 5.1, such as enhancement by increasing the physical difference between the standard and the deviant tones or by reducing the probability of the rare stimulus. These similarities have led to the suggestion that SSA underlies MMN generation; however, existing differences between these two phenomena, the most relevant of which are in latency and anatomical location, indicate that SSA may well lie upstream of MMN generation (Nelken and Ulanovsky 2007). In other words, other intervening processes may occur between SSA and MMN as discussed in Sect. 5.4.

Neurons exhibiting SSA were first described in the primary auditory cortex of the cat (Ulanovsky et al. 2003; Nelken 2014), but they were subsequently discovered in subcortical stations of the auditory pathway, such as the inferior colliculus (IC) of the midbrain (Pérez-González et al. 2005; Malmierca et al. 2009) and medial geniculate body (MGB) of the thalamus (Antunes et al. 2010). Importantly, SSA is present in primary auditory cortex (i.e., the target of the ascending lemniscal pathway), whereas SSA in subcortical stations is stronger in non-lemniscal regions of these nuclei, such as the dorsal and rostral parts of IC (Malmierca et al. 2009) and the dorsal and medial subdivisions of MGB (Antunes et al. 2010), in agreement with seminal intracranial recordings in guinea pigs (Kraus et al. 1994). Thus, while subcortical SSA was originally suggested to originate in auditory cortex and then transmitted via the corticofugal pathway (Nelken and Ulanovsky 2007) to lower auditory centers, studies of transient deactivation of the auditory cortex suggest that SSA may emerge de novo in subcortical stations (Antunes and Malmierca 2011; Anderson and Malmierca 2013), demonstrating the genuine role of the IC and MGB in regularity encoding. The current view accepts the existence of two relatively independent systems for SSA: one lemniscal that is linked to cortex and another non-lemniscal that is linked to the subcortical auditory pathway (Nelken 2014; Malmierca et al. 2015). The functional relationship between these two systems remains to be established. Importantly, these results pave the way toward more fine-grained research of auditory regularity encoding in humans.

5.4 Regularity Encoding in Thalamocortical Networks

The existence of SSA neurons distributed along the animal auditory pathway and the fact that the latency of the novelty responses in these neurons (e.g., their responses to the rare stimuli) is about 100 ms shorter than the typical MMN latency suggest that earlier correlates of deviant detection could be found in humans. Indeed, a series of recent experiments in humans using oddball sequences that set the stimulation, recording, and analysis parameters to measure earlier auditory evoked potentials, such as the auditory brainstem response (ABR) and the middle latency response (MLR), supported the emerging view that regularity encoding and deviance detection are pervasive properties of the auditory system as a whole.

The MLR is a well-characterized sequence of waveforms in the human auditory evoked potential peaking between 12 and 50 ms from sound onset. They are labeled as N_0, P_0, Na, Pa, Nb, and Pb, with earliest components (N_0 and P_0) generated in auditory thalamocortical loops (Picton 2011) and later ones generated in primary auditory cortex (Na and Pa) or beyond (Yvert et al. 2001). Importantly, in oddball experiments aimed at recording MLRs and ABRs, it is necessary to control for stimulus characteristics as these brain responses are sensitive to the specific features of the eliciting stimuli (for recent evidence see Althen et al. 2011); also, controlling for probability factors is important. Indeed, to disclose genuine regularity encoding and disregard adaptation effects yielded by the lower probability of the rare sounds within the oddball sequence, a specific "control" condition needs to be implemented (Fig. 5.2A) (Schröger and Wolff 1996; Ruhnau et al. 2012).

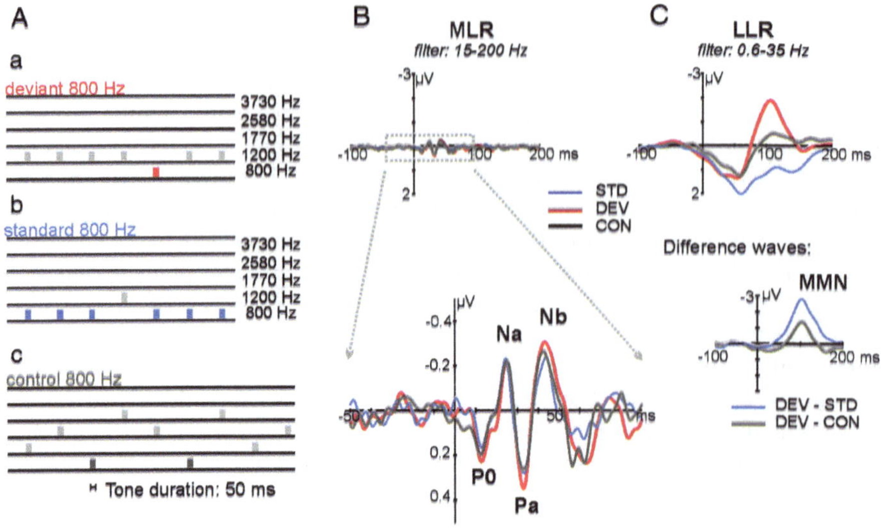

Fig. 5.2 Middle-latency response (*MLR*) correlates of deviance detection. (**A**) Experimental design: A deviant (*DEV*) tone of 800 Hz was delivered with a p = 2.0 among a series of standard (*STD*) tones of 1200 Hz as in (*a*); a reversed-oddball block as in (*b*) allowed for the comparison of responses elicited to the same physical tones in the role of standard; critically, a controlled block (*c*) allowed attributing the effects to true regularity encoding. (**B**) When the brain response was filtered in the MLR range (15–200 Hz), a clear Nb enhancement was elicited to the deviant stimulus in comparison to both the standard and the control tones (same physical stimuli in separate blocks). (**C**) When filtered in the long-latency response (*LLR*) range (0.6–35 Hz), the auditory evoked potential disclosed a remarkable amplitude enhancement to the deviant stimulus, which was generated (shown in the difference waveforms, *lower right*) by both refractoriness (comparison to the standard; *DEV-STD*) and true deviance detection (comparison to the control stimulus; *DEV-CON*). MMN, mismatch negativity; Na, Nb, PO, Pa, MLR waveforms. (Modified with permission from Grimm et al. 2011)

Using such a methodological approach, a number of recent studies revealed correlates of genuine regularity encoding and deviance detection in early thalamocortical networks or early cortical regions (Recasens et al. 2014). Indeed, several MLR waveforms were enhanced for changes in tone frequency (Fig. 5.2) (Grimm et al. 2011; Alho et al. 2012), location (Cornella et al. 2012; Grimm et al. 2012), intensity (Althen et al. 2011), and the spectral content of sound (Slabu et al. 2010). Also, the temporal dynamics of stimulus presentation were tracked by these early thalamocortical networks, as a stimulus occurring earlier than expected elicited clear enhancements of MLR waveforms (Leung et al. 2013).

Considering the time frame of these effects, 20–40 ms from change onset (in neural networks anatomically lower and processing stages about 100 ms earlier than those involved in MMN generation), it has been suggested recently that the deviance-related effects seen at the MLR range might be a better correlate of SSA than the MMN (Escera et al. 2014; Grimm et al. 2016). This is also supported by the fact that MMN is N-methyl-D-aspartate (NMDA) dependent (Umbricht et al. 2000), whereas SSA is not (Farley et al. 2010). In fact, a recent animal study was able to identify two stages of SSA in auditory cortex, the latter of them spanning 200–400 ms and being sensitive to NMDA blockade (Chen et al. 2015), thereby supporting the dissociation between regularity encoding and deviance detection at early (MLR) and later (MMN) processing stages. Moreover, the two stages of regularity encoding and deviance detection, early and late, have also been dissociated with regard to their functional implication in these processes (Cornella et al. 2013; Aghamolaei et al. 2016), indicating that the early thalamocortical networks of the auditory pathway are capable of coding for regularities, but that it takes a further processing step in the cerebral cortex to encode the deviant status. Taken together, the existence of SSA neurons along the animal auditory pathway, the evidence for MMN, and the early correlates of deviance detection in humans support the notion that the encoding of acoustic regularities and the detection of related deviance is a pervasive property of the entire auditory system, spanning from lower levels in the auditory pathway up to higher-order levels of the auditory cortex (Grimm and Escera 2012; Escera and Malmierca 2014; Escera et al. 2014).

5.5 Regularity Encoding in Human Auditory Brainstem

In humans, the involvement of subcortical stations in regularity encoding was recently demonstrated in a functional magnetic resonance imaging (fMRI) study that used the appropriate control and oddball sequences (Cacciaglia et al. 2015). In this study, the oddball trains were composed of two sounds, 500–1000 Hz and 1000–1500 Hz as standard and deviant, respectively, the latter occurring with a 20% probability in the second half of the train. This way, activation to the first part of the sequence that did not contain any deviant sound served as the standard condition for comparison with the second half of the sequence, where deviant sounds occurred. The fMRI acquisition parameters were set to capture activations in structures of the ascending auditory pathway, specifically, the orientation angle was set to 45° with respect to the longitudinal axis of the brainstem. Results yielded significant activations in both the left IC and bilateral MGB when contrasting the standard versus deviant conditions as well as in the contrast of deviant versus control (Fig. 5.3C). These results provide the first demonstration of the involvement of subcortical structures in genuine regularity encoding and deviance detection in humans. However, fMRI lacks the sufficient temporal resolution to disclose whether these activations occurred early in the processing chain or resulted from top-down modulations of the ascending pathway.

Additionally, the involvement of the subcortical auditory pathway in genuine early regularity encoding was examined in another study in which the frequency-following response (FFR) was recorded (Slabu et al. 2012). The FFR is a

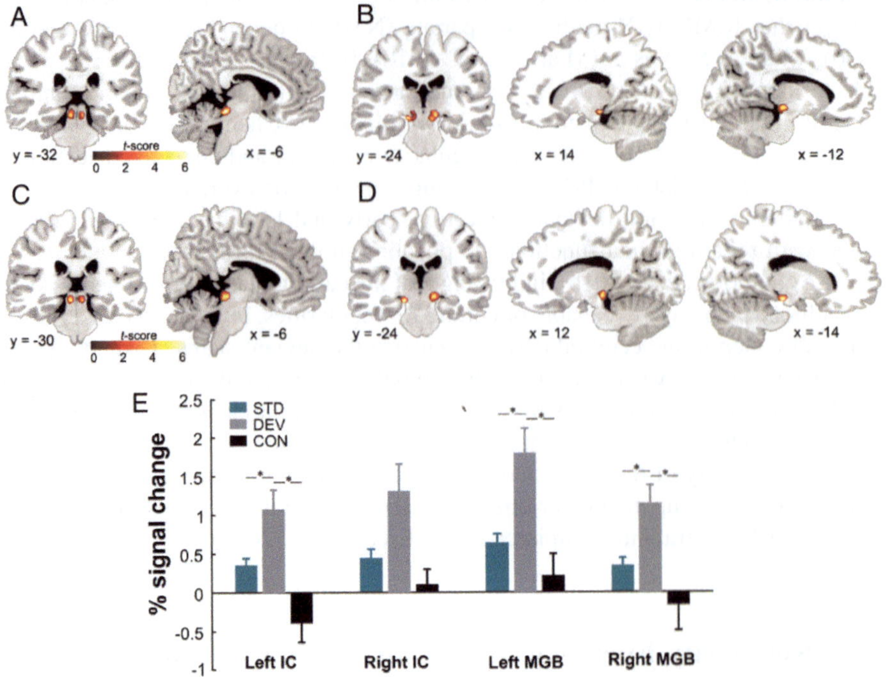

Fig. 5.3 Neuroimaging evidence for the involvement of the subcortical auditory system in regularity encoding and deviance detection in humans. Broadband noise bursts spanning 500 Hz were presented in trains of 20 tokens with a stimulus onset asynchrony of 150 ms. The oddball sequences presented deviant tokens (1000–1500 Hz) from positions 12 onward among standard tokens of 500–1000 Hz. The control sequence presented five different tokens randomly. The figure shows the activations that survived correction for multiple testing using the family-wise error (FWE). The *upper row* shows the deviant > standard contrast whereas the *middle row* shows the deviant > control contrast, disclosing activations in the inferior colliculus (IC) and the medial geniculate body (*MGB*). The *lower row* plots the percent signal change in the bilateral IC and MGB. CON, control; DEV, deviant; STD, standard. *, $P > 0.05$. (Reprinted with permission from Cacciaglia et al. 2015)

sustained part of the ABR typically elicited to periodic and complex auditory stimuli such as speech sounds or music. It emerges at circa 7–15 ms from sound onset after the transient waves V and A of the phasic ABR, therefore reflecting the tonic brainstem response that is phase locked to the spectral and temporal components of the acoustic signal (Chandrasekaran and Kraus 2010; Skoe and Kraus 2010a; Kraus, Anderson, and White-Schwoch, Chap. 1). The FFR has gained recent interest in cognitive auditory neuroscience because it provides a noninvasive measure of the tracking accuracy of periodic sound characteristics in the auditory brainstem. The FFR also allows investigation of the environmental conditions and the biological mechanisms that modulate the representation of incoming sounds at this level of the auditory hierarchy by experience-dependent plasticity (Chandrasekaran et al. 2014), including language experience (Krizman et al. 2012), musical training (Parbery-Clark et al. 2011), short-term auditory training (Anderson et al. 2013), context-dependent encoding (Chandrasekaran et al. 2009), and even sensitivity to statistical properties of the stimulus in real time (Skoe and Kraus 2010b; Skoe et al. 2014).

In the Slabu and colleagues (2012) study mentioned previously, the FFR was recorded in response to a consonant-vowel stimulus /ba/ presented with a low probability ($p = 0.2$) amid a repetitive context set by the repetition of a different syllable (/wa/). To control for the stimulus characteristics, a reversed oddball sequence was used where the /ba/ and /wa/ syllables swapped their deviant/standard status. An additional block featuring four different tokens of the /wa/ syllable (differing in the transition duration of their first and second formants) controlled for probability to preclude mere adaptation effects (Fig. 5.4A). A significant amplitude attenuation in the response to the second and fourth harmonics of the F_0 (Fig. 5.4C) of the deviant syllable compared to the standard and to the control conditions revealed genuine regularity encoding and deviance detection in the human auditory brainstem (Slabu et al. 2012).

The results obtained by Slabu et al. (2012) were replicated and expanded by a recent study that investigated the interaction between stimulus probability (deviant status) and auditory learning (Skoe et al. 2014). In this study, instead of using stimuli pertaining to the phonetic inventory of the listeners, the authors presented an identical syllable (/mi/) that was varied on its pitch trajectory to form two different sounds that were minimally contrastive and with no lexical meaning for the English-speaking participants. Pitch tracking accuracy was measured by autocorrelograms. By using this approach, the authors revealed that pitch tracking was more accurate for frequent (standard) than for infrequent (deviant) stimuli (Skoe et al. 2014), thus supporting the role of the auditory brainstem in extracting statistical information from the acoustic background (i.e., regularity encoding). Moreover, the authors found that probability-dependent plasticity—the encoding of the statistical regularity—interacted with behavioral-relevance plasticity. The relationships between the deviant and standard responses varied when the participants learned to discriminate the minute pitch changes differentiating the standard and deviant stimuli during a training program.

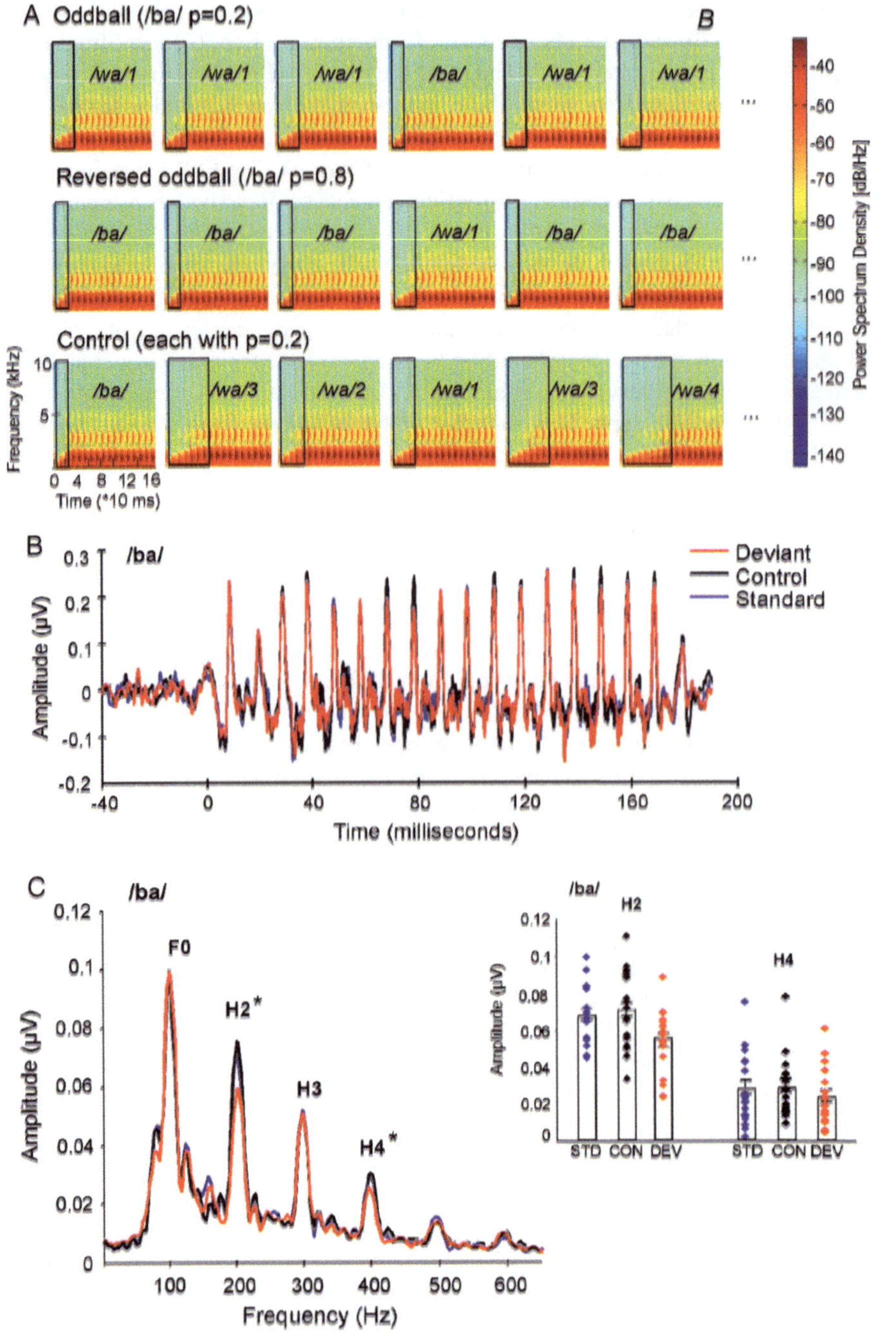

A Oddball (/ba/ p=0.2)
B
/wa/1
/ba/
Reversed oddball (/ba/ p=0.8)
Control (each with p=0.2)
/wa/3
/wa/2
/wa/4
Frequency (kHz)
Time (*10 ms)
Power Spectrum Density [dB/Hz]
B
/ba/
Deviant
Control
Standard
Amplitude (µV)
Time (milliseconds)
C
F0
H2*
H3
H4*
Frequency (Hz)
H2
H4
STD CON DEV STD CON DEV

◀ **Fig. 5.4** Human auditory brainstem correlates of deviance detection. (**A**) Experimental design: a consonant-vowel /ba/ was presented randomly with low probability (p = 0.2) amongst a repetitive /wa/1 stimulus (the difference being in the transition duration of F1 and F2: 20 ms for /ba/, 35 ms for /wa/1; longer durations from 50 to 85 ms for /wa/2, /wa3/ and /wa/4 stimuli in the control block). (**B**) The FFR elicited to the same physical stimulus (/ba/) in the role of standard (*STD*), deviant (*DEV*) and control (*CON*) in different blocks. **C** Amplitude spectrum of the FFR elicited to /ba/ in the different conditions. Notice that compared to the standard and control conditions, the amplitude of the deviant response was attenuated in the second (*H2*) and fourth (*H4*) harmonic of the F_0 (FO). The *inset* shows the individual FFR amplitudes at H2 and H4. *, $P > 0.025$. (Modified with permission from Slabu et al. 2012)

It is interesting to note that the two studies mentioned above (Slabu et al. 2012; Skoe et al. 2014), which addressed regularity encoding and deviant detection in the human auditory brainstem, found that responses to deviant stimuli were attenuated rather than enhanced. This is in agreement with studies that found that a behaviorally relevant stimulus, such as a consonant-vowel of the linguistic repertoire of the participants, elicits larger FFRs when occurring in a repetitive context than when occurring amongst varying stimuli (Chandrasekaran et al. 2009; Parbery-Clark et al. 2011; Strait et al. 2011). However, when the nature of the eliciting stimulus lacks behavioral relevance, such as in an amplitude modulated tone (AM), the occurrence of a change in the AM frequency elicits an enhancement, rather than an attenuation, of the amplitude of the deviant-related FFR compared to the standard (Shiga et al. 2015). This is compatible with the view that the auditory brainstem prioritizes behaviorally relevant stimuli, which had been proposed to originate from mechanisms different from repetition suppression (Skoe and Kraus 2010b; Parbery-Clark et al. 2011); however, repetition suppression is the keystone of regularity encoding proposed here.

The capability of the auditory brainstem to encode for acoustic regularities has been observed not only for auditory objects (the "what" in the auditory scene), but also for the temporal, dynamic component of the auditory background, that is, to "when" a particular object is expected to occur. This is supported by a preliminary study that showed that FFR amplitude to rare delayed stimuli occurring in an otherwise regular sequence (isochronous) was enhanced compared to expected stimuli (e.g., those occurring at the regular intervals; Zarnowiec et al. 2014). It is also supported by a study showing that temporal predictability interacts with stimulus repetitions in shaping brainstem responses (Gorina-Careta et al. 2016). Interestingly, these results complement those observed for thalamocortical networks generating the MLR (Leung et al. 2013) and seminal observations for cortical responses (i.e., the MMN) (Ford and Hillyard 1981) thus indicating that encoding of the temporal dynamics in the acoustic scene also is carried out along the entire auditory system.

5.6 Relationships to Other Forms of On-Line Plasticity

The studies reviewed so far have addressed a particular form of auditory plasticity that can be regarded as on-line; it has to do with the way in which the auditory system captures the ongoing stimulation "on the fly," hence adapting to the moment on a temporal scale that spans a few seconds (although this adaptation to the moment may also interact with longer time scales; e.g., Ulanovsky et al. 2004; Costa-Faidella et al. 2011b). From the evidence discussed, this adaptation to the moment appears as a pervasive property of the entire auditory system, from higher-order regions of the auditory cortex down to the IC at least (Ayala et al. 2012). Correlates of this adaptation to the moment, in forms of neural traces for ongoing statistical regularities, have been described for long-latency (i.e., the MMN), middle latency, and even brainstem responses of the human AEP. Moreover, neurophysiological mechanisms for this kind of on-line encoding of regularities have been associated with SSA, for which the basic strategy for testing, as with human AEP studies, consists of challenging the neural representation of the regularity with a stimulus that does not fulfill the expectation, hence measuring deviance-related responses.

These two processes, regularity encoding and deviance detection, have been considered as two faces of the same coin (Parbery-Clark et al. 2011; Escera and Malmierca 2014), yet they can be dissociated (Taaseh et al. 2011; Aghamolaei et al. 2016). However, this form of plasticity, adapting in the moment, is only one case of the multiple forms of plasticity that the auditory system and, in particular, the subcortical ascending pathway can undergo. In fact, a large series of studies conducted with the FFR to measure responses to complex auditory stimuli (with behavioral relevance) have shown that the human auditory brainstem can experience plasticity over temporal spans ranging from minutes to the entire life (Chandrasekaran et al. 2014), and that these different auditory experiences "layer" along the course of one's own life to shape the individual's auditory subcortical function (Skoe and Chandrasekaran 2014). In particular, three forms of on-line plasticity relate to the studies considered previously in this section.

First, a seminal study by Chandrasekaran et al. (2009) showed that the brainstem encoding of the F_0 of speech sounds is context-dependent and, furthermore, that the capacity to benefit from contextual information (stimulus repetition) correlates with hearing speech-in-noise abilities. In their study, a consonant vowel /da/ was presented repetitively or among other syllables that varied in a number of acoustic features, such as formant structure, duration, voice-onset time, or F_0. The results showed that the second and fourth harmonics of the F_0 of the response were enhanced in the constant context compared to the variable context but only in good compared to poor readers (Chandrasekaran et al. 2009). These results indicate that the human auditory brainstem is sensitive to the ongoing stimulus context. Studies that used the same approach confirmed this effect and went a step beyond to show that the capability of the auditory brainstem to benefit from contextual information underlies enhanced speech-in-noise perception in musicians (Parbery-Clark et al. 2011) and reading and music aptitude in children (Strait et al. 2011).

A further capability of the auditory brainstem is that of performing statistical calculations of the discrete sounds occurring in the soundscape. This was demonstrated by a study that presented a series of musical notes arranged in random or in patterned sequences. In the patterned sequences, the individual sounds were constrained so that a particular note was followed by a fixed one, forming a doublet. With this method, the occurrence of an individual sound predicted a subsequent event with certainty, a prediction that did not occur in the random condition. In other words, the patterned condition set precise local statistics. Recordings revealed attenuated brainstem responses for the patterned condition compared to the random condition (Skoe et al. 2013). More striking was the finding that the capability of the auditory brainstem to extract the local statistics within the sequence predicted the individual capability to learn the implicit syntax of the sequence. These results demonstrate again the capability of the human auditory brainstem to perform computational operations on discrete auditory events and highlight the auditory brainstem's involvement in driving behavioral outcomes.

Finally, two other studies relate more closely to those reviewed in Sect. 5.5 on regularity encoding. In one of them, a piano melody composed of five notes was presented repeatedly for a long recording session. Critically, the melody featured a note repetition (the first and second notes were identical). The results yielded two sets of effects. First, the amplitude of the brainstem responses increased for each note between the first and the second halves of the recording session (Skoe and Kraus 2010b). Second, the note repetition resulted in a repetition enhancement, that is, the amplitude of the brainstem response was larger for the second than for the first note of the melody across the entire recording session, which was analyzed in four separate quarters (Skoe and Kraus 2010b). These results suggest that the human auditory brainstem can encode for both local and global statistics.

The second of these studies (Skoe et al. 2014), as mentioned in Sect. 5.5, confirmed that a rarely occurring stimulus among a series of repetitive ones (a deviant) can be detected by the human auditory brainstem. Moreover, the results of this study also showed that probability-dependent plasticity interacts with another form of plasticity that is behavior dependent through the processes of learning. Hence, the results of this study indicate that behavioral learning can alter the way in which on-line probabilities are computed in the auditory brainstem, thereby highlighting the role of the ascending auditory pathway as a powerful computational network. Interestingly, the authors concluded that by means of long-term experience (e.g., training, but possibly other forms of exposure, such as bilingualism) (Krizman et al. 2012), learning related top-down feedback can override the local brainstem mechanisms that subserve probability detection (Skoe et al. 2014). Although somewhat tentative, this interpretation paves the way for a number of predictions that should guide future research.

The studies considered so far in this section, together with those regarding the auditory brainstem reviewed in Sect. 5.5, indicate that the ascending auditory pathway, beyond being a passive relay of auditory information toward the auditory cortex, possesses complex computational capabilities eventually contributing to auditory cognition. In particular, it is tempting to hypothesize that specific key anatomical structures of the auditory pathway, such as the IC and the MGB, may have the ability to encode for auditory regularities in the acoustic background that go beyond simple stimulus repetition, thus encompassing the relationship between successive discrete auditory stimuli and thereby supporting "primitive intelligence" (Näätänen et al. 2001, 2010).

This idea was preliminarily tested with a sequence of four different tones combining two features (duration: short, long; and pitch: high, low) that was arranged so that the duration of a particular tone predicted the pitch of the next (e.g., high-pitch tones followed short tones). After this contingency was repeated for a number of times to set the regularity, a stimulus that did not follow this implicit contingency was presented. In agreement with former studies (Bendixen et al. 2008), deviant events elicited clear cortical deviance-related responses (MMN). More interesting, however, was the observation of an enhanced amplitude of the FFR elicited to the deviant event compared to that of the standard (Schaefer et al. 2015), suggesting that the auditory brainstem was able to encode for such complex stimulus contingencies. Although preliminary, these results are encouraging and strongly suggestive of the complex and powerful computational capabilities of the human auditory brainstem.

5.7 Summary

This chapter has summarized studies that show that in humans, auditory deviance detection based on regularity encoding occurs at latencies and in neural networks comparable to those revealed in animal studies of single-neuron activity. These studies demonstrate that encoding simple acoustic-feature regularities and the detection of corresponding deviance, such as an infrequent change in frequency or location, occur in thalamocortical networks, giving rise to the MLR in separate auditory cortical regions from those generating the MMN, and occur even at the level of human auditory brainstem, as indicated by the FFR and fMRI. Taken together, these studies give support to the emerging view that regularity encoding is a basic principle of the functional organization of the auditory system, which is organized in ascending levels of complexity along the auditory pathway from the brainstem up to higher-order areas of the cerebral cortex.

Moreover, ongoing studies have started to suggest that subcortical structures in the auditory pathway can implement complex computational operations, mimicking the "primitive intelligence" attributed originally to auditory cortex (Näätänen et al. 2001, 2010) and, therefore, challenging corticocentric views of cognition (Parvizi 2009). Remarkable, for example, are the preliminary results that suggest subcortical structures can support predictive coding, as revealed by enhanced FFRs to individual stimuli that do not accomplish a rule pre-established by the dynamic ongoing sequence (such as the duration of a particular tone determining the pitch of the next)

(Schaefer et al. 2015). In combination with results showing that the auditory brainstem can undergo plasticity at multiple time scales (Chandrasekaran et al. 2014), it is tempting to suggest that the inferior colliculus is a hub for primitive intelligence in audition. However, before the field can reach that far, a series of caveats and research questions need appropriate answers.

First, the FFR is a very small signal generated in very deep cerebral structures so that a large number of trials ($\sim$2000 or more) need to be recorded to reach a sufficient signal-to-nose ratio (Jeng et al. 2011). This limitation becomes stringent when one plans to address changes or effects in these minute responses as a function of complex relationships among discrete auditory stimuli (e.g., differential or contrasting effects). Some improvements have been suggested based on "optimal" paradigms (Bidelman 2015) or multichannel recordings (Bellier et al. 2015), but there is still a need for further improvement.

Second, the specific contribution of discrete subcortical structures to the FFR, and particularly the role of IC in its generation, is still to be disentangled. In fact, most of the evidence about FFR sources come from seminal observations in human patients with brainstem lesions (Sohmer et al. 1977), human intracranial recordings (Møller et al. 1988), analogies from animal studies (Smith et al. 1975), or are based on the electrode montage-dependency of the response (Davis and Britt 1984) or the phase-locking capabilities of subcortical neuronal assemblies compared to cortical ones (Joris et al. 2004). However, direct evidence is lacking. Therefore, approaches that would apply inverse solution methods capable of disclosing putative EEG subcortical sources (Trujillo-Barreto et al. 2004), magnetoencephalography (Parkkonen et al. 2009; Coffey et al. 2016), or an approach that combines FFR recordings with fMRI (Chandrasekaran et al. 2012) may provide compelling evidence for the specific subcortical generation of the FFR.

A more critical issue is the debated contribution of the corticofugal pathway to subcortical encoding of sound and, specifically, of ongoing regularity, particularly if one wants to claim the contribution of the subcortical auditory pathway to auditory cognition. Animal studies have largely demonstrated that the corticofugal pathway plays a critical role in long-term and even short-term plasticity (Suga et al. 2002; Suga 2008; Bajo et al. 2010). However, recent pharmacological (Pérez-González et al. 2012; Ayala and Malmierca 2015) and cortical transient inactivation studies in animals (Antunes and Malmierca 2011; Anderson and Malmierca 2013) suggested that on-line plasticity, or adaptation to the ongoing input statistics, may rely strictly on bottom-up processes. Yet another possibility is that subcortical cognition results from the interplay between bottom-up and top-down interactions (Skoe et al. 2013; Chandrasekaran et al. 2014). In humans, a potential approach to disentangle the interaction of top-down and bottom-up contributions to subcortical auditory cognition may be through temporarily inactivating the auditory cortex by means of transcranial magnetic stimulation (TMS) (Ahveninen et al. 2013). Another potential approach may involve interfering with cortical processing during tailored experiments to address the encoding of acoustic regularities at multiple levels of complexity by means of transcranial direct current stimulation (tDCS) (Riecke et al. 2015). Any of these approaches, used on their own or in combination, will contribute substantial progress in our understanding of the cognitive neuroscience of audition.

Acknowledgments Thanks to all the members of the Brainlab-Cognitive Neuroscience Research Group, past and present, for their work contributing to this research and for being a continuous source of inspiration. Special thanks are for Katarzyna Zarnowiec for her esteemed contribution to set up the FFR experiments and to her and Natàlia Gorina-Careta for their comments on an earlier version of the manuscript. This work was supported by grants from the Fundaçao Bial (Porto, Portugal: 30/12), Fundación Alicia Koplowitz (Madrid, Spain), the Spanish Ministry of Economy and Competitiveness (MINECO: PSI2012-37174, PSI2013-49348-EXPLORA, PSI2015-63664-P), the Catalan Government (SGR2014-177), and the ICREA Acadèmia Distinguished Professorship award.

Compliance with Ethics Requirements Carles Escera declared that he had no conflict of interest.

References

Aghamolaei, M., Zarnowiec, K., Grimm, S., & Escera, C. (2016). Functional dissociations between regularity encoding and deviance detection along the auditory hierarchy. *European Journal of Neuroscience, 43*(4), 529–535.

Ahveninen, J., Huang, S., Nummenmaa, A., Belliveau, J. W., et al. (2013). Evidence for distinct human auditory cortex regions for sound location versus identity processing. *Nature Communications, 4*, 2585. Doi:10.1038/ncomms3585

Alho, K., Grimm, S., Mateo-León, S., Costa-Faidella, J., & Escera, C. (2012). Early processing of pitch in the human auditory system. *European Journal of Neuroscience, 36*, 2972–2978.

Althen, H., Grimm, S., & Escera, C. (2011). Fast detection of unexpected sound intensity decrements as revealed by human evoked potentials. *PLoS ONE, 6*(12), e28522. Doi:10.1371/journal.pone.0028522

Anderson, L. A., & Malmierca, M. S. (2013). The effect of auditory cortical deactivation on stimulus-specific adaptation in the inferior colliculus of the rat. *European Journal of Neuroscience, 37*, 52–62.

Anderson, S., White-Schwoch, T., Parbery-Clark, A., & Kraus, N. (2013). Reversal of age-related neural timing delays with training. *Proceedings of the National Academy of Sciences of the USA, 110*(11), 4357–4362.

Antunes, F. M., & Malmierca, M. S. (2011). Effect of auditory cortex deactivation on stimulus-specific adaptation in the medial geniculate body. *The Journal of Neuroscience, 31*, 17306–17316.

Antunes, F. M., Nelken, I., Covey, E., & Malmierca, M. S. (2010). Stimulus-specific adaptation in the auditory thalamus of the anesthetized rat. *PLoS ONE, 5*, e14071.

Ayala, Y. A., & Malmierca, M. S. (2015). Cholinergic modulation of stimulus-specific adaptation in the inferior colliculus. *The Journal of Neuroscience, 35*(35), 12261–12272.

Ayala, Y. A., Pérez-González, D., Duque, D., Nelken, I., & Malmierca, M. S. (2012). Frequency discrimination and stimulus deviance in the inferior colliculus and cochlear nucleus. *Frontiers in Neural Circuits, 6*, 119. Doi:10.3389/fncir.2012.00119

Bajo, V. M., Nodal, F. R., Moore, D. R., & King, A. J. (2010). The descending corticocollicular pathway mediates learning-induced auditory plasticity. *Nature Neuroscience, 13*(2), 253–260.

Bellier, L., Bouchet, P., Jeanvoine, A., Valentin, O., et al. (2015). Topographic recordings of auditory evoked potentials to speech: Subcortical and cortical responses. *Psychophysiology, 52*(4), 549–594.

Bendixen, A., Prinz, W., Horváth, J., Trujillo-Barreto, N. J., & Schröger, E. (2008). Rapid extraction of auditory feature contingencies. *NeuroImage, 41*(3), 1111–1119.

Bidelman, G. M. (2015). Towards an optimal paradigm for simultaneously recording cortical and brainstem auditory evoked potentials. *The Journal of Neuroscience Methods, 15*(241), 94–100.

Bregman, A. (1990). *Auditory scene analysis. The perceptual organization of sound.* Cambridge, MA: MIT Press.

Cacciaglia, R., Escera, C., Slabu, L. S., Grimm, S., et al. (2015). Involvement of the human midbrain and thalamus in auditory deviance detection. *Neuropsychologia, 68*, 51–58.

Chandrasekaran, B., Hornickel, J., Skoe, E., Nicol, T., & Kraus, N. (2009). Context-dependent encoding in the human auditory brainstem relates to hearing speech in noise: Implications for developmental dyslexia. *Neuron, 64*(3), 311–319.

Chandrasekaran, B., & Kraus, N. (2010). The scalp-recorded brainstem response to speech: Neural origins and plasticity. *Psychophysiology, 47*, 236–246.

Chandrasekaran, B., Kraus, N., & Wong, P. C. (2012). Human inferior colliculus activity relates to individual differences in spoken language learning. *Journal of Neurophysiology, 107*(5), 1325–1336.

Chandrasekaran, B., Skoe, E., & Kraus, N. (2014). An integrative model of subcortical auditory plasticity. *Brain Topography, 27*(4), 539–552.

Chen, I. W., Helmchen, F., & Lütcke, H. (2015). Specific early and late oddball-evoked responses in excitatory and inhibitory neurons of mouse auditory cortex. *The Journal of Neuroscience, 35*(36), 12560–12573.

Coffey, E. B. J., Herholz, S. C., Cepesiuk, A. M. P., Baillet, S., & Zatorre, R. J. (2016). Cortical contributions to the auditory frequency-following response revealed by MEG. *Nature Communications, 7*, 11070. Doi:10.1038/ncomms11070

Cornella, M., Leung, S., Grimm, S., & Escera, C. (2012). Detection of simple and pattern regularity violations occurs at different levels of the auditory hierarchy. *PLoS ONE, 7*(8), e43604.

Cornella, M., Leung, S., Grimm, S., & Escera, C. (2013). Regularity encoding and deviance detection of frequency modulated sweeps: Human middle- and long-latency auditory evoked potentials. *Psychophysiology, 50*, 1275–1281.

Costa-Faidella, J., Baldeweg, T., Grimm, S., & Escera, C. (2011a). Interactions between "what" and "when" in the auditory system: Temporal predictability enhances repetition suppression. *The Journal of Neuroscience, 31*, 18590–18597.

Costa-Faidella, J., Grimm, S., Slabu, L., Díaz-Santaella, F., & Escera, C. (2011b). Multiple time scales of adaptation in the auditory system as revealed by human evoked potentials. *Psychophysiology, 48*, 774–783.

Davis, R. L., & Britt, R. H. (1984). Analysis of the frequency following response in the cat. *Hearing Research, 15*(1), 29–37.

Deouell, L. Y. (2007). The frontal generator of the mismatch negativity revisited. *Journal of Psychophysiology, 21*, 188–203.

Desimone, R. (1996). Neural mechanisms for visual memory and their role in attention. *Proceeding of the National Academy of Sciences of the U S A, 93*(24), 13494–13499.

Escera, C., Alho, K., Winkler, I., & Näätänen, R. (1998). Neural mechanisms of involuntary attention to acoustic novelty and change. *Journal of Cognitive Neuroscience, 10*, 590–604.

Escera, C., & Corral, M. J. (2007). Role of mismatch negativity and novelty-P3 in involuntary auditory attention. *Journal of Psychophysiology, 21*, 251–264.

Escera, C., Leung, S., & Grimm, S. (2014). Deviance detection based on regularity encoding along the auditory hierarchy: Electrophysiological evidence in humans. *Brain Topography, 27*(4), 527–538.

Escera, C., & Malmierca, M. S. (2014). The auditory novelty system: An attempt to integrate human and animal research. *Psychophysiology, 51*, 111–123.

Farley, B. J., Quirk, M. C., Doherty, J. J., & Christian, E. P. (2010). Stimulus-specific adaptation in auditory cortex is an NMDA-independent process distinct from the sensory novelty encoded by the mismatch negativity. *The Journal of Neuroscience, 30*, 16475–16484.

Ford, J. M., & Hillyard, S. A. (1981). Event-related potentials (ERPs) to interruptions of a steady rhythm. *Psychophysiology, 18*(3), 322–330.

Friston, K. (2005). A theory of cortical responses. *Philosophical Transactions of the Royal Society of London B: Biological Sciences, 360*(1456), 815–836.

Friston, K., Kilner, J., & Harrison, L. (2006). A free energy principle for the brain. *Journal of Physiololgy (Paris), 100*(1–3), 70–87.

Gorina-Careta, N., Zarnowiec, K., Costa-Faidella, C., & Escera, C. (2016). Timing predictability enhances regularity encoding in the human subcortical auditory pathway. *Scientific Reports, 6*, 37405. doi:10.1038/srep37405.

Grill-Spector, K., Henson, R., & Martin, A. (2006). Repetition and the brain: Neural models of stimulus-specific effects. *Trends in Cognitive Sciences, 10*, 14–23.

Grimm, S., & Escera, C. (2012). Auditory deviance detection revisited: Evidence for a hierarchical novelty system. *International Journal of Psychophysiology, 85*, 88–92.

Grimm, S., Escera, C., & Nelken, I. (2016). Early indices of deviance detection in humans and animal models. *Biological Psychology, 116*, 23–27.

Grimm, S., Escera, C., Slabu, L. M., & Costa-Faidella, J. (2011). Electrophysiological evidence for the hierarchical organization of auditory change detection in the human brain. *Psychophysiology, 48*, 377–384.

Grimm, S., Recasens, M., Althen, H., & Escera, C. (2012). Ultrafast tracking of sound location changes as revealed by human auditory evoked potentials. *Biological Psychology, 89*, 232–239.

Haenschel, C., Vernon, D. J., Dwivedi, P., Gruzelier, J. H., & Baldeweg, T. (2005). Event-related brain potential correlates of human auditory sensory memory-trace formation. *Journal of Neuroscience, 25*, 10494–10501.

Jeng, F. C., Chung, H. K., Lin, C. D., Dickman, B., & Hu, J. (2011). Exponential modeling of human frequency-following responses to voice pitch. *International Journal of Audiology, 50*(9), 582–593.

Joris, P. X., Schreiner, C. E., & Rees, A. (2004). Neural processing of amplitude-modulated sounds. *Physiological Review, 84*(2), 541–577.

Kraus, N., McGee, T., Littman, T., Nicol, T., & King, C. (1994). Nonprimary auditory thalamic representation of acoustic change. *Journal of Neurophysiology, 72*, 1270–1277.

Krizman, J., Marian, V., Shook, A., Skoe, E., & Kraus, N. (2012). Subcortical encoding of sound is enhanced in bilinguals and relates to executive function advantages. *Proceedings of the National Academy of Sciences of the U S A, 109*(20), 7877–7881.

Leung, S., Recasens, M., Grimm, S., & Escera, C. (2013). Electrophysiological index of acoustic temporal regularity violation in the middle latency range. *Clinical Neurophysiology, 124*, 2397–2405.

Malmierca, M. S., Anderson, L. A., & Antunes, F. M. (2015). The cortical modulation of stimulus-specific adaptation in the auditory midbrain and thalamus: A potential neuronal correlate for predictive coding. *Frontiers in Systems Neuroscience, 9*, 19. Doi:10.3389/fnsys.2015.00019

Malmierca, M. S., Cristaudo, S., Perez-Gonzalez, D., & Covey, E. (2009). Stimulus-specific adaptation in the inferior colliculus of the anesthetized rat. *The Journal of Neuroscience, 29*, 5483–5493.

Møller, A. R., Jannetta, P. J., & Sekhar, L. N. (1988). Contributions from the auditory nerve to the brain-stem auditory evoked potentials (BAEPs): Results of intracranial recording in man. *Electroencephalography and Clinical Neurophysiology, 71*, 198–211.

Näätänen, R., Astikainen, P., Ruusuvirta, T., & Huotilainen, M. (2010). Automatic auditory intelligence: An expression of the sensory-cognitive core of cognitive processes. *Brain Research Reviews, 64*(1), 123–136.

Näätänen, R., Paavilainen, P., Rinne, T., & Alho, K. (2007). The mismatch negativity (MMN) in basic research of central auditory processing: A review. *Clinical Neurophysiology, 118*, 2544–2590.

Näätänen, R., Tervaniemi, M., Sussman, E., Paavilainen, P., & Winkler, I. (2001). "Primitive intelligence" in the auditory cortex. *Trends in Neurosciences, 24*, 283–288.

Näätänen, R., & Winkler, I. (1999). The concept of auditory stimulus representation in cognitive neuroscience. *Psychological Bulletin, 6*, 826–859.

Nelken, I. (2014). Stimulus-specific adaptation and deviance detection in the auditory system: Experiments and models. *Biological Cybernetics, 108*, 655–663.

Nelken, I., & Ulanovsky, N. (2007). Mismatch negativity and stimulus-specific adaptation in animal models. *Journal of Psychophysiology, 21*, 214–223.

Opitz, B., Mecklinger, A., Von Cramon, D. Y., & Kruggel, F. (1999). Combining electrophysiological and hemodynamic measures of the auditory oddball. *Psychophysiology, 36*(1), 142–147.

Paavilainen, P. (2013). The mismatch-negativity (MMN) component of the auditory event-related potential to violations of abstract regularities: A review. *International Journal of Psychophysiology, 88*, 109–123.

Parbery-Clark, A., Strait, D. L., & Kraus, N. (2011). Context-dependent encoding in the auditory brainstem subserves enhanced speech-in-noise perception in musicians. *Neuropsychologia, 49*(12), 3338–3345.

Parkkonen, L., Fujiki, N., & Mäkelä, J. P. (2009). Sources of auditory brainstem responses revisited: Contribution by magnetoencephalography. *Human Brain Mapping, 30*(6), 1772–1782.

Parvizi, J. (2009). Corticocentric myopia: Old bias in new cognitive sciences. *Trends in Cognitive Sciences, 13*(8), 354–359.

Pérez-González, D., Covey, E., & Malmierca, M. S. (2005). Novelty detector neurons in the mammalian auditory midbrain. *European Journal of Neuroscience, 22*, 2879–2885.

Pérez-González, D., Hernandez, O., Covey, E., & Malmierca, M. S. (2012). GABA(A)-mediated inhibition modulates stimulus-specific adaptation in the inferior colliculus. *PLoS ONE, 7*, e34297.

Picton, T. W. (2011). *Human auditory evoked potentials*. San Diego: Plural Publishing.

Recasens, M., Grimm, S., Capilla, A., Nowak, R., & Escera, C. (2014). Two sequential processes of change detection in hierarchically ordered areas of the human auditory cortex. *Cerebral Cortex, 24*, 143–153.

Recasens, M., Leung, S., Grimm, S., Nowak, R., & Escera, C. (2015). Repetition suppression and repetition enhancement underlie auditory memory-trace formation in the human brain: An MEG study. *NeuroImage, 108*, 75–86.

Riecke, L., Formisano, E., Herrmann, C. S., & Sack, A. T. (2015). 4-Hz Transcranial alternating current stimulation phase modulates hearing. *Brain Stimulation, 8*(4), 777–783.

Ruhnau, P., Herrmann, B., & Schröger, E. (2012). Finding the right control: The mismatch negativity under investigation. *Clinical Neurophysiology, 123*, 507–512.

Sabri, M., Kareken, D. A., Dzemidzic, M., Lowe, M. J., & Melara, R. D. (2004). Neural correlates of auditory sensory memory and automatic change detection. *NeuroImage, 21*(1), 69–74.

Schaefer, J., Zarnowiec, K., SanMiguel, I., Malmierca, M. S., & Escera, C. (2015). Predicting complex acoustic contingencies in the human auditory brainstem. In A. Widmann, J. Steinberg, A. Bendixen, A. D. Friederici, et al. (Eds.), *Error Signals from the Brain: 7th Mismatch Negativity Conference* (pp. 61–62). Leipzig: University of Leipzig Press.

Schröger, E., & Wolff, C. (1996). Mismatch response to changes in sound location. *NeuroReport, 7*, 3005–3008.

Shiga, T., Althen, H., Cornella, M., Zarnowiec, K., et al. (2015). Deviance-related responses along the auditory hierarchy: Combined FFR. *MLR and MMN evidence. PLOS ONE, 10*(9), e0136794. Doi:10.1371/journal.pone.0136794

Skoe, E., & Chandrasekaran, B. (2014). The layering of auditory experiences in driving experience-dependent subcortical plasticity. *Hearing Research, 311*, 36–48.

Skoe, E., Chandrasekaran, B., Spitzer, E. R., Wong, P. C., & Kraus, N. (2014). Human brainstem plasticity: The interaction of stimulus probability and auditory learning. *Neurobiology of Learning and Memory, 109*, 82–93.

Skoe, E., & Kraus, N. (2010a). Auditory brainstem response to complex sounds: A tutorial. *Ear and Hearing, 31*, 302–324.

Skoe, E., & Kraus, N. (2010b). Hearing it again and again: On-line subcortical plasticity in humans. *PLoS ONE, 5*(10), e13645. Doi:10.1371/journal.pone.0013645

Skoe, E., Krizman, J., Spitzer, E., & Kraus, N. (2013). The auditory brainstem is a barometer of rapid auditory learning. *Neuroscience, 243*, 104–114.

Slabu, L. M., Escera, C., Grimm, S., & Costa-Faidella, J. (2010). Early change detection in humans as revealed by auditory brainstem and middle-latency evoked potentials. *European Journal of Neuroscience, 32*, 859–865.

Slabu, L., Grimm, S., & Escera, C. (2012). Novelty detection in the human auditory brainstem. *The Journal of Neuroscience, 32*(4), 1447–1452.

Smith, J. C., Marsh, J. T., & Brown, W. S. (1975). Far-field recorded frequency-following responses: Evidence for the locus of brainstem sources. *Electroencephalography and Clinical Neurophysiology, 39*(5), 465–472.

Sohmer, H., Pratt, H., & Kinarti, R. (1977). Sources of frequency following responses (FFR) in man. *Electroencephalography and Clinical Neurophysiology, 42*(5), 656–664.

Strait, D. L., Hornickel, J., & Kraus, N. (2011). Subcortical processing of speech regularities underlies reading and music aptitude in children. *Behavioral and Brain Functions, 7*, 44. Doi:10.1186/1744-9081-7-44

Suga, N. (2008). Role of corticofugal feedback in hearing. *Journal of Comparative Physiology A: Neuroethology, Sensory, Neural, and Behavioral Physiology, 194*, 169–183.

Suga, N., Xiao, Z., Ma, X., & Li, W. (2002). Plasticity and corticofugal modulation for hearing in adults animals. *Neuron, 36*, 9–18.

Taaseh, N., Yaron, A., & Nelken, I. (2011). Stimulus-specific adaptation and deviance detection in the rat auditory cortex. *PLoS ONE, 6*, e23369. Doi:10.1371/journal.pone.0023369

Trujillo-Barreto, N. J., Aubert-Vázquez, E., & Valdés-Sosa, P. A. (2004). Bayesian model averaging in EEG/MEG imaging. *NeuroImage, 21*, 1300–1319.

Ulanovsky, N., Las, L., Farkas, D., & Nelken, I. (2004). Multiple time scales of adaptation in auditory cortex neurons. *The Journal of Neuroscience, 24*, 10440–10453.

Ulanovsky, N., Las, L., & Nelken, I. (2003). Processing of low-probability sounds by cortical neurons. *Nature Neuroscience, 6*, 391–398.

Umbricht, D., Schmid, L., Koller, R., Vollenweider, F. X., et al. (2000). Ketamine-induced deficits in auditory and visual context-dependent processing in healthy volunteers: Implications for models of cognitive deficits in schizophrenia. *Archives of General Psychiatry, 57*, 1139–1147.

Winkler, I. (2007). Interpreting the mismatch negativity. *Journal of Psychophysiology, 27*, 147–163.

Winkler, I., Denham, S. L., & Nelken, I. (2009). Modeling the auditory scene: Predictive regularity representations and perceptual objects. *Trends in Cognitive Sciences, 13*, 532–540.

Yvert, B., Crouzeix, A., Bertrand, O., Seither-Preisler, A., & Pantev, C. (2001). Multiple supratemporal sources of magnetic and electric auditory evoked middle latency components in humans. *Cerebral Cortex, 11*, 411–423.

Zarnowiec, K., Costa-Faidella, J., & Escera, C. (2014). Fine-grained violations in rhythmic auditory stimulation modulate the human frequency-following response. *International Journal of Psychophysiology, 94*, 167.

Chapter 6
The Janus Face of Auditory Learning: How Life in Sound Shapes Everyday Communication

Travis White-Schwoch and Nina Kraus

Abstract Human communication is rooted in sound. The frequency-following response (FFR) reveals the integrity of sound processing in the brain and characterizes sound processing skills in a diverse group of listeners spanning a gamut from expertise to disorder. These neurophysiological processes are shaped by life experiences, for better or worse. Here two lifelong experiences are juxtaposed—music training (a model of expertise) and low socioeconomic status (a model of disorder)—with an emphasis on how FFR studies reveal their combined influences on everyday communication skills. A view emerges wherein the auditory system is an integrated, but distributed, processing circuit that interacts with other modalities, cognitive systems, and the limbic system to shape auditory physiology. Objective indices of auditory physiology, such as the FFR, therefore can be thought of as measures of "whole brain" auditory processing that recapitulate the past by revealing the legacy of prior experiences and predict the future by elucidating their consequences for everyday communication.

Keywords Auditory neurophysiology · Auditory processing · Communication disorders · Language · Learning and memory · Music training · Neuroplasticity · Poverty · Socioeconomic status

T. White-Schwoch
Auditory Neuroscience Laboratory, Department of Communication Sciences, Northwestern University, 2240 Campus Dr., Evanston, IL 60208, USA
e-mail: whiteschwoch@northwestern.edu

N. Kraus (✉)
Auditory Neuroscience Laboratory, Departments of Communication Sciences, Neurobiology and Otolaryngology, Northwestern University, 2240 Campus Dr., Evanston, IL 60208, USA
e-mail: nkraus@northwestern.edu

N. Kraus et al. (eds.), *The Frequency-Following Response*,
Springer Handbook of Auditory Research 61,
DOI 10.1007/978-3-319-47944-6_6

6.1 Introduction

Janus, the Roman god of beginnings, transitions, and endings, is classically depicted with two faces: one that looks forward and one that looks back (Fig. 6.1). In this regard he symbolizes two interrelated principles of auditory learning that are manifest in the frequency-following response (FFR). First, auditory learning is inherently a double-edged sword that shapes neural function for better or for worse. Second, auditory physiology recapitulates the past by reflecting the legacy of these experiences and predicts the future by delineating their consequences for everyday communication. Thus, a single FFR is a snapshot of an individual's auditory processing past and potential.

These two principles are reviewed in this chapter with an emphasis on the enduring biological legacy imparted by life in sound and the presumed consequences of auditory experiences for everyday communication. It is important to keep in mind that these principles apply to the auditory system in general; the FFR simply provides an experimental avenue to explore them in humans and to relate these findings to animal models.

Here, FFR studies of everyday communication skills and how these skills are shaped by experience are reviewed. Two long-term experiences are contrasted: music, a case of enrichment, and low socioeconomic status (SES), a background associated with disorders of brain and cognition functions. Juxtaposing these discrepant experiences illustrates how overlapping circuitry is involved in states of expertise and disorder and suggests they arise through similar mechanisms. Importantly, the idea that similar, malleable neural pathways are involved in states

Fig. 6.1 The Janus face of auditory learning. Janus, the Roman god of beginnings, transitions, and endings, is classically depicted with two faces. His two faces symbolize two principles of auditory learning. First, learning is a double-edged sword, which can shape neural function for better or worse. Second, auditory physiology recapitulates the past by reflecting the biological legacy of a life in sound and predicts the future by delineating the consequences of those experiences for everyday communication. (Photograph from www.wikipedia.org)

of expertise and disorder motivates experience-dependent interventions to improve listening (Carcagno and Plack, Chap. 4). This idea cuts across clinical populations, including individuals with auditory processing disorder (Schochat, Rocha-Muniz, and Filippini, Chap. 9), individuals with reading impairment (Reetzke, Xie, and Chandrasekaran, Chap. 10), and older adults (Anderson, Chap. 11).

6.2 The Auditory System: Distributed but Integrated

Classic models of the auditory system emphasize its hierarchical nature, that is, how each station along the neuraxis is specialized (Webster 1992). Despite the theoretical contributions of these models to auditory neuroscience, they risk failing to account for the remarkable integration the auditory system achieves. After all, sound catalyzes activity *throughout* the auditory neuraxis, from cochlea to cortex and back again. A complete understanding of how the brain makes sense of sound therefore has to consider not just what each auditory relay uniquely contributes to auditory processing, but how they integrate (Kraus and White-Schwoch 2015). This integration occurs along with complementary brain networks, such as cognitive and limbic circuits.

Thus, a view emerges wherein the auditory system is *distributed but integrated.* Any measure of auditory physiology, including the FFR, can be thought to index "whole brain" auditory processing, insofar as it reflects activity that occurs in concert with the rest of the system and that has been shaped in the context of that system. This view does not ignore the fact that different measures of auditory function have different generators: otoacoustic emissions (OAEs) are generated by the outer hair cells, and the FFR is generated primarily by the auditory midbrain. But both hair cell and midbrain physiology are shaped by experience, meaning OAEs and FFRs reflect this experience. This network-based account of auditory physiology relates to emerging models of sensory systems that emphasize the "push and pull" among sensory organs, subcortical networks, and cortical networks (Behrmann and Plaut 2013; Pafundo et al. 2016).

The FFR encapsulates this view. As a simple case in point, consider the extraordinary diversity of research conducted with the FFR, spanning lifelong language experience (Krishnan and Gandour, Chap. 3) to in-the-moment adaptation (Escera, Chap. 5). Consider also the broad clinical applications studied with the FFR for auditory processing disorders (Schochat, Rocha-Muniz, and Filippini, Chap. 9), reading impairment (Reetzke, Xie, and Chandrasekaran, Chap. 10), and age-related hearing impairment (Anderson, Chap. 11). Needless to say, these three clinical populations do not share a common lesion in the auditory midbrain; rather, the midbrain functions in the context of bottom-up and top-down networks that shape its response properties, making the FFR a viable tool in the laboratory and clinic. This diversity of research employing the FFR highlights how it is a measure of integrative auditory function (Coffey et al. 2016).

6.3 Conceptual Frameworks for Auditory Learning

The premise that motivates studies of auditory experience is that sound provides a fundamental medium for learning about the world. Thus, life in sound (particularly during early childhood) shapes both auditory and nonauditory social and cognitive brain processes.

6.3.1 Maturation by Experience

The *maturation-by-experience hypothesis* posits that sensory experience is necessary for the development of sensory circuitry. This idea has been studied extensively, especially with respect to language development. For example, Kuhl et al. (1992) have shown that newborns are sensitive to sound contrasts that are not meaningful in their native language, but this sensitivity narrows by approximately six months-of-age, suggesting that ongoing exposure to the linguistic environment shapes basic perception. These studies are complemented by physiological studies that show how language experiences shape sound processing (Jeng, Chap. 2; Krishnan and Gandour, Chap. 3).

Additional evidence for maturation by experience comes from studies of cochlear implantation. Gordon et al. (2012) have shown that the auditory brainstem matures as a function of auditory experience by measuring electrically evoked auditory brainstem responses (eABRs) in children with cochlear implants. They have documented a systematic relationship between eABR latency and years since implantation, suggesting that ongoing auditory experience modulates brainstem function.

The idea that sensory experience is tied to sensory development is also supported by evidence from the visual system. For example, van Rheede et al. (2015) show that visual input rapidly drives formerly dormant neurons in the tadpole optic tectum (a midbrain visual system structure) to spike in response to future visual input. In other words, visual experience creates the conditions for visual neurons to fire. Noteworthy is that van Rheede and colleagues show that experience is required for a fundamental unitary phenomenon—firing an action potential—thus highlighting how experience and perception are intrinsically intertwined.

The maturation-by-experience hypothesis motivates a strong prediction: if sensory experience is required for the development of sensory circuitry, then altering said experiences should alter the default state of said circuitry. In other words, two individuals reared in different sensory environments should exhibit distinct profiles in adulthood. This prediction is borne out by investigations of language experience (Intartaglia et al. 2016; Krishnan and Gandour, Chap. 3), music experience, and socioeconomic status. This prediction is also supported by studies in animals that have been raised in enriched environments and exhibited more sophisticated sensory processing (audition: Engineer et al. 2004; vision: Wang et al. 2010) compared to those that have been raised in toxic environments and exhibited less mature processing (Chang and Merzenich 2003).

These investigations do not rule out the clear role that predispositions, such as genetics, play in shaping adult brain circuitry. Nevertheless, they highlight how predispositions interact with experience in childhood and adulthood to sculpt a sensory phenotype. It should also be emphasized that early experiences influence future learning and plasticity in addition to what becomes automatic sensory processing.

6.3.2 The Cognitive-Sensorimotor-Reward Framework for Auditory Learning

Kraus and White-Schwoch (2015) propose a *cognitive-sensorimotor-reward* framework for auditory learning. Cognitive factors optimize learning because listening is an active process, and repeated engagement of cognitive systems sculpts sensory infrastructure to process sound more effectively. Sensorimotor systems optimize learning through the convergence of processing stations within and across modalities, suggesting that sensory neurons can be thought of as a form of "memory storage" because their intrinsic response properties retune to reflect past experience. Limbic systems optimize learning via neuromodulatory input throughout the auditory neuraxis.

It should be mentioned that these three ingredients may not be necessary for learning. For example, listeners are sensitive to the statistics of a soundscape and the auditory system rapidly adapts to exploit these regularities (Escera, Chap. 5). The focus of the cognitive-sensorimotor-reward framework, however, is long-term experience that reshapes fundamental sensory infrastructure. Noteworthy is that these enduring changes are reflected by the FFR, including in studies of the effects of music experience and poverty (Sects. 6.5 and 6.6).

6.3.3 The Afferent/Efferent, Primary/Non-primary Trade-off Model

A major view in auditory neuroscience is that the mechanisms responsible for in-the-moment plasticity, such as attending to one sound among many, are also responsible for large-scale functional remodeling following prolonged and repeated experiences (Fritz et al. 2003; Weinberger 2004). The idea is that plastic experiences, such as a repeated evanescent change in a cell's receptive field, eventually are engrained in that cell's intrinsic response properties.

The *afferent/efferent, primary/non-primary trade-off model* (Kraus and White-Schwoch 2015) contends that in-the-moment changes are qualitatively distinct from long-term changes to neural function. This view does not discount clear evidence that both selective attention and long-term experiences engender forms of neuroplasticity, nor does it discount evidence that cellular consequences of these experiences are similar. Rather, it distinguishes between an active *learning* phase and a *learned* phase (e.g., Reed et al. 2011).

Kraus and White-Schwoch (2015) propose this evanescent-enduring dichotomy is rooted in the parallel primary and non-primary streams of the auditory neuraxis. The primary (lemniscal) pathway consists of "core" stations (such as the central nucleus of the inferior colliculus, the ventral division of the medial geniculate, and the primary auditory cortex). The lemniscal pathway is strongly tonotopic and characterized by short-latency, stimulus-dependent, unitary responses with a bias to tonal stimuli. The non-primary pathway (paralemniscal) consists of "belt" or "parabelt" stations (such as the lateral cortex of the inferior colliculus, the shell nucleus of the medial geniculate, or peripheral auditory fields). The paralemniscal pathway is characterized by less tonotopic response profiles and connectivity, more context-dependent coding and, in human cortex, a tendency to specialize for certain sounds such as speech (reviewed in Abrams et al. 2011). Additionally, the primary pathway is biased to process faster sounds, whereas the non-primary pathway is biased to process slower sounds (Ahissar et al. 2000; Abrams et al. 2011). Finally, it has been speculated that these parallel pathways may underlie the remarkable balance struck between stability and flexibility in auditory processing (Kraus and White-Schwoch 2015).

Both primary and non-primary divisions of the auditory neuraxis contain afferent (ear-to-brain) and efferent (brain-to-ear) projections. Kraus and White-Schwoch (2015) propose a trade-off between the relative weights of these projections between the two streams (Fig. 6.2). Specifically, they argue that the primary stream is biased to afferent processing that reflects deeply engrained response profiles with relatively fewer resources dedicated to efferent processing. In contrast, they argue that the non-primary stream is biased to efferent processing that facilitates rapid task-dependent plasticity with relatively fewer resources dedicated to afferent processing.

This framework motivates the prediction that rapid task-dependent plasticity (i.e., the learning phase) should originate in non-primary auditory fields. Indeed, in ferret models, active attending drives spectrotemporal receptive field plasticity in posterior auditory fields prior to (and to a much greater degree than) in primary auditory cortex (Atiani et al. 2014). Moreover, recordings from rat auditory cortex show that stimulus-specific adaptation (SSA)—rapid plasticity occurring as a function of stimulus predictability (see Escera, Chap. 5)—occurs more quickly and robustly in non-primary auditory fields than in primary auditory cortex (Nieto-Diego and Malmierca 2016). This non-primary SSA is the presumed neural substrate of the mismatch negativity response (MMN, a scalp-recorded response reflecting detection of stimulus deviants). Similarly, recordings from guinea pigs show that non-primary auditory thalamus contributes to the MMN (Kraus et al. 1994a, b). Also noteworthy is that SSA-sensitive neurons in IC are strongly biased

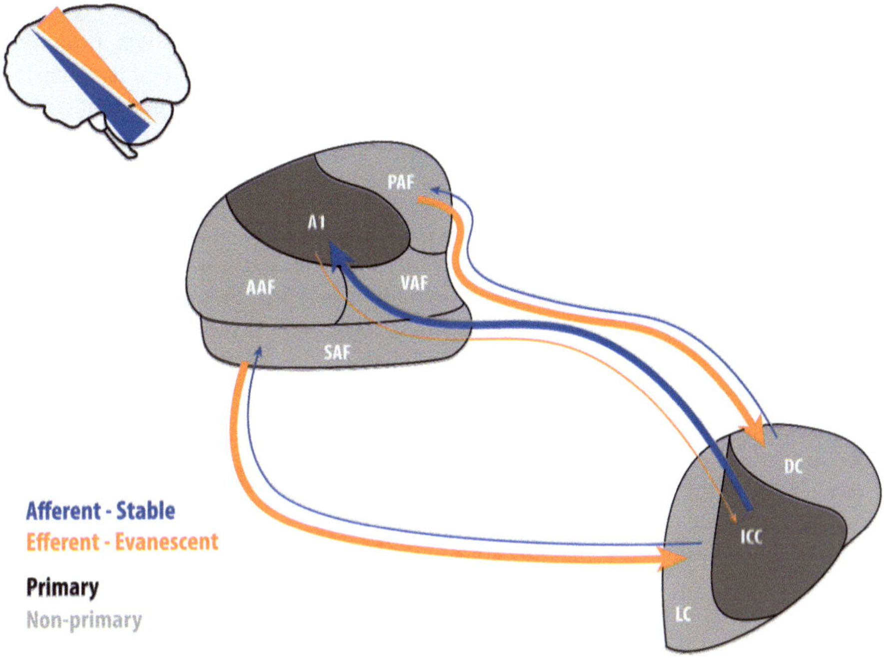

Fig. 6.2 The afferent/efferent, primary/non-primary trade-off model. The auditory neuraxis contains primary and non-primary divisions, each of which are innervated by afferent (*blue*) and efferent (*orange*) projections. Kraus and White-Schwoch (2015) propose a model wherein the primary division is biased to stable, afferent processing that is only remodeled over time. In contrast, the non-primary division is biased to efferent processing that is constantly undergoing evanescent plasticity. This idea is illustrated by schematized projections between the auditory cortex and inferior colliculus, showing primary (*darker*) and non-primary (*lighter*) fields, but applies throughout the system. (*A1*, primary auditory cortex; *AAF*, anterior auditory field; *DC*, dorsal cortex; *ICC*, central nucleus of inferior colliculus; *LC*, lateral cortex; *PAF*, posterior auditory field; *SAF*, superior auditory field; *VAF*, ventral auditory field. (Original figure by White-Schwoch and Kraus; auditory cortex view after Nieto-Diego and Malmierca 2016; inferior colliculus view after Loftus et al. 2008)

toward non-primary fields and receive input from non-primary corticocollicular fibers (Ayala et al. 2015).

This framework motivates a second prediction, namely, that primary auditory fields should only be resculpted following prolonged experiences (i.e., in the learned phase). This prediction is more difficult to test, especially given the challenges of following a single neuron over time. There is, however, evidence from the somatosensory system that cortical maps are reshaped following partial amputation (and restoration) of a finger (Merzenich et al. 1983). Studies in congenitally deaf humans and cats show that altered, daily sensory experience leads to drastic reorganization of primary auditory cortex (Kral and Sharma 2012). Finally, primary corticollicular fibers are necessary for recalibrating sound localization abilities to accommodate an earplug, a form of auditory learning (Bajo et al. 2010). Together, this shows that primary remodeling accumulates following intensive, prolonged experiences, such as a permanent alteration of sensory input. This idea is consistent with Kilgard's (2012) *expansion-renormalization model*, which argues that cortical map plasticity is a transient phenomenon, not necessary for changes in long-term performance.

Thus, it appears that rapid plasticity is qualitatively distinct from long-term functional remodeling. Evidence for this principle is found in several sites along the auditory neuraxis, reinforcing the review of a distributed-but-integrated circuit. With respect to the FFR, this model can explain a paradox that pervades the field. Although the FFR profoundly illustrates the legacy of prolonged and repeated experiences (see Sects. 6.5–6.7), evanescent changes due to attentional modulation are difficult to capture (see Shinn-Cunningham, Varghese, Wang, and Bharadwaj, Chap. 7). The FFR may be biased to index the more enduring and automatic processing route in the auditory system, which is only reshaped following experience. Noteworthy is the observation that online changes in inferior colliculus receptive fields are relatively small and in the opposite direction of auditory cortex receptive field plasticity induced by the same task (Slee and David 2015).

6.3.4 Selective Modulation of Sensory Function

An important principle of auditory neuroplasticity is that learning occurs along the dimensions that are trained, that is, the specific acoustic parameters that a listener attends to and connects to meaning are those whose neural coding is modulated. The *selective modulation* occurs in contrast to an overall "volume knob" effect that could be thought of as a broad gain (or decrease) in neural activity.

This principle was demonstrated elegantly in an experiment by Polley et al. (2006), who trained two groups of rats on a single stimulus set. One group trained on frequency contrasts between the stimuli, whereas another group trained on intensity contrasts. Perceptual acuity increased along the specific dimensions on which the rats were trained. Additionally, receptive fields in auditory cortex improved only along the trained dimension (frequency versus intensity). This experiment highlights how cognitive factors interact with sensory factors during learning. It also illustrates how plasticity is constrained to the factors that a listener is taught to care about as opposed to a product of the trained stimuli per se.

The FFR profoundly illustrates the principle of selective modulation of sensory function. A single response presents a plethora of information about how discrete components of sound are processed, and a central hypothesis behind much FFR research is that these reflect non-overlapping processes (see Kraus, Anderson, and White-Schwoch, Chap. 1). That is to say, discrete FFR parameters reflect distinct mechanisms that are modulated by divergent auditory experiences. This concept is broadly consistent with the idea that neuroplasticity occurs along dimensions that are behaviorally relevant to a listener (Recanzone et al. 1993). Additionally, it is consistent with the idea of the auditory system as a distributed, but integrated, circuit insofar as a single index of neurophysiological function reveals multiple aspects of auditory processing.

6.3.5 Experimental Models of Auditory Learning in Humans

A major challenge in auditory neuroscience has been to bridge research in animal models and research on humans. Animal models offer clear advantages, including careful control over learning, environments, and genetics, in addition to the ability to make invasive recordings of neural activity. There are also several limitations with animal models, however, the largest being an inability to evaluate outcomes of learning for complex behaviors (such as communication) and the fact that laboratory conditions are rarely naturalistic. The FFR fills this gap by providing a robust index of auditory function across species (e.g., Warrier et al. 2011; White-Schwoch et al. in press).

Both music training and socioeconomic status (SES) provide models for auditory learning in humans because they both encapsulate a constellation of factors that drive neural remodeling. In other words, they provide "best case scenarios" for considering what is possible in terms of neuroplasticity. Music training provides a model of *adaptive plasticity*—gains in perceptual and cognitive skills exhibited in the learned phase. In contrast, SES provides a model of *maladaptive plasticity*—poorer perceptual and cognitive skills in the learned phase. Additional models include language experience (Krishnan and Gandour, Chap. 3) and perceptual learning (Carcagno and Plack, Chap. 4). An understanding of this learning, however, needs to be grounded in an understanding of everyday communication, its impairments, and their relationships to the FFR.

6.4 FFR Studies of Everyday Communication

FFR studies provide a foundation for informing the biology of auditory experience by elucidating not just the neural remodeling engendered by these experiences but the consequences of this remodeling for communication. This literature specifically lays the groundwork for understanding FFR studies of music training and SES. On the one hand, seeing that the same neurophysiological processes implicated in communication disorders are boosted by music training makes a strong case for therapeutic interventions. On the other hand, seeing that the neurophysiological profile of communication disorders partially overlaps that of SES can help understand the linguistic and cognitive sequelae of growing up in low SES environments. Finally, showing how FFR parameters do—and do not—overlap between different populations reinforces the idea that plasticity is a double-edged sword that occurs along dimensions of behavioral relevance. Thus, an understanding of FFR studies of everyday communication provides a context for studies of long-term auditory experience by highlighting their consequences for everyday behavior.

6.4.1 *Listening and Language Are Connected*

FFR studies of communication skills are rooted in the idea that listening, language, and literacy are connected. Many models of communication impairments, including reading impairment and autism spectrum disorder, identify abnormal auditory processing as a core deficit (e.g., Tallal 2004). A child who cannot efficiently connect sounds to meaning will create a poor infrastructure on which to develop skills such as reading. Although it remains debated whether this abnormal auditory perception is a cause or consequence of poor language skills (Rosen 2003), objective measures (such as the FFR) prove advantageous in evaluating communication skills in these populations (Reetzke, Xie, and Chandrasekaran, Chap. 10). Specifically, because the FFR does not require a behavioral response from listeners, it may be applied to difficult-to-test populations such as infants (see Jeng, Chap. 2) or children with attention problems (see Schochat, Rocha-Muniz, and Filippini, Chap. 9).

The FFR has clarified that listening skills (such as understanding speech in noisy environments) and language skills (such as reading) only partially overlap. Listening in noise pulls on the ability to group talkers and auditory objects, whereas reading pulls on the ability to categorize sounds into phonemes. Both rely on accurate transcription of speech sounds and sensitivity to the auditory environment. Thus, individuals with communication impairments can struggle in one or both domains. Importantly, FFR studies of auditory enrichment and deprivation have also shown that the same facets of sound processing important for communication are malleable through experience (Kraus and White-Schwoch 2016), thereby motivating therapeutic approaches to improve the neural foundations of everyday communication.

6.4.2 *Everyday Communication Skills are Revealed by the FFR*

6.4.2.1 Phase One: FFRs are Distinct in Children with Learning Problems

The first set of FFR studies of communication skills involved a broadly defined group of children with auditory-based learning problems (LPs) such as dyslexia, specific language impairment (SLI), auditory processing disorder (APD), and attention-deficit/hyperactivity disorder. The emphasis of this set of studies was to establish group differences between LP children and their typically developing peers. Together, these studies provide a proof of concept that the FFR characterizes auditory-neurophysiological processing in children with and without communication impairments.

Cunningham et al. (2001) conducted the first FFR study of LP children. They compared FFRs to the consonant-vowel (CV) syllable [da] in quiet and background noise. The LP children had slower responses than their peers and less robust representation of the speech harmonics that conveyed speech formant features. When the FFRs were correlated to the evoking stimulus, the LP children had smaller scores, suggesting a less faithful representation of sound. However, group differences were only apparent when the [da] was played in noise. Consistent with the integrated-but-distributed model of the auditory system (see Sect. 6.2), these children also had smaller cortical responses to speech in noise. Finally, they required larger acoustic differences to distinguish two similar syllables presented in noise.

King et al. (2002) followed up on this work and replicated the finding that LP children have slower responses to CV syllables concomitant to less robust cortical responses to speech. Wible et al. (2004) showed LP children had less-refined onset responses to speech, suggesting a loss of neural synchrony, especially for rapidly presented streams of stimuli. Again, consistent with the distributed-but-integrated view of the auditory system, Abrams et al. (2006) found that FFRs to speech relate to the integrity with which rapid temporal cues are processed in auditory cortex; however, this coherence is disrupted in LP children (Wible et al. 2005).

Together, these studies show that children with communication problems broadly spanning language, literacy, and auditory processing, have poorer speech-evoked FFRs than their peers. Specifically, their FFRs tend to be slower, smaller, and to reproduce stimulus features less accurately. When sounds are presented in noise or rapid succession, group differences become more apparent, suggesting that listening conditions that tax the system emphasize neurophysiological processing bottlenecks in LP children.

6.4.2.2 Phase Two: FFRs Reveal Mechanistic Bottlenecks in Sound Processing

The second phase aimed to elucidate direct relationships between FFR parameters and communication behaviors such as literacy skills or the ability to understand speech in noise. This phase is noteworthy for introducing more diverse stimuli and complex paradigms and techniques. This work reinforces the relationship between the ability to understand speech in everyday environments and literacy skills and points to potential auditory-neurophysiological mechanisms underlying this link. Finally, this phase of work introduced the idea of partially overlapping “neural signatures” for communication—patterns of enhancements and diminutions to sound processing manifest by the FFR that characterize groups of listeners (Kraus and Nicol 2014).

Banai et al. (2009) investigated phonological processing (knowledge of the sound structure of spoken language; see Reetzke, Xie, and Chandrasekaran, Chap. 10) in a group of children with a wide range of reading skills. Children with poor phonological processing had slower FFRs than their peers with better phonological processing. Additionally, these children had smaller responses to speech harmonics. Notably, Banai et al. (2009) showed that groups of children had similar responses to the fundamental frequency (F_0), which is consistent with the idea of selective modulation of neurophysiological functions (see Sect. 6.3.4). Thus, children with poor phonological skills have poor coding of speech formant features but intact coding of the F_0 (Kraus and Nicol 2005).

Chandrasekaran et al. (2009) introduced a new paradigm to measure FFRs: instead of eliciting responses to a single stimulus, they presented eight CV syllables in a random order, one of which was also presented in a continuous stream in a second recording session. The hypothesis was that the auditory system would be sensitive to stimulus context, motivated by evidence that neural coding is modulated by the statistics of sounds in the environment (see Escera, Chap. 5). Indeed, Chandrasekaran et al. (2009) found a stimulus adaptation effect—responses to harmonics in the stimulus were amplified when it was played in a predictable condition. Interestingly, the extent of this amplification was related to children's ability to understand sentences in noise, and it was attenuated in children with dyslexia. This finding ties into Ahissar's (2007) argument that a primary deficit in dyslexia is an inability to exploit regularities in a sensory stream.

Hornickel et al. (2009) measured FFRs to the contrastive CV syllables [ba], [da], and [ga]. The three stimuli are expected to elicit a stereotypical response pattern based on their acoustics (see Fig. 6.3). The extent of FFR timing differences related both to the ability to understand sentences in noise and to phonological processing. Additionally, poor readers did not exhibit the expected timing pattern between responses to the three sounds, suggesting they have "blurry" neurophysiological encoding of these speech features (Fig. 6.3).

Together, these studies reinforce the idea that accurate representation of fine-grained speech features is crucial for literacy and language development and the idea that everyday listening skills support reading development (Tallal 2004).

Hornickel and Kraus (2013) investigated a different aspect of neural coding with respect to literacy. Instead of asking how quickly or accurately stimulus features are coded, they quantified the across-trial variability of the FFR. They found a systematic relationship between reading fluency and FFR variability: the best readers had the most consistent responses to speech and the poorest readers had the most variable responses. This suggests that children with poor literacy skills do not process sensory input reliably, which presumably hampers the development of a robust knowledge of language. Noteworthy is a similar phenomenon in auditory cortex of rats with a dyslexia candidate gene knockdown (Centanni et al. 2014).

Anderson et al. (2010a) focused on children's abilities to understand sentences in noise. They compared FFRs to [da] presented in quiet and background noise and quantified the extent to which noise delayed timing in response to the consonant-vowel transition. Children with better sentence perception in noise

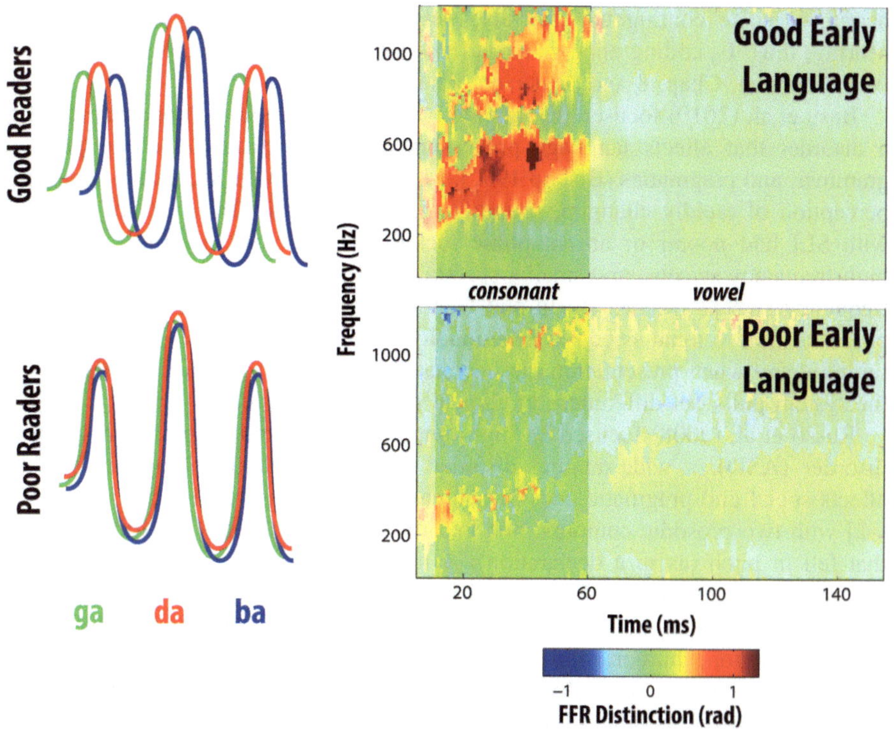

Fig. 6.3 FFR distinction of contrastive sounds. *Left* FFRs to the sounds [ba], [da], and [ga] are expected to have different timing based on the acoustics of the sounds. /G/ is the highest frequency sound, and so should elicit the earliest response, followed by /d/, followed by /b/. Good readers show this pattern, but poor readers do not. (Adapted from Hornickel et al. 2009). *Right* The cross-phaseogram measures the extent to which these sounds are distinguished in time-frequency space. Stronger distinctions (deeper red colors) are found in preschoolers with better early language skills. (Adapted from White-Schwoch and Kraus 2013)

exhibited a smaller FFR timing delay. Additionally, children with stronger reading fluency (the ability to quickly and accurately parse text) exhibited a smaller timing delay in noise. This study reinforced the link between FFR timing and communication skills and the link between literacy and perception in noise. In a follow-up study, however, they showed that children with better sentence perception in noise had stronger representation of the F_0 in speech (Anderson et al. 2010b). Thus, while the neural signatures for literacy and hearing in noise seemed to overlap with respect to neural response timing, they differed in that hearing in noise was associated with representation of the F_0, whereas literacy was associated with representation of speech harmonics.

The link between F_0 coding and the ability to understand sentences in noise was also observed in young adults whose F_0 coding and sentence-in-noise perception are both strengthened following auditory training (Song et al. 2012). This F_0

hearing-in-noise contingency has also been found in older adults with normal hearing; thus, F_0 coding appears to underlie listening in noise across the life span (see Bidelman, Chap. 8 and Anderson, Chap. 11).

Basu et al. (2010) focused on children with specific language impairment (SLI), a disorder that affects language processing and production, including phonics, grammar, and pragmatics (Bishop 1997). Many children with SLI exhibit abnormal perception of rapidly changing sounds (e.g. Benasich and Tallal 2002). Children with SLI had poorer or absent phase locking in response to rapidly changing, high-frequency sweeps, and smaller spectral amplitudes in response to fast-moving frequency sweeps (see Krishnan and Gandour, Chap. 3 for a discussion of FFR phase locking to tonal sweeps). Additionally, children with SLI exhibited a longer timing delay when presentation rate increased, consistent with the idea that taxing the system pulls out differences in clinical populations (Sect. 6.4.2.1).

Russo et al. (2008) focused on high-functioning children with autism spectrum disorder (ASD), a wide-ranging disorder of language and communication that affects social and pragmatic cues in language. They measured FFRs to the sound [ya] with two prosodic contours: one that rose in pitch (as in a question) and one that fell in pitch (as in a declaration). A majority of children with ASD did not accurately track the pitch of the sound in their FFRs. This provides neurophysiological evidence that the auditory coding of prosodic cues in speech, which often convey emotion, humor, and intention, is disrupted in children with ASD. Additionally, children with ASD had slower responses to the onset and formant features of the CV syllable [da] (Russo et al. 2009).

Together, these studies show how FFR signatures evoke the hallmark behavioral phenotype of the communication disorder. In other words, the coding of specific acoustic and acoustic-phonetic parameters that challenge distinct groups of listeners is disrupted (e.g., formant transitions in dyslexia, prosody in ASD, etc.). Additionally, this literature illustrates how neural signatures of distinct communication abilities and disabilities only partially overlap. If different communication impairments have distinct neural signatures, then the FFR may be a viable tool for evaluating auditory processing in clinical populations (Anderson and Kraus 2016).

6.4.2.3 Phase Three: FFRs Predict the Future

Recent attention in FFR studies has turned to the potential for the FFR to predict the development of communication skills. As reviewed by Jeng (Chap. 2), FFRs are robust during infancy and in preschoolers (also see Anderson et al. 2015). The long-term goal of this line of research is therefore to identify FFR features that predict communication impairments before children struggle with language development. Paradoxically, many disorders cannot be identified until a child has struggled and failed. For example, a diagnosis of dyslexia requires that a child has received prolonged formal instruction, setting opportunities for remediation back several years (Ozernov-Palchik and Gaab 2016). This goal is motivated by the intuitive fact that early interventions are extremely effective. Bishop and Adams (1990) show that if a child's language problems are resolved by age 5.5 years, literacy development proceeds smoothly; otherwise, literacy challenges may be anticipated. This is not to say that interventions later in life are ineffective (see Carcagno and Plack, Chap. 4); rather, because they cannot piggyback a sensitive period for auditory development (see Sect. 6.3.1) interventions likely have to be more intense (but see Sect. 6.7.1.2).

White-Schwoch and Kraus (2013) replicated Hornickel and colleagues' (2009) study of speech sound distinction in the FFR (see Sect. 6.4.2.2) in preschool children who had not yet learned to read. They found that pre-readers with stronger early phonological processing had stronger neurophysiological distinctions of contrastive sounds (Fig. 6.3). This shows a neural correlate of a core literacy skill that precedes formal instruction in reading, suggesting poor neurophysiological processing is a bottleneck in literacy development (as opposed to a consequence of limited reading experience). Additionally, this suggests that preschool FFRs identify children with early constraints on literacy development.

Woodruff Carr et al. (2014) conducted a study in a similar population focused on rhythm skills, motivated by evidence that individuals with dyslexia struggle to maintain a steady beat (Goswami 2011). Preschoolers who could maintain a steady beat had more accurate FFR coding of the temporal envelope of a speech syllable. Consistent with the hypothesis that sensitivity to envelope cues is crucial for literacy development (Goswami 2011), these children also had more advanced early literacy skills than their peers. Thus, it once again seems that neural correlates of literacy are tied to early language development in children who have not yet learned to read.

White-Schwoch et al. (2015) conducted a second study of phonological processing in preschoolers. Unlike previous experiments that focused on one aspect of the FFR, they combined three facets of neurophysiological processing, each of which is tied to literacy in older children (see Sect. 6.4.2.2): timing, representation of formant features, and consistency. They developed a statistical model incorporating these features in response to a CV transition in noise and found that the model was strongly predictive of children's phonological processing. Additionally, they could forecast how children perform on an array of early literacy tests one year later, providing the first longitudinal evidence that the FFR predicts the development of language skills. Finally, they applied this model to school-aged children and found it reliably identifies which children have a learning disability, supporting the idea that the FFR could be a clinical tool to identify children at risk for learning disabilities (see Anderson, Chap. 11 for a discussion of FFR clinical applications).

6.4.3 Interim Summary

Together, the studies discussed in the previous sections show how FFRs are distinct in children with language-based LPs, how FFRs elucidate mechanisms underlying communication impairments in those populations, and how FFRs predict future success and struggle in everyday communication skills. A common thread through these studies is that listening and language are connected. Specifically, the ability to understand speech in noise is partially connected to language skills, and FFR signatures of those domains partially overlap (Fig. 6.4). Hearing in noise pulls distinctly on the ability to group talkers and auditory objects and is characterized by FFR markers of those skills. Language and literacy skills pull distinctly on the ability to categorize phonemes and are characterized by FFR markers of those skills. Both rely on a resiliency to difficult listening situations, accurate distinctions of sounds, and sensitivity to stimulus context. This work sets the stage for studies of long-term auditory experience, because these same neurophysiological processes are affected by that experience.

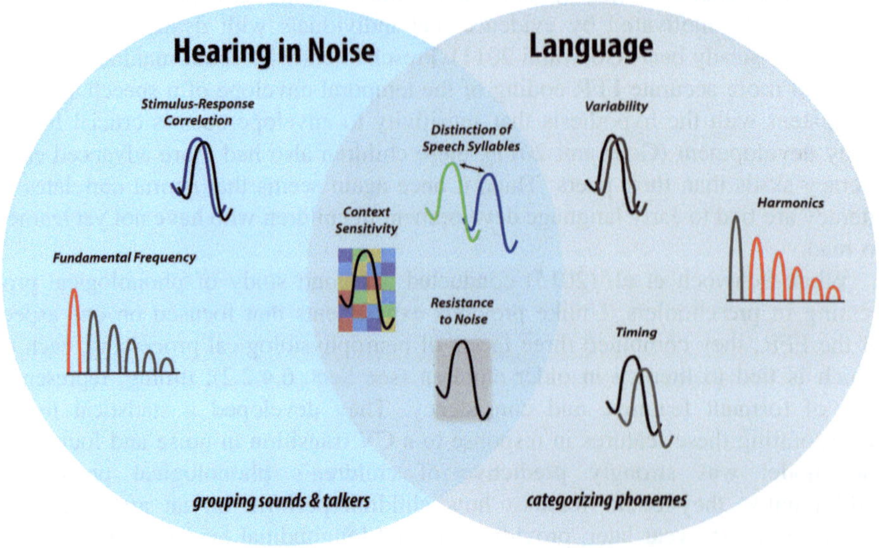

Fig. 6.4 Neural signatures of everyday communication: the listening-language overlap. Language skills are tied to listening, particularly the ability to understand speech in noise, but these neural signatures only partially overlap. The ability to understand speech in noise requires grouping sounds and talkers, facilitated by strong representation of the fundamental frequency and processing of meaningful stimulus ingredients. Language development (particularly language skills important for literacy) requires accurate categorization of incoming phonemes, facilitated by strong harmonic representation, fast timing, and a healthy base of consistent auditory processing. Both domains are supported by a sensitivity to stimulus context, accurate distinctions of contrastive speech sounds, and resiliency to the degrading effects of background noise on speech processing. (Original figure by White-Schwoch and Kraus)

6.5 FFR Studies of Music Training

The literature on music training vis-à-vis the FFR is large, and many converging experiments reinforce the evidence discussed below. Strait and Kraus (2014) provide a comprehensive review on the topic, couched in a broader literature on music and brain plasticity.

6.5.1 Music Training as a Model for Auditory Learning

Music engages a rich and diverse series of brain networks. The literature on music training and brain plasticity is vast and diffuse with respect to sensory modalities, cognitive functions, and methodology. Although the emphasis here is on FFR studies of music training, these should be considered in the broader context of changes in auditory anatomy and physiology attributed to music training, ranging from the cochlea to cortex. In these experiments, "musicians" are individuals who regularly (at least 2–3 times/week) produce music by instrument or voice. In general, they exclude individuals who engage solely in music listening or music appreciation, both of which appear insufficient to spark neural remodeling (Kraus et al. 2014a). And while many studies are correlational, random-assignment experiments reinforce a causal role for music training in conferring this plasticity (Chobert et al. 2012; Kraus et al. 2014b). Importantly, these neurophysiological enhancements are not merely parlor tricks: as summarized in Fig. 6.5, musicians enjoy a host of advantages for auditory cognition and perception over their peers.

The cognitive-sensorimotor-reward model of auditory learning (Sect. 6.3.2) is encapsulated by music training. The *sensorimotor* component of music making is perhaps the most self-evident, but what is thought to be important in the context of music and learning is the integration across sensory modalities (Limb and Braun 2008). Imagine that a piano player must continuously listen to sounds that are produced, use auditory feedback to modulate motor control, and use visual information, such as cues from other musicians. The *cognitive* component includes directed attention that frequently shifts between auditory streams (for example, rapidly switching listening from violins to trombones) and strong engagement of auditory working memory systems. The *reward* component comes into play given the extraordinary activation of limbic systems during music performance and listening (Blood et al. 1999). Classic studies on auditory cortex plasticity show that pairing limbic system stimulation with auditory cues catalyzes neuroplasticity (Bakin and Weinberger 1996; Kilgard and Merzenich 1998). Thus, the auditory-reward coupling inherent in music making, combined with top-down cognitive modulation, creates ideal circumstances for auditory learning (Patel 2011).

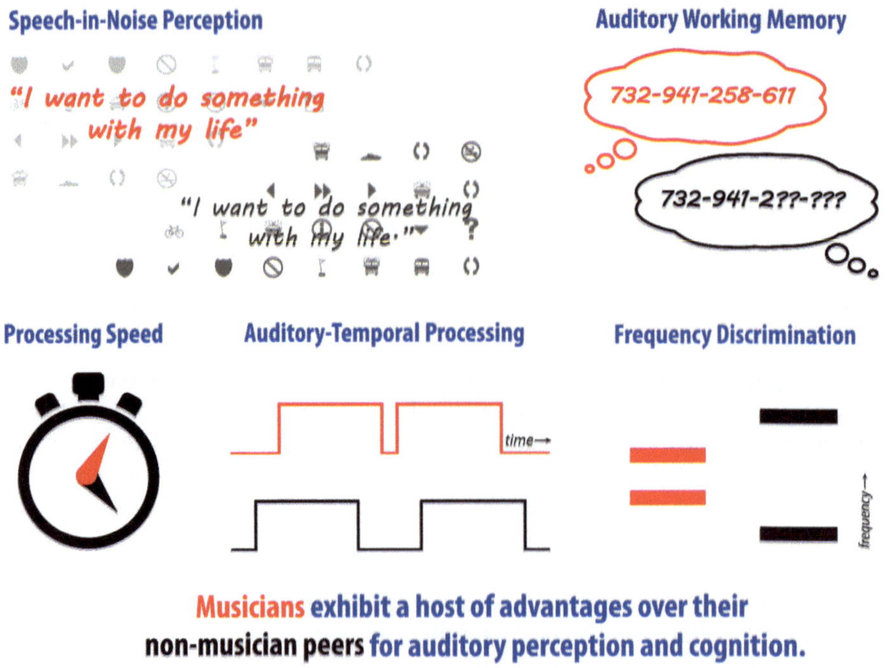

Fig. 6.5 Music training strengthens auditory perception and cognition. Evidence across their life spans shows that musicians outperform their nonmusician peers on tests of auditory function. These include (*clockwise from top left*) speech perception in noise, auditory working memory, frequency discrimination, auditory-temporal processing, and processing speed. (Original figure by White-Schwoch and Kraus)

6.5.2 FFRs are Distinct in Listeners with Music Experience

Musacchia et al. (2007) conducted the first study of music training and the FFR. They measured FFRs to a CV syllable and to a cello sound in audio-only and audiovisual conditions. Across stimuli and listening conditions, musicians had faster and larger responses than their peers. Additionally, FFR amplitudes were correlated to the extent of music training in both audio-only and audiovisual conditions, reinforcing the idea that neurophysiological enhancements observed in musicians are products of experience, not preexisting differences.

The second landmark study of music training and the FFR was inspired by Krishan and colleagues' seminal work showing that experience with a tone language enhances FFR tracking of pitch contours (Krishnan and Gandour, Chap. 3). Wong et al. (2007) asked if music training affects this linguistic processing as well. They found that amateur musicians had more accurate FFR tracking of pitch contours than nonmusicians. Additionally, the extent of music training correlated to the accuracy of pitch tracking, suggesting that additional music training heightens neurophysiological acuity for linguistically relevant sounds, which is consistent with the idea that neural circuits for music and language overlap (Patel 2008).

Bidelman et al. (2011) replicated and extended this work, comparing musicians to tone language speakers (native Mandarin speakers). Although musicians and Mandarin speakers both had stronger tracking of tone contours and iterated rippled noise than English-speaking nonmusicians, musicians specialized for more musical sounds and Mandarin speakers specialized for more linguistic sounds. Thus, although there is a cross-domain transfer between music and language, specialization still occurs along dimensions that are the most behaviorally salient for a listener (see Sect. 6.3.4). Strait et al. (2012a) tested a similar concept by comparing two groups of musicians—pianists and non-pianists—all with a similar degree of training. Pianists had stronger FFR coding of the timbre of a piano than of bassoon or tuba sounds, whereas non-pianists had equivalent coding across instruments. This supports the notion of "specialization among the specialized" in experience-dependent plasticity: auditory expertise is a continuum that is amplified by further training.

Strait et al. (2009) pursued both the questions of selective modulation of auditory function and of the generalization of music training to other kinds of sounds by eliciting FFRs to a baby cry (an emotional sound) in musicians and nonmusicians. The stimulus was important because it contained spectrally simple (periodic) and complex (containing acoustic transients) segments. Musicians had smaller responses to the simple segment, but larger responses to the complex segment. This was interpreted to reflect both attenuation of neural processing for simple sounds (a neural economy of resources) and enhancement of neural processing for challenging sounds (neural recruitment to process difficult stimuli). This study reinforces that experience-dependent plasticity is not an "all-or-none" phenomenon (see Sect. 6.3.4).

It should be noted that the FFR literature on music training reinforces the view of the auditory system as a distributed, but integrated, circuit (see Sect. 6.2). It has been argued that music training influences subcortical processing through the corticofugal system and that this integration heightens neural acuity throughout the system (Kraus and Chandrasekaran 2010). This is evident in studies that consider the interactions between different levels of auditory processing. For example, Lee et al. (2009) show that musically relevant cochlear nonlinearities are represented more robustly in the midbrain FFRs of musicians. Additionally, music training strengthens the coherence between FFRs and cortical responses in younger (Musacchia et al. 2008) and older adults (Bidelman and Alain 2015).

Do FFR signatures of music training generalize to everyday communication? With respect to auditory cognition and perception, there is evidence that musicians have superior speech perception in noise, processing speed, auditory working memory, auditory-temporal processing (such as backward masking thresholds), and frequency discrimination (Fig. 6.5). Longitudinal and cross-sectional studies argue that these cognitive and communicative gains are intimately tied to augmented auditory processing (Strait and Kraus 2014). Literacy skills and the ability to listen to speech in noise are emphases in the music-communication overlap (Kraus and White-Schwoch 2016).

Are these enhancements linked to the FFR? Weiss and Bidelman (2015) addressed this question in the context of speech perception in an elegant experiment. They measured FFRs to sounds along a continuum from the vowels /u/ to /a/. They "sonified" the FFRs—transformed the waveforms into audible sound files, capitalizing on the spectotemporal richness of FFRs—then used the sonified FFRs as stimuli in a simple categorical perception task. Listeners categorized the FFRs of musicians more quickly and more accurately than the FFRs of nonmusicians. This suggests that the stronger auditory processing by musicians, particularly of speech in noise (Parbery-Clark et al. 2009b; Zendel and Alain 2012), may be grounded, in part, in processes reflected by the FFR.

6.5.3 Music Training Affects Speech Processing Across the Life Span

6.5.3.1 Preschoolers

Strait et al. (2013) investigated the auditory-neurophysiological impact of preschool music classes by eliciting FFRs to a consonant-vowel syllable in quiet and noise. Preschoolers (ages 3–5 years) engaged in music classes had faster responses to the onset and consonant transition of speech in quiet and noise, but the groups had identical response timing for a vowel. Additionally, these children had stronger stimulus-response correlations, indicating more accurate neural coding of speech, and their responses were more resilient to background noise. Moreover, preschoolers had stronger neurophysiological distinctions of contrastive syllables (Strait et al. 2014); because these FFR metrics are linked to literacy (see Sect. 6.4.2.3), including in preschoolers, these results support the idea that music training boosts communication skills. In contrast to older children and adults engaged in music lessons, preschoolers did not exhibit enhanced processing of speech harmonics (see Sects. 6.5.3.2 and 6.5.3.3).

A subset of children returned after one additional year of music training, providing the first longitudinal study of music training vis-à-vis the FFR. Preschoolers exhibited even stronger neurophysiological resiliency to the degrading effects of background noise on response timing after the additional year of training. These results corroborate a causal role for music training in engendering brain plasticity. In the context of studies on music training across the life span, these results suggest that the musician's neural signature emerges progressively (Fig. 6.6).

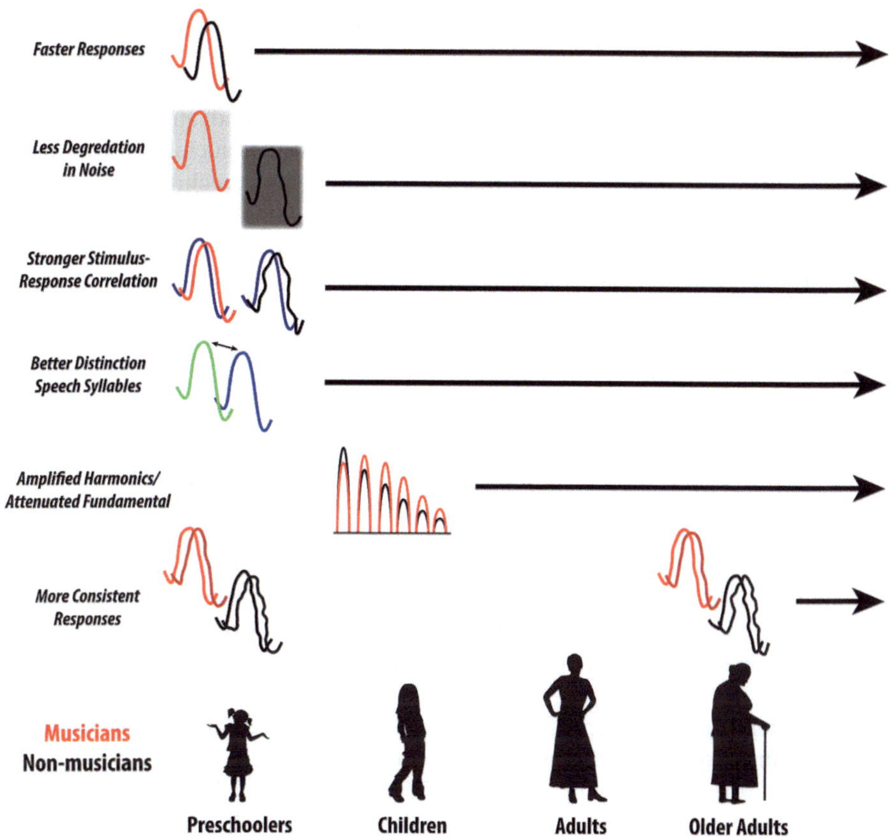

Fig. 6.6 Musicians exhibit a neural signature of their experience that is manifest in the FFR and emerges progressively across the life span. Preschool musicians have faster responses, more resiliency to background noise, stronger stimulus-response correlations, and better neurophysiological distinctions of contrastive speech syllables (see Fig. 6.3). Older children begin to show stronger representation of speech harmonics. The last element of the musician's neural signature emergences in senescence, when musicians have more consistent responses than their nonmusician peers. (Original figure by White-Schwoch and Kraus)

6.5.3.2 Children

Strait et al. (2012b) conducted a complementary series of studies in older children (ages 7–13 years), employing identical techniques as in preschoolers (see Sect. 6.5.3.1). Children with music training had faster responses to the onset and CV transition of speech presented in quiet and noise; they also exhibited more resiliency to background noise and stronger stimulus-response correlations. Similar to preschoolers, these older children with music training exhibited stronger neurophysiological distinctions of contrastive speech syllables (Strait et al. 2014). Additionally, children with music training had stronger processing of speech harmonics, suggesting this aspect of the musician's neural signature emerges following additional music training.

Strait and colleagues also investigated auditory behaviors, including sentence-in-noise perception, auditory working memory, and auditory attention. Consistent with evidence from adults (Fig. 6.5), they found that children with music training outperformed their peers on these tests. Additionally, a child's performance on these tests related to the integrity of neural processing (including response timing and harmonic representation). This again corroborates the idea that strengthening neurophysiological processing boosts skills important for everyday communication.

6.5.3.3 Young Adults

Parbery-Clark et al. (2009a) conducted a study of young adults (ages 19–30 years) and measured FFRs to speech in quiet and noise. Young adult musicians had faster responses to the onset and CV transition, stronger stimulus-response correlations, more resiliency to background noise, stronger representation of speech harmonics, and stronger neurophysiological distinctions of speech syllables. All of these facets of neurophysiological processing related to the ability to understand sentences in noise, again corroborating the idea that music training engenders neurophysiological enhancements to speech processing that generalize to everyday communication skills.

In a later study, Parbery-Clark et al. (2013) evaluated the extent to which young adults with music training take advantage of binaural hearing, which has a recognized role for listening in noise. They measured FFRs to /d/ in quiet monaurally to each ear and diotically (same sound to both ears). They chose a sound for which a musician advantage had not been observed and found that while the groups had similar responses to monaural input, musicians had faster and less variable responses to the diotic input (Fig. 6.7). Moreover, musicians had a larger binaural "benefit" in terms of the acceleration of FFR timing and decrease of response variability, suggesting a musician advantage may be rooted in binaural processing.

6.5.3.4 Middle-Aged and Older Adults

Parbery-Clark et al. (2012a, b) adopted similar techniques to study auditory processing in middle-aged and older adults with music training (ages 45–65 years). A hallmark of auditory aging is loss of temporal precision that disrupts the ability to understand rapidly changing speech sounds and the ability to understand speech in noisy environments (for a review see Gordon-Salant 2014). In the FFR, aging manifests as a diverse and wide-ranging neural signature, including a prolongation of response timing in response to consonant transitions (see Anderson, Chap. 11). Parbery-Clark et al. (2012a) showed that older adult musicians have identical

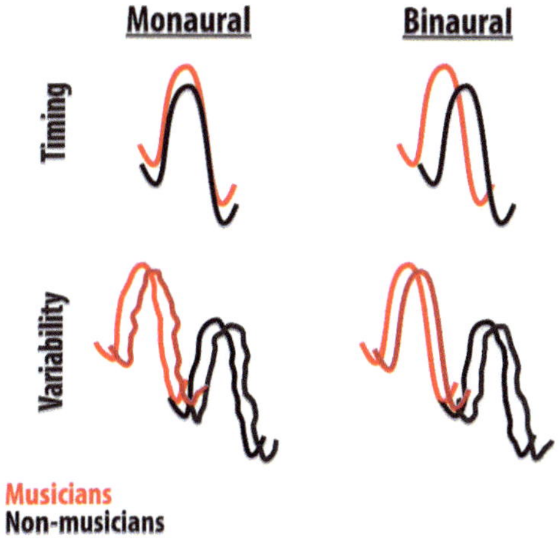

Fig. 6.7 Young adult musicians have stronger binaural responses to speech. Parbery-Clark et al. (2013) compared monaural and binaural FFRs in young adult musicians and nonmusicians. The two groups had similar *monaural* responses with respect to timing and consistency. However, musicians had faster and more consistent *binaural* responses tied to superior speech perception in noise, suggesting music experience may strengthen binaural processing. (Original figure by White-Schwoch and Kraus)

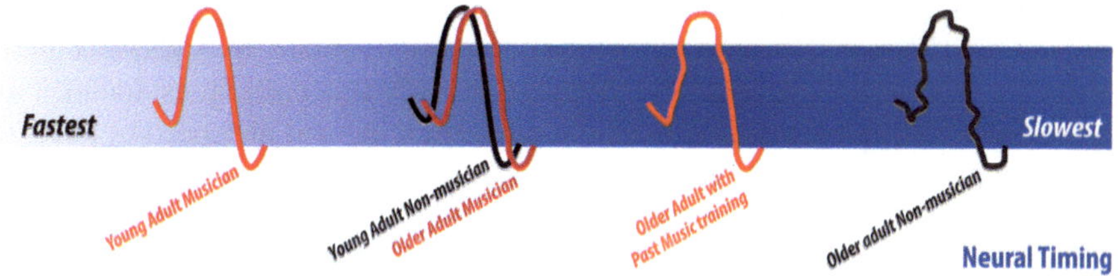

Fig. 6.8 Music training affects neural timing. FFRs to consonant-vowel transitions are faster in musicians than in nonmusicians. The fastest responses are observed in young adult musicians. Older adults with a life of music training have identical response timing as young adult nonmusicians. Older adults with music training early in life (but not for 50+ years) still have faster responses than older adult nonmusicians. (Original figure by White-Schwoch and Kraus)

response timing to young adult nonmusicians, suggesting that lifelong music training offsets this signature of aging (see Fig. 6.8). Bidelman and Alain (2015) also reported that older adult musicians exhibited fewer age-related declines in neural processing, and that older musicians had tighter integration between cortical and subcortical processing than their peers.

In a subsequent study, Parbery-Clark et al. (2012b) compared older adults with music training to older adult nonmusicians, and found that the musician's signature persists in this older population. Older adults with continuous, lifelong music training had faster responses, more robust responses to speech harmonics, stronger stimulus-response correlations, and more resiliency to background noise. Again, neurophysiological benefits were concomitant to gains in speech perception in noise and auditory working memory. Interestingly, a new aspect of the musician signature arose: older adult musicians had less variable neurophysiological responses, offsetting an additional element of the FFR signature for aging (Anderson et al. 2012).

6.5.4 Music Training Imparts an Enduring Biological Legacy

Does music training have to be lifelong to engender biological enhancements to sound processing? The simple answer is no, although the neurophysiological profile in listeners with past music experience is more circumscribed.

Skoe and Kraus (2012) tackled this question in young adults with a history of music training who did not currently play an instrument. They measured FFRs to a series of triangle tones and found that young adults with the most prior music training had the largest response to the F_0 of the tones, followed by a group with a small amount of music training, followed by a group with no music training. Thus, it appears that past auditory experiences, even if they are not maintained, still influence auditory function.

White-Schwoch et al. (2013) addressed a similar question in older adults. They also compared three groups (one with a modest amount of training, one with a little, and one with none). However, they considered individuals who had not played music for 50+ years. Nevertheless, White-Schwoch and colleagues found a systematic relationship between the extent of music training and the speed of FFRs to a consonant transition: older adults with the most music training had the fastest responses (Fig. 6.8). This is noteworthy because it ties into one of the hallmarks of aging (see Sect. 6.5.3.4 and Anderson Chap. 11). However, no other aspects of the aging musician signature were apparent, suggesting that while music training during childhood imparts a lifelong biological legacy, that legacy is not as pronounced as in lifelong musicians.

Three hypotheses emerge to explain these findings. The first hypothesis is that innate differences in auditory abilities drive some individuals to pursue music for longer than others. For example, an individual with faster FFRs may experience more success playing an instrument. It is unlikely, however, that there is a direct relationship between FFR timing and the number of years an individual feels driven to make music, suggesting a causal role for music training in engendering these changes. Additionally, longitudinal studies show a causal role for music training in shaping auditory physiology (Chobert et al. 2012; Kraus et al. 2014b).

The second hypothesis is that early experience tunes the auditory system during childhood and these changes are fixed. This possibility is also unlikely, however, given the potential for learning to shape auditory functions throughout life (see Carcagno and Plack, Chap. 4 for a discussion of short-term auditory training in adults). Thus, a third hypothesis emerges: music training teaches a listener to interact more meaningfully with sound, making everyday auditory activities learning experiences. By directing attention to sound in cognitively demanding and rewarding contexts (see Sect. 6.3.2), music training may facilitate sound-to-meaning connections during daily listening, an idea evocative of meta-plasticity: neuroplasticity that is a consequence of neuroplasticity (Abraham 2008).

6.6 FFR Studies of Socioeconomic Status

The literature on SES and the FFR is in its infancy. Nevertheless, early evidence supports the contention that growing up in poverty negatively affects auditory brain function in ways that are both similar and distinct to neural profiles of populations with communication impairments (see Sect. 6.4).

6.6.1 Socioeconomic Status as a Model for Auditory Learning

The idea that SES provides a model for auditory learning is rooted in the concept that a child's everyday sensory milieu shapes brain structure and function (see Sect. 6.3.1). The maturation-by-experience hypothesis predicts divergent outcomes for children raised in relatively high versus low socioeconomic environments. Much like music, the socioeconomic environment includes a large number of factors that are candidates for driving this plasticity and, much like music, it is thought that SES engenders plasticity through this confluence. Socioeconomic status is a broad term reflecting factors that are difficult to quantify. Chief among these is an impoverished linguistic repertoire that may impede children's mapping of sounds in the environment to meaning. Many studies (including those using the FFR) use maternal education as a proxy for SES (Hoff et al. 2012).

Hart and Risley (1995) conducted a landmark study of toddlers' home language environments. Children in low SES households heard 30 million fewer words by three years of age than their peers. Additionally, these children heard 40% the number of different words as their peers. Thus, these children grow up in homes that are linguistically impoverished with respect to both the quantity and quality of their daily exposure. The latter is often forgotten in popular accounts of this seminal finding but is likely to be a major factor. After all, a mother's voice is one of the most rewarding stimuli available to a child (Abrams et al. 2016) and may facilitate learning a mother's linguistic repertoire even if said repertoire is impoverished. This scenario reinforces an important idea of neuroplasticity: the conditions may be correct to facilitate learning, but what is learned may be maladaptive (Kilgard 2012).

The physical environments of low SES neighborhoods provide another cluster of factors that may shape auditory function. Low SES neighborhoods run the risk of higher levels of toxins such as poor water quality and air pollution. Additionally, low SES neighborhoods may be noisier (imagine housing near an airport); background noise interferes with auditory cortical map development (Chang and Merzenich 2003) and degenerates afferent synapses at the inner hair cells (Sergeyenko et al. 2013; also see Shinn-Cunningham, Varghese, Wang, and Bharadwaj, Chap. 9 for a discussion of environmental noise exposure and the FFR). Finally, poor nutrition may impede brain development. Although there are no direct relationships established between any of these factors and the FFR, the integrative model of the auditory system supports the hypothesis that this confluence is reflected by the FFR.

Together, the constellation of factors associated with SES cover cognitive domains (such as the sophistication of language input and educational opportunities) and sensorimotor domains (such as environmental noise exposure and the total amount of auditory input) in rewarding settings (such as hearing mothers' voices). As in the previous discussions of the role of music, the emphasis of this chapter is studies of SES vis-à-vis the FFR. Note this only provides a glimpse into a large literature concerned with the development of brain structure and function in low SES populations (Stevens et al. 2009; Noble et al. 2015) and educational and socioemotional outcomes (Leventhal and Brooks-Gunn 2000).

6.6.2 The Neural Signature of SES

Skoe et al. (2013) established the neural signature of SES (Fig. 6.9) by studying adolescents, whom they dichotomized into groups with lower levels of maternal education (high school or less) and higher levels of maternal education (any amount of postsecondary schooling). Children in the low maternal education group had less stable FFRs and poorer representations of speech harmonics. Additionally, this group had "noisier" neural activity; that is, amplitudes were higher during the brief intervals of silence between each presentation of the [da] in the recording session. This increase in spontaneous neural activity may be likened to static on the telephone, and Skoe et al. (2013) suggested it may interfere with precise processing. Moreover, the low maternal education group performed more poorly than their peers on tests of reading and auditory working memory.

A noteworthy aspect of this study is that differences in academic achievement and neurophysiological processing were evident despite a great deal of homogeneity between the groups, which were matched for age, sex, intelligence, hearing, and even came from the same schools. Although the groups differed in the extent of maternal education, this difference was small ($\approx$3.5 years). Nevertheless, there were clear group differences. This suggests that modest disparities in maternal education, and by extension SES, cascade to tangible differences in auditory neurophysiology and skills important for everyday communication.

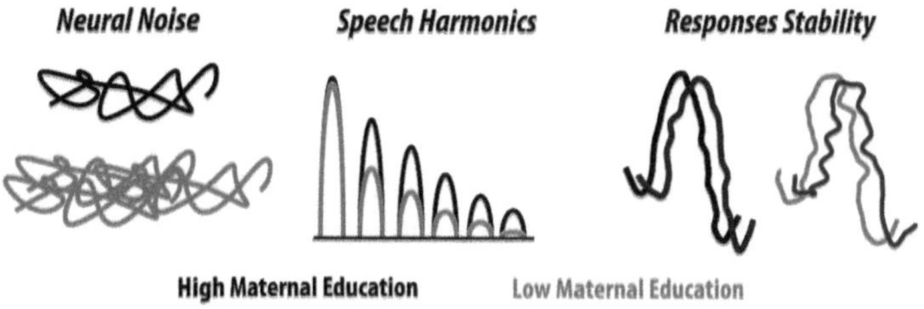

Fig. 6.9 The neural signature of socioeconomic status (SES). Children whose mothers have low levels of maternal education (a proxy for low SES) have distinct neural processing from their peers. Their FFRs have increased levels of neural noise, exhibit poor representation of speech harmonics, and are less stable across trials. (Original figure by White-Schwoch and Kraus)

6.7 Music Meets SES

There is broad interest in the use of community music programs to enrich perceptual, cognitive, and emotional development in underserved populations, including low SES children. Two groups of studies have used the FFR as an outcome measure to determine if music sparks neuroplasticity in low SES populations.

6.7.1 Music Training Catalyzes Neuroplasticity in Low SES Populations

6.7.1.1 Children

The first group of studies involved the Harmony Project (www.harmony-project.org), a mentorship organization in Los Angeles that provides free music instruction to children from gang-reduction zones. Children (ages 6–9 years at study onset) were randomly assigned to engage in instrumental music lessons or to spend a year on a waitlist, with guaranteed admission to music lessons one year later. Kraus et al. (2014b) found that the neurophysiological distinction of contrastive syllables (see Fig. 6.3) was stronger following two years of music training, but one year was insufficient. In a follow-up study, they pitted instrumental music training against music appreciation classes and found that children who actively made music had faster responses and stronger representation of speech harmonics (Kraus et al. 2014a).

This cluster of neurophysiological measures is evocative of the neural signature of literacy and language disorders (see Sect. 6.4.2.2) and provides additional evidence that music training may support the development of these skills or, at a minimum, their neural substrates (for reviews see Tierney and Kraus 2014;

Kraus and White-Schwoch 2016). This work extends this idea to low SES populations, which are known to exhibit lower levels of academic achievement (see Sect. 6.6.2). In fact, children engaged in instrumental music instruction performed better on tests of rhythmic aptitude (Slater et al. 2013) and rapid automatized naming (Slater et al. 2014)—two skills thought to support reading. Additionally, these children had better perception of sentences in noise (Slater et al. 2015), providing longitudinal evidence supporting the aforementioned correlational findings (see Sect. 6.5.1).

These experiments illustrate two principles of auditory learning. First, it takes time to change the brain. One year of music training was insufficient to spark neural remodeling, suggesting that the auditory system has some resistance to changing its automatic response properties. This is not to say that auditory processing cannot adapt to new contexts (see Escera, Chap. 5); rather, the deeply ingrained response properties evident during passive listening situations reflect the consequences of prolonged and repeated auditory experience (see Sect. 6.3.3). Second is that *making music matters*, that is, active engagement during the learning phase is necessary to engender remodeling evident in the learned phase.

6.7.1.2 Adolescents

The second group of experiments was conducted in collaboration with the Chicago Public Schools and involved a group of adolescents (starting age $\approx$ 14 years). The study focused on children in charter schools that offered co-curricular enrichment programs, either incorporating music training (instrumental lessons or choir) or paramilitary training (Junior Reserve Officer's Training Corps, JROTC, emphasizing discipline and athletics). Although these experiments did not employ random assignment (children chose their training), including the JROTC group provides an active control matched for the amount of time spent engaged in enriching activates. Thus, this work may arbitrate between potential effects of music training sensu stricto and childhood enrichment more generally.

Once again a single year of music training was insufficient to prompt neurophysiological changes, but after two years adolescents engaged in music training had faster responses to speech in noise (Tierney et al. 2013), consistent with the musician's neural signature (see Sect. 6.5.2). After a third year of training, the music group showed less variable FFRs than their peers (Tierney et al. 2015). Additionally, cortical responses to speech were more mature in children with music training. These results harken back to the maturation-by-experience hypothesis by showing that auditory enrichment piggybacks on neurophysiological processes still undergoing a phase of late maturation. In addition, the results are consistent with the view that altering a sensory milieu cascades to changes in sound processing. Indeed, adolescents engaged in music training also showed stronger phonological processing than their peers, a key literacy skill that is linked to FFR variability (White-Schwoch et al. 2015).

6.7.2 SES Shapes Auditory Processing Directly and Indirectly

Anderson et al. (2013) considered SES in a study of speech perception in noise. They studied 120 older adults and used statistical modeling to determine the influences of auditory-neurophysiological processing (FFRs to [da] in noise), cognition (auditory working memory and attention), hearing (air-conduction thresholds and distortion product otoacoustic emissions), and life experience (SES, past music training, and physical activity) on the ability to understand speech in noise. Life experience (including SES, measured by maternal education and the subjects' own educational attainment) influenced the ability to understand speech in noise with an additional indirect influence via neurophysiological processing. In other words, life experiences shaped the integrity of the FFR, which in turn shaped the ability to understand speech in noise. Moreover, like Skoe et al. (2013), they found a direct link between SES and formant processing (see Sect. 6.6.2).

Additionally, Anderson et al. (2013) considered whether their subjects underwent music training at any point during their lives. Older adults with music experience—regardless of whether or not they still played—relied more heavily on cognitive factors to understand speech in noise, consistent with the idea that music training shapes a listener's everyday relationship with sound (see Sect. 6.5.4). However, the influence of SES belied the apparent effects of music training (Fig. 6.10). This is not surprising, as families with more resources may be more likely to engage in music training.

The compounding influence of SES and music training highlights how SES is a complex and multifactorial phenomenon with both direct and indirect effects of auditory neurophysiology. For example, noise toxicity in impoverished neighborhoods may directly affect the auditory system, but the lack of opportunities for auditory enrichment, which is itself a consequence of low SES, also likely affects auditory processing.

6.7.3 SES and Bilingualism: A Complementary Case of Auditory Enrichment

The majority of studies discussed above asked how one sort of life experience, such as music or low SES, affects auditory function. Krizman et al. (2016) asked how experience with sound—bilingualism—interacts with SES. They tested a population of Spanish-English bilingual adolescents and compared them to a matched group of English-monolingual adolescents, building on work establishing that Spanish-English bilinguals have enhanced FFRs to the F_0 and more consistent FFRs to vowels (Krizman et al. 2012). Next, they split the two groups according to SES, using a similar strategy as Skoe and colleagues (see Sect. 6.6.2).

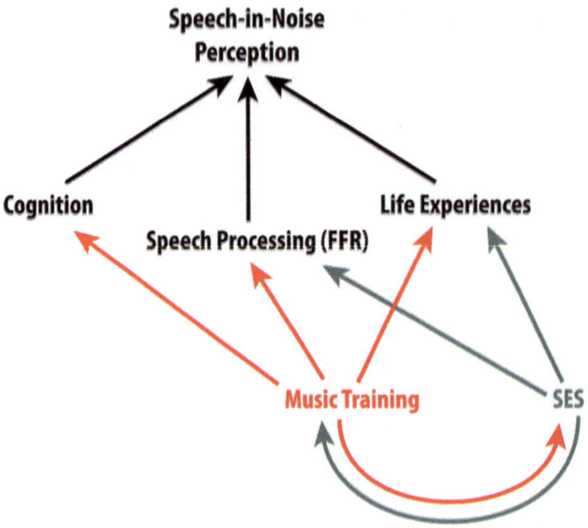

Fig. 6.10 Socioeconomic status (SES) influences auditory functions directly and indirectly. The ability of older adults to understand speech in noise is shaped by cognitive factors, speech processing (FFR), and life experience (music training and SES). Music training enhances cognition and speech processing, and SES affects speech processing. However, the apparent influence of music training could be attributed to SES. Individuals from lower-SES backgrounds were less likely to have received music training at all, pointing to the complex interactions between multiple types of auditory experience in predicting communication skills. Thus, music and SES interact to shape auditory processing and everyday communication. (Adapted from Anderson et al. 2013)

Bilingualism buffered the effects of SES on FFR stability. Specifically, the responses of low SES bilinguals were as consistent as high-SES monolinguals. Thus, enriching an auditory environment (by learning a second language) may counteract some of the consequences of growing up in impoverished environments. Presumably this buffer specifically addresses the linguistic impoverishment endemic to low SES children (see Sect. 6.6.1) and shows how enrichment and deprivation interact to shape auditory function. A similar finding comes from Zhu et al. (2014), who raised a group of rats in environmental deprivation but found that intense perceptual training as adults restored auditory cortical processing.

This also illustrates that the selective modulation(s) conferred through auditory learning may interact with modulations following other experiences. Practically, this suggests that therapeutic interventions might reverse some of the neural signatures of deprivation and communication disorders (see Carcagno and Plack, Chap. 4). Together, the bilingualism and musician models of auditory learning provide complementary evidence that auditory enrichment counteracts auditory deprivation.

6.8 Summary: Time Traveling Through the Auditory System

The auditory system is inherently plastic but strikes a remarkable balance between a propensity for, and a resiliency against, change. Everyday experiences in and through sound push the auditory system to reorganize, and the FFR provides an approach to explore this reorganization in humans. This reorganization bears on communication skills. An individual's FFR provides a window into his or her communication abilities, and signature patterns of FFR properties characterize several groups of listeners. These include individuals with language problems (such as dyslexia, SLI, APD, and ASD), and listeners who have undergone extensive experiences, such as musicians or children from low SES backgrounds. The latter have been the emphasis of this chapter and provide two experimental models of auditory learning in humans.

An intriguing notion emerges when considering this work as a whole: like Janus, an individual's FFR is a window to the past and doorway to the future, and thus offers a form of time travel through the auditory system. Each person's life in sound is a push and pull between enhancements and diminutions to sound processing, meaning FFR properties indicate the imprint of past experiences, good or bad (Fig. 6.11). The same response properties predict what the future holds for that individual's ability to communicate. The FFR provides this remarkable window to the past, present, and future because it is a measure of integrative auditory function. But diverse research on the FFR shows that the auditory future is not set in stone: auditory experiences that boost a relationship to sound through cognitive,

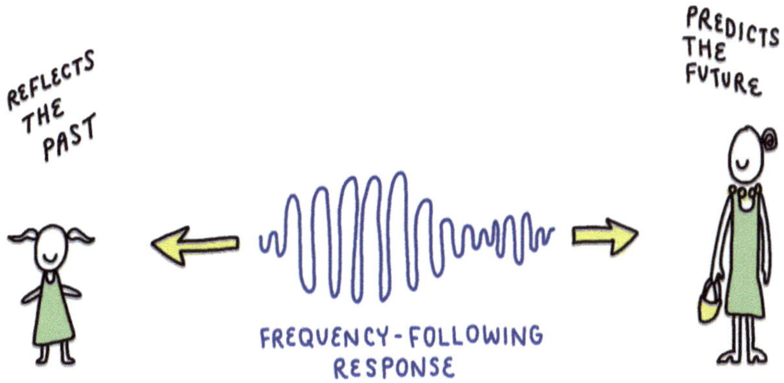

Fig. 6.11 Time traveling through the auditory system. Auditory anatomy and physiology reflect the legacy of a life in sound and predict the consequences of these experiences for future communication skills. Thus, the FFR can be thought of as a vehicle for time travel. An individual's FFR, measured today, is a biological mosaic that reflects the past and can objectively predict future success and challenges in everyday communication such as language and listening in noise. (Image by Katie Shelly)

sensorimotor, and reward networks can ameliorate poor communication skills and set up a new path for auditory processing.

This idea represents the next step for FFR research in learning and communication. How do multiple sorts of experiences interact to shape communication? Can adaptive learning completely offset maladaptive learning, or does it at least compensate for reduced function (e.g., Chambers et al. 2016)? Given promising evidence that the FFR can predict future communication impairment, can it also be used to motivate targeted early interventions? These questions can be explored as FFR techniques advance. As new neural signatures of communication crystallize (e.g., concussion; Kraus et al. 2016), these signatures can be used to fit individuals to different patterns. For example, one child may show a pattern that motivates music training and another may show a pattern that motivates second language training. As the FFR becomes more widely available, and its underlying mechanisms more thoroughly understood, it has exciting potential as a clinical tool to aid human communication. Moreover, it has the potential to bridge auditory neuroscience with education and social policy, both by highlighting the importance of everyday auditory experiences in development and well-being and by providing strategies in their support.

Acknowledgments Many thanks to Samira Anderson, Trent Nicol, and Sebastian Otto-Meyer for their comments on this chapter, and members of the Auditory Neuroscience Laboratory, past and present, for their contributions. The authors are supported by NIH (R01 DC010016 and R01 HD069414), NSF (BCS 1430400), the National Association of Music Merchants, the Dana Foundation, and the Knowles Hearing Center.

Compliance with Ethics Requirements Travis White-Schwoch declared that he had no conflict of interest.

Nina Kraus is Chief Scientific Officer of Synaural, a company working to develop a user-friendly measure of auditory processing.

References

Abraham, W. C. (2008). Metaplasticity: Tuning synapses and networks for plasticity. *Nature Reviews Neuroscience, 9*(5), 387.

Abrams, D. A., Nicol, T., Zecker, S. G., & Kraus, N. (2006). Auditory brainstem timing predicts cerebral asymmetry for speech. *The Journal of Neuroscience, 26*(43), 11131–11137.

Abrams, D. A., Nicol, T., Zecker, S., & Kraus, N. (2011). A possible role for a paralemniscal auditory pathway in the coding of slow temporal information. *Hearing Research, 272*(1), 125–134.

Abrams, D. A., Chen, T., Odriozola, P., Cheng, K. M., et al. (2016). Neural circuits underlying mother's voice perception predict social communication abilities in children. *Proceedings of the National Academy of Sciences of the U S A, 113*(22), 6295–6300.

Ahissar, M. (2007). Dyslexia and the anchoring-deficit hypothesis. *Trends in Cognitive Sciences, 11*(11), 458–465.

Ahissar, E., Sosnik, R., & Haidarliu, S. (2000). Transformation from temporal to rate coding in a somatosensory thalamocortical pathway. *Nature, 406*(6793), 302–306.

Anderson, S., & Kraus, N. (2016). Auditory processing disorder: Biological basis and treatment efficacy. In C. G. Le Prell, E. Lobarinas, A. N. Popper, & R. R. Fay (Eds.), *Translational research in audiology, neuro-otology, and the hearing sciences*. Springer Science + Business Media: Berlin.

Anderson, S., Parbery-Clark, A., White-Schwoch, T., & Kraus, N. (2012). Aging affects neural precision of speech encoding. *The Journal of Neuroscience, 32*(41), 14156–14164.

Anderson, S., Skoe, E., Chandrasekaran, B., & Kraus, N. (2010a). Neural timing is linked to speech perception in noise. *The Journal of Neuroscience, 30*(14), 4922–4926.

Anderson, S., Skoe, E., Chandrasekaran, B., Zecker, S., & Kraus, N. (2010b). Brainstem correlates of speech-in-noise perception in children. *Hearing Research, 270*(1–2), 151–157.

Anderson, S., White-Schwoch, T., Parbery-Clark, A., & Kraus, N. (2013). A dynamic auditory-cognitive system supports speech-in-noise perception in older adults. *Hearing Research, 300*, 18–32.

Anderson, S., Parbery-Clark, A., White-Schwoch, T., & Kraus, N. (2015). Development of subcortical speech representation in human infants. *The Journal of the Acoustical Society of America, 137*(6), 3346–3355.

Atiani, S., David, S. V., Elgueda, D., Locastro, M., et al. (2014). Emergent selectivity for task-relevant stimuli in higher-order auditory cortex. *Neuron, 82*(2), 486–499.

Ayala, Y. A., Udeh, A., Dutta, K., Bishop, D., et al. (2015). Differences in the strength of cortical and brainstem inputs to SSA and non-SSA neurons in the inferior colliculus. *Scientific Reports, 5*(10383), 2016. Doi:10.1038/srep10383

Bajo, V. M., Nodal, F. R., Moore, D. R., & King, A. J. (2010). The descending corticocollicular pathway mediates learning-induced auditory plasticity. *Nature Neuroscience, 13*(2), 253–260.

Bakin, J. S., & Weinberger, N. M. (1996). Induction of a physiological memory in the cerebral cortex by stimulation of the nucleus basalis. *Proceedings of the National Academy of Sciences of the U S A, 93*(20), 11219–11224.

Banai, K., Hornickel, J., Skoe, E., Nicol, T., et al. (2009). Reading and subcortical auditory function. *Cerebral Cortex, 19*, 2699–2707.

Basu, M., Krishnan, A., & Weber-Fox, C. (2010). Brainstem correlates of temporal auditory processing in children with specific language impairment. *Developmental Science, 13*(1), 77–91.

Behrmann, M., & Plaut, D. C. (2013). Distributed circuits, not circumscribed centers, mediate visual recognition. *Trends in Cognitive Sciences, 17*(5), 210–219.

Benasich, A. A., & Tallal, P. (2002). Infant discrimination of rapid auditory cues predicts later language impairment. *Behavioural Brain Research, 136*(1), 31–49.

Bidelman, G. M., & Alain, C. (2015). Musical training orchestrates coordinated neuroplasticity in auditory brainstem and cortex to counteract age-related declines in categorical vowel perception. *The Journal of Neuroscience, 35*(3), 1240–1249.

Bidelman, G. M., Gandour, J. T., & Krishnan, A. (2011). Cross-domain effects of music and language experience on the representation of pitch in the human auditory brainstem. *Journal of Cognitive Neuroscience, 23*(2), 425–434.

Bishop, D. V. (1997). *Uncommon understanding (classic edition): Development and disorders of language comprehension in children*. New York: Psychology Press.

Bishop, D. V., & Adams, C. (1990). A prospective study of the relationship between specific language impairment, phonological disorders and reading retardation. *Journal of Child Psychology and Psychiatry, 31*(7), 1027–1050.

Blood, A. J., Zatorre, R. J., Bermudez, P., & Evans, A. C. (1999). Emotional responses to pleasant and unpleasant music correlate with activity in paralimbic brain regions. *Nature Neuroscience, 2*(4), 382–387.

Centanni, T. M., Booker, A., Sloan, A., Chen, F., et al. (2014). Knockdown of the dyslexia-associated gene *Kiaa0319* impairs temporal responses to speech stimuli in rat primary auditory cortex. *Cerebral Cortex, 24*(7), 1753–1766.

Chambers, A. R., Resnik, J., Yuan, Y., Whitton, J. P., et al. (2016). Central gain restores auditory processing following near-complete cochlear denervation. *Neuron, 89*(4), 867–879.

Chandrasekaran, B., Hornickel, J., Skoe, E., Nicol, T., & Kraus, N. (2009). Context-dependent encoding in the human auditory brainstem relates to hearing speech in noise: Implications for developmental dyslexia. *Neuron, 64*(3), 311–319.

Chang, E. F., & Merzenich, M. M. (2003). Environmental noise retards auditory cortical development. *Science, 300*(5618), 498–502.

Chobert, J., François, C., Velay, J.-L., & Besson, M. (2012). Twelve months of active musical training in 8- to 10-year-old children enhances the preattentive processing of syllabic duration and voice onset time. *Cerebral Cortex, 24*(4), 956–967.

Coffey, E. B., Herholz, S. C., Chepesiuk, A. M., Baillet, S., & Zatorre, R. J. (2016). Cortical contributions to the auditory frequency-following response revealed by MEG. *Nature Communications, 7*(11070), 2016. Doi:10.1038/ncomms11070

Cunningham, J., Nicol, T., Zecker, S. G., Bradlow, A., & Kraus, N. (2001). Neurobiologic responses to speech in noise in children with learning problems: Deficits and strategies for improvement. *Clinical Neurophysiology, 112*(5), 758–767.

Engineer, N. D., Percaccio, C. R., Pandya, P. K., Moucha, R., et al. (2004). Environmental enrichment improves response strength, threshold, selectivity, and latency of auditory cortex neurons. *Journal of Neurophysiology, 92*(1), 73–82.

Fritz, J., Shamma, S., Elhilali, M., & Klein, D. (2003). Rapid task-related plasticity of spectrotemporal receptive fields in primary auditory cortex. *Nature Neuroscience, 6*(11), 1216–1223.

Gordon, K. A., Salloum, C., Toor, G. S., van Hoesel, R., & Papsin, B. C. (2012). Binaural interactions develop in the auditory brainstem of children who are deaf: Effects of place and level of bilateral electrical stimulation. *The Journal of Neuroscience, 32*(12), 4212–4223.

Gordon-Salant, S. (2014). Aging, hearing loss, and speech recognition: Stop shouting, I can't understand you. In A. N. Popper & R. R. Fay (Eds.), *Perspectives on auditory research* (pp. 211–228). New York: Springer Science + Business Media.

Goswami, U. (2011). A temporal sampling framework for developmental dyslexia. *Trends in Cognitive Sciences, 15*(1), 3–10.

Hart, B., & Risley, T. R. (1995). *Meaningful differences in the everyday experience of young American children*. Baltimore: Paul H Brookes Publishing.

Hoff, E., Laursen, B., & Bridges, K. (2012). Measurement and model building in studying the influence of socioeconomic status on child development. In L. Mayes & M. Lewis (Eds.), *The Cambridge handbook of environment in human development* (pp. 590–606). Cambridge: Cambridge University Press.

Hornickel, J., & Kraus, N. (2013). Unstable representation of sound: A biological marker of dyslexia. *The Journal of Neuroscience, 33*(8), 3500–3504.

Hornickel, J., Skoe, E., Nicol, T., Zecker, S., & Kraus, N. (2009). Subcortical differentiation of stop consonants relates to reading and speech-in-noise perception. *Proceedings of the National Academy of Sciences of the U S A, 106*(31), 13022–13027.

Intartaglia, B., White-Schwoch, T., Meunier, C., Roman, S., et al. (2016). Native language shapes automatic neural processing of speech. *Neuropsychologia, 89*, 57–65.

Kilgard, M. P. (2012). Harnessing plasticity to understand learning and treat disease. *Trends in Neurosciences, 35*(12), 715–722.

Kilgard, M. P., & Merzenich, M. M. (1998). Cortical map reorganization enabled by nucleus basalis activity. *Science, 279*(5357), 1714–1718.

King, C., Warrier, C. M., Hayes, E., & Kraus, N. (2002). Deficits in auditory brainstem pathway encoding of speech sounds in children with learning problems. *Neuroscience Letters, 319*(2), 111–115.

Kral, A., & Sharma, A. (2012). Developmental neuroplasticity after cochlear implantation. *Trends in Neurosciences, 35*(2), 111–122.

Kraus, N., & Chandrasekaran, B. (2010). Music training for the development of auditory skills. *Nature Reviews Neuroscience, 11*(8), 599–605.

Kraus, N., & Nicol, T. (2005). Brainstem origins for cortical "what" and "where" pathways in the auditory system. *Trends in Neurosciences, 28*(4), 176–181.

Kraus, N., & Nicol, T. (2014). The cognitive auditory system. In R. R. Fay & A. N. Popper (Eds.), *Perspectives on auditory research* (pp. 299–319). Heidelberg: Springer.

Kraus, N., & White-Schwoch, T. (2015). Unraveling the biology of auditory learning: A cognitive-sensorimotor-reward framework. *Trends in Cognitive Sciences, 19*(11), 642–654.

Kraus, N., & White-Schwoch, T. (2016). Neurobiology of everyday communication: What have we learned from music? *The Neuroscientist.* Doi:10.1177/1073858416653593

Kraus, N., McGee, T., Carrell, T., King, C., Littman, T., & Nicol, T. (1994a). Discrimination of speech-like contrasts in the auditory thalamus and cortex. *The Journal of the Acoustical Society of America, 96*(5), 2758–2768.

Kraus, N., McGee, T., Littman, T., Nicol, T., & King, C. (1994b). Nonprimary auditory thalamic representation of acoustic change. *Journal of Neurophysiology, 72*(3), 1270–1277.

Kraus, N., Slater, J., Thompson, E. C., Hornickel, J., et al. (2014a). Auditory learning through active engagement with sound: Biological impact of community music lessons in at-risk children. *Frontiers in Neuroscience, 8*(351), 2016. Doi:10.3389/fnins.2014.00351

Kraus, N., Slater, J., Thompson, E. C., Hornickel, J., et al. (2014b). Music enrichment programs improve the neural encoding of speech in at-risk children. *The Journal of Neuroscience, 34* (36), 11913–11918.

Kraus, N., Thompson, E. C., Krizman, J., Cook, K., White-Schwoch, T., & LaBella, C. R. (2016). Auditory biological marker of concussion in children. *Nature, Scientific Reports, 6*(39009).

Krizman, J., Marian, V., Shook, A., Skoe, E., & Kraus, N. (2012). Subcortical encoding of sound is enhanced in bilinguals and relates to executive function advantages. *Proceedings of the National Academy of Sciences of the U S A, 109*(20), 7877–7881.

Krizman, J., Skoe, E., & Kraus, N. (2016). Bilingual enhancements have no socioeconomic boundaries. *Developmental Science*. Doi:10.1111/desc.12347

Kuhl, P. K., Williams, K. A., Lacerda, F., Stevens, K. N., & Lindblom, B. (1992). Linguistic experience alters phonetic perception in infants by 6 months of age. *Science, 255*(5044), 606–608.

Lee, K. M., Skoe, E., Kraus, N., & Ashley, R. (2009). Selective subcortical enhancement of musical intervals in musicians. *The Journal of Neuroscience, 29*(18), 5832–5840.

Leventhal, T., & Brooks-Gunn, J. (2000). The neighborhoods they live in: The effects of neighborhood residence on child and adolescent outcomes. *Psychological Bulletin, 126*(2), 309–337.

Limb, C. J., & Braun, A. R. (2008). Neural substrates of spontaneous musical performance: An fMRI study of jazz improvisation. *PLoS ONE, 3*(2), e1679. Doi:10.1371/journal.pone.0001679

Loftus, W. C., Malmierca, M. S., Bishop, D. C., & Oliver, D. L. (2008). The cytoarchitecture of the inferior colliculus revisited: A common organization of the lateral cortex in rat and cat. *Neuroscience, 154*(1), 196–205.

Merzenich, M. M., Kaas, J., Wall, J., Nelson, R., et al. (1983). Topographic reorganization of somatosensory cortical areas 3b and 1 in adult monkeys following restricted deafferentation. *Neuroscience, 8*(1), 33–55.

Musacchia, G., Sams, M., Skoe, E., & Kraus, N. (2007). Musicians have enhanced subcortical auditory and audiovisual processing of speech and music. *Proceedings of the National Academy of Sciences of the U S A, 104*(40), 15894–15898.

Musacchia, G., Strait, D., & Kraus, N. (2008). Relationships between behavior, brainstem and cortical encoding of seen and heard speech in musicians and nonmusicians. *Hearing Research, 241*(1), 34–42.

Nieto-Diego, J., & Malmierca, M. S. (2016). Topographic distribution of stimulus-specific adaptation across auditory cortical fields in the anesthetized rat. *PLoS Biology, 14*(3), e1002397.

Noble, K. G., Houston, S. M., Brito, N. H., Bartsch, H., et al. (2015). Family income, parental education and brain structure in children and adolescents. *Nature Neuroscience, 18*(5), 773–778.

Ozernov-Palchik, O., & Gaab, N. (2016). Tackling the “dyslexia paradox”: Reading brain and behavior for early markers of developmental dyslexia. *Wiley Interdisciplinary Reviews: Cognitive Science, 7*(2), 156–176.

Pafundo, D. E., Nicholas, M. A., Zhang, R., & Kuhlman, S. J. (2016). Top-down-mediated facilitation in the visual cortex is gated by subcortical neuromodulation. *The Journal of Neuroscience, 36*(10), 2904–2914.

Parbery-Clark, A., Skoe, E., & Kraus, N. (2009a). Musical experience limits the degradative effects of background noise on the neural processing of sound. *The Journal of Neuroscience, 29*(45), 14100–14107.

Parbery-Clark, A., Skoe, E., Lam, C., & Kraus, N. (2009b). Musician enhancement for speech-in-noise. *Ear and Hearing, 30*(6), 653–661.

Parbery-Clark, A., Anderson, S., Hittner, E., & Kraus, N. (2012a). Musical experience offsets age-related delays in neural timing. *Neurobiology of Aging, 33*, 1483.

Parbery-Clark, A., Anderson, S., Hittner, E., & Kraus, N. (2012b). Musical experience strengthens the neural representation of sounds important for communication in middle-aged adults. *Frontiers in Aging Neuroscience, 4*(30), 2016. Doi:10.3389/fnagi.2012.00030

Parbery-Clark, A., Strait, D. L., Hittner, E., & Kraus, N. (2013). Musical training enhances neural processing of binaural sounds. *The Journal of Neuroscience, 33*(42), 16741–16747.

Patel, A. D. (2008). *Music, language, and the brain*. New York, NY: Oxford University Press.

Patel, A. D. (2011). Why would musical training benefit the neural encoding of speech? The OPERA hypothesis. *Frontiers in Psychology, 2*(142), 2016. Doi:10.3389/fpsyg.2011.00142

Polley, D. B., Steinberg, E. E., & Merzenich, M. M. (2006). Perceptual learning directs auditory cortical map reorganization through top-down influences. *The Journal of Neuroscience, 26* (18), 4970–4982.

Recanzone, G. H., Schreiner, C. E., & Merzenich, M. M. (1993). Plasticity in the frequency representation of primary auditory cortex following discrimination training in adult owl monkeys. *The Journal of Neuroscience, 13*(1), 87–103.

Reed, A., Riley, J., Carraway, R., Carrasco, A., et al. (2011). Cortical map plasticity improves learning but is not necessary for improved performance. *Neuron, 70*(1), 121–131.

Rosen, S. (2003). Auditory processing in dyslexia and specific language impairment: Is there a deficit? What is its nature? Does it explain anything? *Journal of Phonetics, 31*(3), 509–527.

Russo, N., Skoe, E., Trommer, B., Nicol, T., et al. (2008). Deficient brainstem encoding of pitch in children with autism spectrum disorders. *Clinical Neurophysiology, 119*(8), 1720–1731.

Russo, N., Nicol, T., Trommer, B., Zecker, S., & Kraus, N. (2009). Brainstem transcription of speech is disrupted in children with autism spectrum disorders. *Developmental Science, 12*(4), 557–567.

Sergeyenko, Y., Lall, K., Liberman, M. C., & Kujawa, S. G. (2013). Age-related cochlear synaptopathy: An early onset contributor to auditory functional decline. *The Journal of Neuroscience, 33*(34), 13686–13694.

Skoe, E., & Kraus, N. (2012). A little goes a long way: How the adult brain is shaped by musical training in childhood. *The Journal of Neuroscience, 32*(34), 11507–11510.

Skoe, E., Krizman, J., & Kraus, N. (2013). The impoverished brain: Disparities in maternal education affect the neural response to sound. *The Journal of Neuroscience, 33*(44), 17221–17231.

Slater, J., Tierney, A., & Kraus, N. (2013). At-risk elementary school children with one year of classroom music instruction are better at keeping a beat. *PLoS ONE, 8*(10), e77250.

Slater, J., Strait, D. L., Skoe, E., O’Connell, S., et al. (2014). Longitudinal effects of group music instruction on literacy skills in low-income children. *PLoS ONE, 9*(11), e113383.

Slater, J., Skoe, E., Strait, D. L., O’Connell, S., et al. (2015). Music training improves speech-in-noise perception: Longitudinal evidence from a community-based music program. *Behavioural Brain Research, 291*, 244–252.

Slee, S. J., & David, S. V. (2015). Rapid task-related plasticity of spectrotemporal receptive fields in the auditory midbrain. *The Journal of Neuroscience, 35*(38), 13090–13102.

Song, J. H., Skoe, E., Banai, K., & Kraus, N. (2012). Training to improve hearing speech in noise: Biological mechanisms. *Cerebral Cortex, 122*, 1890–1898.

Stevens, C., Lauinger, B., & Neville, H. (2009). Differences in the neural mechanisms of selective attention in children from different socioeconomic backgrounds: An event-related brain potential study. *Developmental Science, 12*(4), 634–646.

Strait, D. L., & Kraus, N. (2014). Biological impact of auditory expertise across the life span: Musicians as a model of auditory learning. *Hearing Research, 308*, 109–121.

Strait, D. L., Kraus, N., Skoe, E., & Ashley, R. (2009). Musical experience and neural efficiency—effects of training on subcortical processing of vocal expressions of emotion. *European Journal of Neuroscience, 29*(3), 661–668.

Strait, D. L., Chan, K., Ashley, R., & Kraus, N. (2012a). Specialization among the specialized: Auditory brainstem function is tuned into timbre. *Cortex, 48*, 360–362.

Strait, D. L., Parbery-Clark, A., Hittner, E., & Kraus, N. (2012b). Musical training during early childhood enhances the neural encoding of speech in noise. *Brain and Language, 123*, 191–201.

Strait, D. L., Parbery-Clark, A., O'Connell, S., & Kraus, N. (2013). Biological impact of preschool music classes on processing speech in noise. *Developmental Cognitive Neuroscience, 6*, 51–60.

Strait, D. L., O'Connell, S., Parbery-Clark, A., & Kraus, N. (2014). Musicians' enhanced neural differentiation of speech sounds arises early in life: Developmental evidence from ages three to thirty. *Cerebral Cortex, 24*(9), 2512–2521.

Tallal, P. (2004). Improving language and literacy is a matter of time. *Nature Reviews Neuroscience, 5*(9), 721–728.

Tierney, A. T., & Kraus, N. (2014). Music training for the development of reading skills. In M. M. Merzenich, M. Nahum, & T. van Vleet (Eds.), *Applying brain plasticity to advance and recover human ability* (pp. 209–241). London: Elsevier.

Tierney, A., Krizman, J., Skoe, E., Johnston, K., & Kraus, N. (2013). High school music classes enhance the neural processing of speech. *Frontiers in Psychology, 4*(855), 2016. Doi:10.3389/fpsyg.2013.00855

Tierney, A., Krizman, J., & Kraus, N. (2015). Music training alters the course of adolescent auditory development. *Proceedings of the National Academy of Sciences of the U S A, 112*(32), 10062–10067.

van Rheede, J. J., Richards, B. A., & Akerman, C. J. (2015). Sensory-evoked spiking behavior emerges via an experience-dependent plasticity mechanism. *Neuron, 87*(5), 1050–1062.

Wang, B.-S., Sarnaik, R., & Cang, J. (2010). Critical period plasticity matches binocular orientation preference in the visual cortex. *Neuron, 65*(2), 246–256.

Warrier, C. M., Abrams, D. A., Nicol, T. G., & Kraus, N. (2011). Inferior colliculus contributions to phase encoding of stop consonants in an animal model. *Hearing Research, 282*(1–2), 108–118.

Webster, D. B. (1992). An overview of mammalian auditory pathways with an emphasis on humans. In D. B. Webster, A. N. Popper, & R. R. Fay (Eds.), *The mammalian auditory pathway: Neuroanatomy* (pp. 1–22). New York: Springer.

Weinberger, N. M. (2004). Specific long-term memory traces in primary auditory cortex. *Nature Reviews Neuroscience, 5*(4), 279–290.

Weiss, M. W., & Bidelman, G. M. (2015). Listening to the brainstem: Musicianship enhances intelligibility of subcortical representations for speech. *The Journal of Neuroscience, 35*(4), 1687–1691.

White-Schwoch, T., & Kraus, N. (2013). Physiologic discrimination of stop consonants relates to phonological skills in pre-readers: A biomarker for subsequent reading ability? *Frontiers in Human Neuroscience, 7*(899), 2016. Doi:10.3389/fnhum.2013.00899

White-Schwoch, T., Nicol, T., Warrier, C. M., Abrams, D. A., & Kraus, N. (in press). Individaul differences in human auditory processing: Insights from single-trial auditory midbrain activity in an animal model. *Cerebral Cortex*. Doi:10.1093/cercor/bhw29

White-Schwoch, T., Woodruff Carr, K., Anderson, S., Strait, D. L., & Kraus, N. (2013). Older adults benefit from music training early in life: Biological evidence for long-term training-driven plasticity. *The Journal of Neuroscience, 33*(45), 17667–17674.
White-Schwoch, T., Woodruff Carr, K., Thompson, E. C., Anderson, S., et al. (2015). Auditory processing in noise: A preschool biomarker for literacy. *PLoS Biology, 13*(7), e1002196.
Wible, B., Nicol, T., & Kraus, N. (2004). Atypical brainstem representation of onset and formant structure of speech sounds in children with language-based learning problems. *Biological Psychology, 67*(3), 299–317.
Wible, B., Nicol, T., & Kraus, N. (2005). Correlation between brainstem and cortical auditory processes in normal and language-impaired children. *Brain, 128*(2), 417–423.
Wong, P. C. M., Skoe, E., Russo, N. M., Dees, T., & Kraus, N. (2007). Musical experience shapes human brainstem encoding of linguistic pitch patterns. *Nature Neuroscience, 10*, 420–422.
Woodruff Carr, K., White-Schwoch, T., Tierney, A., Strait, D. L., & Kraus, N. (2014). Beat synchronization predicts neural speech encoding and reading readiness in preschoolers. *Proceedings of the National Academy of Sciences of the U S A, 111*(40), 14559–14564.
Zendel, B. R., & Alain, C. (2012). Musicians experience less age-related decline in central auditory processing. *Psychology and Aging, 27*(2), 410–417.
Zhu, X., Wang, F., Hu, H., Sun, X., et al. (2014). Environmental acoustic enrichment promotes recovery from developmentally degraded auditory cortical processing. *The Journal of Neuroscience, 34*(16), 5406–5415.

Chapter 7
Individual Differences in Temporal Perception and Their Implications for Everyday Listening

Barbara Shinn-Cunningham, Leonard Varghese, Le Wang, and Hari Bharadwaj

Abstract Growing evidence shows that individual differences among listeners with normal hearing thresholds reflect underlying differences in how well the auditory system encodes temporal features of sound. In the laboratory, these differences manifest in a range of psychophysical tasks. In everyday life, however, the situations that reveal these differences are often social settings where listeners are trying to understand one talker in the presence of other competing sound sources (the "cocktail party" setting). Physiologically, the brainstem's envelope-following response (a specific form of the frequency-following response) correlates with individual differences in behavior. Motivated by both animal and human studies, this chapter reviews the evidence that behavioral and physiological differences across individual listeners with normal hearing thresholds reflect differences in the number of auditory nerve fibers responding to sound despite normal cochlear mechanical function (cochlear neuropathy). The chapter also points out some of the measurement issues that need to be considered when designing experiments trying to probe these kinds of individual differences in coding of clearly audible, supra-threshold auditory information.

Keywords Auditory brainstem · Binaural hearing · Binaural interaction component · Envelope-following response · FFR · Frequency-following response · Individual differences · Selective attention

B. Shinn-Cunningham (✉) · L. Varghese · L. Wang
Biomedical Engineering, Boston University, 677 Beacon Street, Boston, MA 02215, USA
e-mail: shinn@bu.edu

L. Varghese
e-mail: lennyv@bu.edu

L. Wang
e-mail: lwang@bu.edu

H. Bharadwaj
Martinos Center for Biomedical Imaging, Massachusetts General Hospital, 149 Thirteenth St., Boston, MA 02129, USA
e-mail: hari@nmr.mgh.harvard.edu

N. Kraus et al. (eds.), *The Frequency-Following Response*, Springer Handbook of Auditory Research 61,
DOI 10.1007/978-3-319-47944-6_7

7.1 Introduction

Historically, the majority of psychoacoustic studies have explored how variations in perceptual ability depend on acoustic stimulus parameters. Often in such studies, individual differences across listeners confound interpretations: they are a source of noise and interfere with the differences that are the focus of study. However, a growing number of studies have started to exploit repeatable individual differences that are present across listeners with normal audiometric thresholds.

The envelope-following response (EFR), a specific form of the frequency-following response (FFR), indexes important differences in temporal coding fidelity in listeners with normal hearing thresholds. The finding that individual differences in ability are related to differences in objective physiological measurements supports the idea that sensory coding fidelity differs amongst listeners with normal audiometric thresholds and that this affects hearing in everyday settings.

After briefly describing the EFR in Sects. 7.2, 7.3 reviews evidence that the EFR reflects, in part, differences in the number of auditory nerve fibers (ANFs) responding to sound, which can be reduced (e.g., by noise exposure and by ordinary aging) without affecting detection thresholds (Kujawa and Liberman 2009; Lobarinas et al. 2013). Section 7.4 then considers why these differences manifest in everyday settings when listeners need to direct selective auditory attention. Section 7.5 discusses various factors that should be considered when designing experiments to investigate individual differences in the EFR, and Sect. 7.6 comments on some aspects of brainstem sound processing that *do not* cause consistent signatures in the EFR.

7.2 The Envelope-Following Response

7.2.1 Defining the Envelope-Following Response

The voltage measured on the scalp in response to sound reflects a mixture of brain activity, including both cortical responses (which are relatively large at the scalp) and responses from deeper, subcortical portions of the auditory pathway (roughly two orders of magnitude smaller than cortically generated potentials). The observable voltages on the scalp are the sum of electrical activity from an enormous number of individual neurons and their interconnections, each of which can produce electrical potentials. If they are in opposing directions, the sum of these electrical potentials will cancel. As a result, the only signals that survive to be observable on the scalp are those that cause synchronous potentials with similar polarity at the measuring electrodes (for a review of the physics of neuroelectric and neuromagnetic measurements, see Hamalainen et al. 1993).

Responses phase locked to periodic sounds originating from either the subcortical or cortical portions of the auditory pathway often are collectively referred to as auditory steady-state responses or ASSRs (Galambos et al. 1981; Stapells et al. 1984). The EFR is a specific form of ASSR measured by presenting a periodic input signal (typically with a periodicity in the 80–450 Hz range so that the subcortical portion of the response is emphasized; see Sect. 2.3) in opposite polarities and then averaging the responses (Goblick and Pfeiffer 1969; Aiken and Picton 2008).

To understand what the EFR encodes, one can first consider how a sound pressure wave is represented when it is transduced into neural energy. This transduction takes place in the cochlea, which turns the uni-dimensional sound pressure into a multi-dimensional representation. Specifically, the cochlea breaks the input down into a parallel representation in which each "channel" of the representation responds to a different band of input acoustic frequencies. As a result of this cochlear frequency selectivity, a narrow frequency band of sound drives each ANF. Because the "driving signal" for each ANF is narrowband, the temporal information encoded by each ANF frequency channel can be logically separated into two parts: (1) the temporal fine-structure (TFS) corresponding to nearly sinusoidal *carrier* fluctuations near the center frequency of the narrowband driving signal, and (2) slow fluctuations in the energy of the TFS known as the *envelope* of that carrier, whose frequency content is limited by the bandwidth of the corresponding cochlear filter (Boashash 1992). The (non-negative) envelope of a stimulus is also referred to as the *modulation* in the signal.

The bandwidth of the cochlear filters increases with increasing center frequency; therefore, the temporal fluctuations in the envelope of the driving functions can be more rapid with increasing center frequency (Moore 2003). This is illustrated in Fig. 7.1, which shows the signal exciting three different places along the cochlea for a click train repeating at 100 Hz. Each of the illustrated cochlear channels (100, 1000, and 4000 Hz) responds to the frequency components of the click train falling within its critical band. For the 100 Hz signal, only a single component (at 100 Hz) falls within the cochlear critical band (Fig. 7.1A). As a result, the response in this channel consists of TFS fluctuating at 100 Hz and an envelope that rises to a constant value, beginning from the start of the click train. In contrast, the higher-frequency channels have many harmonics (frequency components that are multiples of the fundamental frequency of 100 Hz) that sum together to produce the driving function at that cochlear place (see Fig. 7.1B, C). Summing harmonics with a common fundamental frequency produces a signal that is periodic with the period of the fundamental frequency. Therefore, the 1000-Hz and 4000-Hz cochlear places have periodic driving functions with a period of 1/100 Hz (10 ms), which is seen in their envelopes. In addition, because the bandwidth of the response from the 4000-Hz cochlear channel is broader than that of the 1000-Hz channel (and thus encompasses more harmonics), the envelope from the 4000-Hz channel has more rapid amplitude fluctuations (the waveform is "sharper" in time; compare envelopes in Fig. 7.1B, C). Finally, the TFS fluctuates at a rate determined by the center

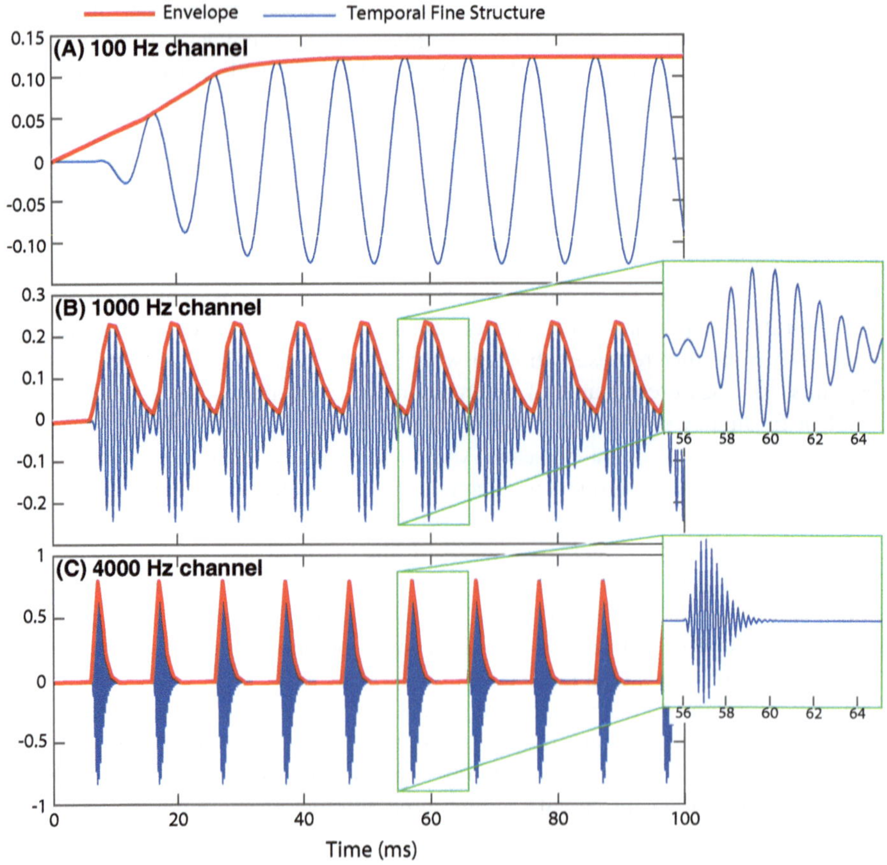

Fig. 7.1 Illustration of the "driving function" at three different places along the cochlea for a click train repeating at 100 Hz. In a low-frequency channel centered at 100 Hz (**A**), the driving signal looks like a sinusoid in the steady state with a flat envelope. In the 1000-Hz (**B**) and 4000-Hz (**C**) channels, the envelope is periodic with a repetition rate equal to the 100 Hz periodicity of the input. However, the 1000-Hz envelope varies more slowly in time than the 1000-Hz channel envelope (compare *red waveforms* in **B** and **C**). The temporal fine structure (the rapidly fluctuation oscillations in *blue*) varies with a periodicity equal to the center frequency of the channel and is four times slower for the 1000-Hz channel than for the 4000-Hz channel (compare *blue waveforms* in the insets of **B** and **C**)

frequency of each critical band (with periods of 1/1000 or 1 ms, and 1/4000 or 0.25 ms for the 1000-Hz and the 4000-Hz channels, respectively; see the insets in Fig. 7.1B, C).

While Fig. 7.1 shows the driving functions at different cochlear places, the actual response coming out of the cochlea cannot track very rapid positive and negative TFS fluctuations. Instead, ANFs at each place along the cochlea respond to some combination of the TFS and the envelope in their driving functions in a proportion that depends on the rate of TFS fluctuations. The ANF firing pattern does not track TFS well above about 2000 Hz; thus, the envelope of the driving function tends to dominate responses of the high-frequency cochlear channels.

Because the EFR is measured by averaging responses to positive and negative polarity acoustic presentations, the EFR cancels out all portions of the measured electrical response that take on opposite values in response to the two types of presentations (see Fig. 3.3 of Krishan and Gandor, Chap. 3). This includes much of the response elicited by a signal's TFS, as well as electromagnetic contamination (e.g., from unshielded or improperly shielded audio transducers) and the cochlear microphonic (see Skoe and Kraus 2010). Conversely, the EFR enhances responses that are the same for positive and negative polarity inputs. Given this, the EFR tends to be driven by periodicities in the envelope of the input acoustic waveform, which explains the name *envelope-following response* (see Sect. 7.4.1 for further discussion). The EFR differs from other forms of FFRs in that it tends to have a higher signal-to-noise ratio (SNR) (Skoe and Kraus 2010) and, therefore, is often a more robust signature of neural activity than other types of FFRs.

One frustration in conducting research on FFRs is the lack of consistent nomenclature. The term "FFR" is now often used to denote both EFRs as well as other forms of ASSRs. Yet historically, ASSRs tracking neural envelope periodicities were known as either "amplitude modulation following responses" or EFRs (Dolphin and Mountain 1992; Kuwada et al. 2002) in order to distinguish them from ASSRs phase locked to the TFS of pure tones, which were called FFRs (Marsh et al. 1975). In the interest of specificity and based on historical precedent, this chapter uses the term EFR to refer to responses derived by summing equal numbers of positive and negative polarity presentations of a periodic auditory input.

7.2.2 Challenges in Localizing the Source of the Envelope-Following Response

EFRs provide a convenient, noninvasive method for measuring some aspects of subcortical neural activity. However, inferring what brain structures produce EFRs is challenging. Any form of electroencephalography (EEG) recorded at the scalp, including EFRs, measures a response that is the sum of all neural activity recorded through multiple layers of tissue and bone. On top of this, electromagnetic noise is ubiquitous, coming from line noise in the power system, electric fields generated by experimental equipment, muscle artifacts (e.g., from eye blinks), and even incidental neural activity that is not associated with the presented stimulus. Even if noise were insignificant, determining what brain region produces a particular voltage on the scalp is an ill-posed problem that is mathematically impossible to solve without independently constraining the solution (Hamalainen et al. 1993).

These challenges are particularly problematic when trying to determine what subcortical activity is reflected in the scalp voltage. The subcortical sites generating neural activity are deep in the brain and far from the recording sites on the scalp. As a result, the SNR at the scalp is low. Moreover, all of these deep, subcortical sources are at roughly the same distance to all electrodes, which means that they contribute nearly the same signal to every electrode on the scalp—even electrodes that are far away from each other. This detail, combined with the fact that activity at each subcortical stage is temporally correlated with activity at the next stage, makes it almost impossible to separate activity from different sources by combining information across multiple electrical sensors (an approach that is used to infer source locations of cortical neural activity from an EEG).

7.2.3 Frequency Content of the Brainstem Envelope-Following Response

Luckily, cortical activity that is observable in scalp voltages tends to be made up of relatively low frequency components, dropping off above about 80 Hz, while synchronous subcortical activity can be observed up to much higher frequencies. One reason for this is that the way temporal information is encoded changes as information ascends the auditory pathway.

In the peripheral parts of the pathway, the temporal pattern of the neural firing encodes both TFS and amplitude modulation. As information passes to higher processing centers, temporal cues are transformed from being represented in temporal structure to being encoded by a neural "place" code (coded by which neurons are firing, rather than the temporal pattern of their firing). For instance, coding of envelope modulation undergoes a transformation from phase-locked changes in firing rate (which would help to drive scalp signals phase locked to the modulation) to a more place-based code as one traverses from the ANF to the midbrain (see Joris et al. 2004). Consistent with this, cortical neurons tend to have longer lasting, slower, and more integrative responses than do subcortical neurons (e.g., Escabi and Read 2003).

Because of this frequency dependence, lower-frequency responses in the EEG signal (below 60 Hz or so) are dominated by cortical responses, while higher-frequency responses (above about 80 Hz) primarily reflect responses from subcortical regions of the brain. Both reversible inactivation studies (Kuwada et al. 2002) and irreversible lesion studies (Sohmer et al. 1977; Kiren et al. 1994) offer additional, direct evidence that a subcortical source is the dominant generator of EFRs above 80 Hz. These studies suggest that across different mammalian species the currents in inferior colliculus (IC) neurons produce the dominant response in ASSRs for frequencies above 80 Hz; lesions at higher stages of the auditory pathway, including primary auditory cortex, do not strongly influence ASSRs in this frequency range.

Practically speaking, in humans, EFR measurements are weak above 500 Hz due to SNR limitations (Cohen et al. 1991), although one study reports measuring significant EFRs up to 1000 Hz (see Purcell et al. 2004). In understanding this limit, it is helpful to recall that the ANF critical bandwidth increases with characteristic frequency (i.e., the frequency of input that causes the greatest response in a particular ANF); moreover, the critical bandwidth also increases with sound level (see Moore 2003). Since the critical bandwidth determines the fastest modulations contained in the signal driving a particular ANF, the modulation rates that might be conveyed increase with characteristic frequency.

In humans, the broadest cochlear filters can support about a 5–600 Hz modulation rate at moderate sound levels. In contrast to humans, in cats, ANFs with high characteristic frequencies (i.e., >10 kHz) phase lock to envelope fluctuations only up to about 1000 Hz; that is, the modulation bandwidth is not limited by the cochlea (Joris and Yin 1992). For the lower-frequency ANFs in cats, the envelope cutoff frequency decreases with the characteristic frequency of the fiber, suggesting that the limitation is imposed by the critical bandwidth of the cochlear filter (Joris and Yin 1992).

Figure 7.2A illustrates the low-pass nature of the EFR in human subjects measured in response to amplitude-modulated noise for modulation frequencies between 100 and 500 Hz. Because the noise in the measurements tends to be proportional to the reciprocal of the frequency, the SNR in the EFR decreases more slowly than does the absolute power. This can be seen in Fig. 7.2B, which plots the phase-locking value (PLV), a measure of response synchrony (see Sect. 7.4.2) in the EFR as a function of frequency. The PLV is a metric that directly reflects the SNR (e.g., Zhu et al. 2013). As shown in the figure, both the absolute strength of the EFR (Fig. 7.2A) and the PLV (Fig. 7.2B) decrease rapidly above 450 Hz. Thus, for all practical purposes, the EFR can be measured effectively in humans for input signals with periodicities falling in the range of roughly 80–500 Hz.

7.2.4 *Using Response Phase to Infer the Source of the Envelope-Following Response*

Analysis of the phase of the EFR as a function of input modulation frequency supports invasive animal studies in suggesting a dominant midbrain source of the EFR. The slope (derivative) of the phase versus stimulus-modulation-frequency function gives the response group delay. In frequency ranges where the group delay is constant (phase is a linear function of frequency), the group delay reflects the neural delay from the input to the dominant response source. If the measured EFR is not dominated by a single source at a given frequency but rather reflects a mixture that contains strong responses from multiple stages of the auditory pathway (each with a different delay), the group delay generally will vary with modulation frequency. By considering how the phase versus frequency slope changes with

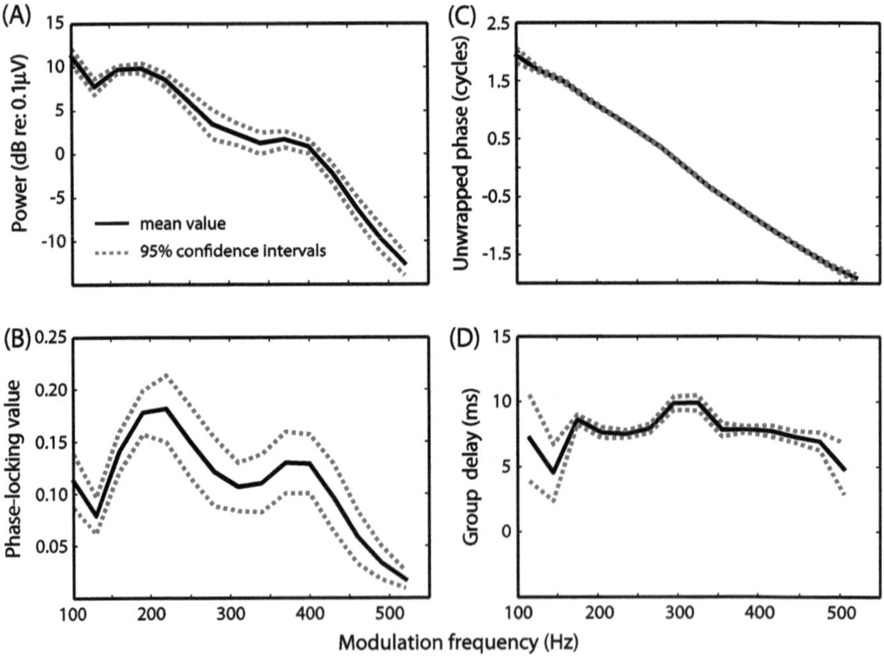

Fig. 7.2 Sample human envelope-following response (EFR) data in response to amplitude modulated noise as a function of modulation frequency. *Solid lines* show the mean response across subjects, while the *dashed lines* surrounding the mean show the 95% confidence intervals. (**A**) Power at the EFR modulation frequency. (**B**) The phase-locking value (measuring the consistency in the phase of the response at the modulation frequency across independent trials in response to the modulated noise). (**C**) The unwrapped EFR phase. (**D**) The estimated group delay (taking the difference of the EFR phase at adjacent frequencies from Fig. 7.1C)

frequency, one can begin to infer the frequency ranges over which the voltage mixture on the scalp reflects one dominant source as well as the latency of the response of this source (Kuwada et al. 2002; Shaheen et al. 2015).

Above 200 Hz, group delay varies substantially across species, probably due to anatomical differences, leading to different mixtures of responses from different parts of the pathway in the scalp-recorded voltage (e.g., Okada et al. 1997). While rabbits and mice have frequency regions of constant phase slopes out to 500 and 700 Hz, respectively (Kuwada et al. 2002; Pauli-Magnus et al. 2007), slopes are constant only up to 200 Hz in gerbils (Dolphin and Mountain 1992). In humans, EFRs exhibit a relatively constant group delay above about 80 Hz. The estimated (unwrapped) phase of the response as a function of modulation frequency is shown in Fig. 7.2C for modulation rates from 100 up to 500 Hz. Across this range, the phase is nearly linear. Figure 7.2D shows the group delay as a function of modulation frequency; the group delay is nearly constant, hovering around a value of about 8 ms, consistent with a midbrain source (likely the IC).

While the IC is likely a dominant source of the EFR, this does not mean that the recorded responses directly reflect the output firing patterns of IC neurons. Instead, the post-synaptic currents flowing in IC neurons, driven by lemniscal inputs, are likely the primary source of the EFR response. In particular, the action potentials (spikes) generated in IC are unlikely to induce observable potentials on the scalp. Spikes are brief and induce both depolarizing and hyperpolarizing currents along the neuronal axis. Consequently, they would have to be significantly better aligned in time across the IC population than the lemniscal inputs to produce an observable net signal. Spikes produce a quadrupolar current pattern, which does not produce large observable voltages beyond a very short distance; in contrast, post-synaptic currents produce a dipole current pattern, yielding voltages that can be observed at greater distances, such as between two scalp electrodes (Hamalainen et al. 1993; Milstein and Koch 2008).

7.3 Individual Differences in Listeners with Normal Cochlear Function

A number of EFR studies have found that the strength of the EFR differs across groups, such as musicians versus nonmusicians (Wong et al. 2007; Strait et al. 2011) or listeners with and without learning disabilities (Wible et al. 2005; Hornickel et al. 2011). These results demonstrate that the strength of the brainstem response is an important index of perceptual abilities. The finding that experience helps to shape subcortical neural responses (e.g., Skoe et al. 2014) helps to explain some of these group differences, such as why trained musicians tend to have stronger brainstem responses than do nonmusicians (e.g., Parbery-Clark et al. 2011). Experience is thus one potential source of individual variation in EFR strength (e.g., Anderson et al. 2013). Such results suggest that training may help ameliorate perceptual deficits indexed by the EFR (e.g., Whitton et al. 2014; Slater et al. 2015).

Other recent studies reveal robust individual differences in the subcortical responses across ordinary individuals selected randomly within a fairly homogenous subject group. Specifically, there are now a number of studies that measured EFRs as well as various perceptual abilities in ordinary listeners, all of whom had no known hearing or learning deficits, normal hearing thresholds, and normal cochlear function. These studies identified differences in EFR strength that were correlated with a range of basic perceptual abilities and the ability to selectively attend to speech in a noisy mixture (e.g., Ruggles et al. 2012; Bharadwaj et al. 2015). These studies suggest that EFRs may be useful for diagnosing subtle hearing differences that come from differences in the fidelity of auditory coding in the brainstem and that have real-world behavioral consequences (Bharadwaj et al. 2014; Shaheen et al. 2015).

This section presents evidence that at least a portion of the individual variation in EFR strength comes from differences in the number of ANFs that respond to sound. This view is motivated by the growing interest in *hidden hearing loss* (more formally known as *cochlear synaptopathy* or *cochlear neuropathy*) in both animals (Kujawa and Liberman 2009; Valero et al. 2016) and humans (Schaette and McAlpine 2011; Plack et al. 2014). Cochlear neuropathy is distinct from *auditory neuropathy*, which is characterized by a profound disruption of auditory brainstem responses (ABRs) with reduced amplitude or even absent ABR wave V responses (see Starr et al. 1996).

7.3.1 What Is Hidden Hearing Loss?

It has long been known that moderate noise exposure can lead to temporary threshold shifts (TTS) (see Quaranta et al. 1998). Immediately after noise exposure, sound detection thresholds can be elevated by as much as 40 dB but then recover back to normal over the course of days. Because clinical "hearing loss" is defined as having elevated hearing thresholds, by definition, listeners with TTS do not have hearing loss. Indeed, until relatively recently, because hearing thresholds recover and there is no loss of hair cells due to TTS (Bohne and Harding 2000), noise exposure of this type was assumed to incur no permanent hearing damage.

A growing number of animal studies have upended this assumption (Kujawa and Liberman 2015; Liberman 2015). Noise exposure that causes TTS has been shown to produce a rapid loss of as many as 40–60% of the ANF synapses driven by cochlear inner hair cells, which are the cells that generate the ascending signal conveying information in the auditory pathway (Kujawa and Liberman 2006, 2009). This loss of synapses subsequently leads to a slow death of ANF cell bodies (spiral ganglion cells) and central axons (Lin et al. 2011; Kujawa and Liberman 2015). Even in cases where the effects on synapses and spiral ganglion cells are pronounced, the effect on cochlear function can be negligible; cochlear mechanical function (including the tuning of the cochlea) can be normal in animals suffering from cochlear neuropathy (Kujawa and Liberman 2009). Most hearing screenings reveal losses associated with damage to inner and outer hair cells by looking for: (1) elevated detection thresholds, (2) reduced amplification in the cochlea, (3) wider-than-normal cochlear tuning, and (4) reduced otoacoustic emissions. Yet, with hidden hearing loss, these measures are normal, making the deficit "hidden" to typical hearing screening.

How can it be that hearing thresholds are normal even though the number of ANFs is significantly reduced? One reason is that synaptopathy causes a diffuse loss, leaving behind ANFs throughout the cochlea (Liberman et al. 1997; Lobarinas et al. 2013). In addition, some evidence suggests that synapse loss preferentially affects ANFs that have high thresholds and low spontaneous firing rates (SR) (Furman et al. 2013; Kujawa and Liberman 2015). As shown in Fig. 7.3, each healthy inner hair cell in the cochlea typically drives multiple ANFs that differ in

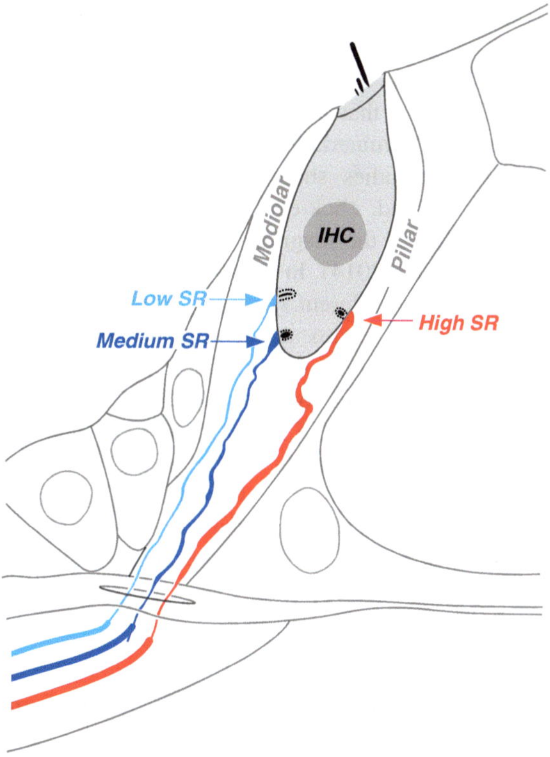

Fig. 7.3 Illustration of how terminals of the cochlear nerve innervate a single inner hair cell. Each inner hair cell typically has synaptic contacts with multiple auditory nerve fibers with *high*, *medium*, and *low spontaneous rates* (*SR*). (Figure from Bharadwaj et al. 2014)

their spontaneous firing rates and thresholds (i.e., the sound level at which the response of the ANF differs from its spontaneous firing pattern) (see Moore 2003). While low-threshold, high-SR ANFs begin to increase their firing rates when sound just exceeds perceptual detection thresholds, mid-threshold and high-threshold ANFs (or "higher-threshold ANFs" for brevity) only contribute to neural activity at supra-threshold sound levels. If noise exposure preferentially damages higher-threshold ANFs, it makes sense that a large number of ANFs may cease to respond without influencing detection thresholds. However, the effect of this loss on the encoding of acoustic temporal details in supra-threshold sound can be substantial (Plack et al. 2014).

Why might higher-threshold fibers be more susceptible to noise-exposure damage than other fibers? Pharmacological studies suggest that cochlear neuropathy is the result of a type of glutamate excitotoxicity, a process in which neurons are damaged and die off through over-activity in response to the neurotransmitter glutamate (e.g., Pujol et al. 1993; Mehta et al. 2013). In the central nervous system, glutamate excitotoxicity is mediated by an increase in intracellular calcium concentration (Szydlowska and Tymianski 2010). Mitochondria within cell bodies comprise an important intracellular calcium buffering system. In inner hair cells within the cochlea, fewer mitochondria are associated with higher-threshold fibers

(Liberman 1980). Given that higher-threshold fibers typically respond with lower firing rates than do low-threshold ANFs, the smaller number of mitochondria may be sufficient to ward off excitotoxicity in ordinary settings; however, in the face of ongoing noise that drives higher-threshold fibers at a continuous, high rate, these cells may be vulnerable to glutamate excitotoxicity (Bourien et al. 2014).

Animal studies show that cochlear neuropathy decreases the magnitude of supra-threshold, click-evoked ABR wave I responses (coming from the auditory nerve) but not the magnitude of wave V (coming from the midbrain; see Hickox and Liberman 2014). In animals with extreme neuropathy (with a loss of >95% of cochlear nerve afferent synapses), plasticity leads to an enhanced neural gain in the brainstem and cortex that compensates for the weak ANF response, producing detection thresholds that are near normal (Chambers et al. 2016). While these changes ameliorate some of the effects of a weak ANF drive, they cannot compensate fully for ANF loss; temporal coding in the denervated animals is poorer than in control animals. These findings further help explain why more subtle cochlear neuropathy may have a big impact on the representation of temporal features of supra-threshold sounds without affecting detection thresholds and why ABR wave V may have a normal magnitude even when ABR wave I amplitude is reduced.

7.3.2 Hidden Hearing Loss in Humans

While there are no data yet to directly support the idea that cochlear neuropathy occurs in humans, a growing number of studies hint that it accounts for some of the individual variability seen in listeners with normal cochlear mechanical function. As noted above, listeners with normal hearing thresholds vary significantly in their ability to utilize precise temporal information (Ruggles et al. 2011; Bharadwaj et al. 2015). This variability correlates with difficulties in using spatial-selective attention to focus on and understand speech in a noisy background (Ruggles and Shinn-Cunningham 2011), underscoring the clinical relevance of these differences.

In one such study, young adult subjects were recruited with no special criteria except that they had normal hearing thresholds and no known auditory deficits (Bharadwaj et al. 2015). Individual differences amongst this cohort were nonetheless large. Perceptual abilities correlated with EFR strength, especially at high sound levels and shallow modulation depths when higher-threshold ANFs are important for coding temporal features. This is illustrated in Fig. 7.4, which demonstrates the consistent relationships between the EFR strength (plotted along the x axes) and perceptual thresholds for amplitude modulation detection (Fig. 7.4A) and for envelope interaural time difference (ITD) discrimination (Fig. 7.4B). Both of these perceptual measures rely on fine temporal information, and both are significantly correlated with the strength of the EFR when a shallow modulation drives the brainstem response.

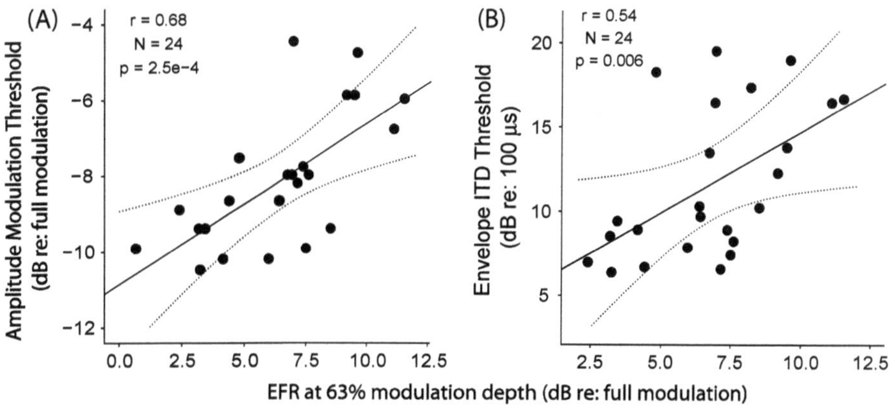

Fig. 7.4 Relationship between envelope-following response (EFR) strength and perceptual thresholds in young adult listeners with normal cochlear function. (**A**) Amplitude modulation detection thresholds are correlated with the EFR strength. (**B**) Discrimination thresholds for envelope ITD are correlated with EFR strength. *r*, correlation coefficient (Data are from the study described in Bharadwaj et al. 2014)

Crucially, listeners had normal compressive growth of cochlear responses (measured by distortion product otoacoustic emissions), normal frequency tuning (measured by psychoacoustic estimation), and pure-tone audiometric thresholds of 15 dB hearing level (HL) or better at octave frequencies between 250 Hz and 8 kHz. In other words, although perceptual differences were correlated with the EFR (an objective measure of the precision of brainstem temporal coding), these differences could not be explained by cochlear mechanical function. These findings suggest that cochlear neuropathy may be quite common, affecting a large percentage of the population, including relatively young listeners.

Other studies in humans also support the view that human listeners with normal cochlear function may suffer from different degrees of cochlear neuropathy. For instance, listeners can vary significantly in their ability to discriminate both frequency modulation and ITDs (see Strelcyk and Dau 2009; Grose and Mamo 2010). The computation of ITDs depends directly on temporal precision in ANF responses and subsequent processing centers (such as neurons in the superior olivary complex). Indeed, sensitivity to ITD cues was one of the perceptual abilities that correlated with EFR strength (Bharadwaj et al. 2015).

On the physiological side, listeners with normal hearing thresholds show large inter-subject variability in the magnitude of ABR wave I (Schaette and McAlpine 2011; Stamper and Johnson 2015) again supporting the view that listeners with normal audiograms suffer from neuropathy to varying degrees. As in animal studies, while ABR wave I amplitude varies significantly across individuals, the magnitude of ABR wave V does not (Schaette and McAlpine 2011; Stamper and Johnson 2015).

One study has shown that perceptual differences correlate with differences in human ABRs: in young adults with no known hearing deficits, wave I magnitude was related to ITD sensitivity (Mehraei et al. 2016). Consistent with previous animal studies, wave V magnitude was unrelated to wave I magnitude or perceptual ability (although effects of noise on wave V timing were correlated with wave I amplitude). Taken together, these results suggest that cochlear neuropathy is common amongst human listeners who have normal audiograms, many of whom do not even realize that they may have communication difficulties.

7.3.3 Effects of Aging on Hidden Hearing Loss

In animal models, natural aging produces cochlear neuropathy (see Anderson, Chap. 11, for a discussion of the effects of aging and hearing loss). Aging mice raised without exposure to any loud sound (and without significant hair cell loss) exhibit a loss of 30–40% of inner hair cell synapses by roughly 3/4 of their lifespan, an age at which thresholds are elevated by less than 10 dB (Sergeyenko et al. 2013). This kind of neurodegeneration may selectively affect higher-threshold fibers (Schmiedt et al. 1996). Counts of spiral ganglion cells in an age-graded series of human temporal bones show degeneration of 30%, on average, from birth to death, even in cases with no hair cell loss (Makary et al. 2011). These anatomical results support the idea that aging alone can produce hidden hearing loss.

Older human listeners with normal hearing thresholds exhibit basic temporal processing deficits (see Fitzgibbons and Gordon-Salant 2010 for a review). Aging degrades temporal modulation sensitivity (Purcell et al. 2004; He et al. 2008) and leads to weaker brainstem responses (Anderson et al. 2012). Temporal deficits correlate with the strength of the EFR in older listeners with normal thresholds (Purcell et al. 2004). The highest modulation frequency to which EFRs exhibit phase locking decreases with age (Purcell et al. 2004; Grose et al. 2009), and temporal processing of both monaural and binaural sound features degrades with age (e.g., Grose and Mamo 2012; Grose et al. 2015). Indeed, even after factoring out effects of elevated hearing thresholds, aging causes degradations in temporal processing that appear well before there is evidence of speech processing deficits (Snell and Frisina 2000; Snell et al. 2002). Aging also interferes with the ability to understand speech in the presence of competing sound (Fullgrabe et al. 2014; Helfer 2015). All of these symptoms implicate deficits in temporal coding in aging listeners. Indeed, a number of researchers have concluded that in older listeners with normal thresholds, difficulties with understanding speech in noise arise because of temporal processing deficits (Helfer and Vargo 2009; Jin et al. 2014). These studies, like anatomical studies, support the view that aging leads to cochlear neuropathy in human listeners.

7.3.4 *Hidden Hearing Loss and Individual Differences in the Envelope-Following Response*

When considering the idea that cochlear neuropathy produces differences in the EFR, one question is how to reconcile the view that individual differences in the EFR come from an irreversible loss of ANFs with the many studies showing that appropriate experience and training can increase the strength of the brainstem response (e.g., Carcagno and Plack 2011; Strait and Kraus 2014). One intriguing possibility is that long-term training and experience can partially, but only partially, compensate for cochlear neuropathy. For instance, training could increase the efficiency with which the information in remaining ANFs is extracted by higher centers of the auditory pathway. In line with this, as noted previously, neuropathy decreases the magnitude of ABR wave I but not in the magnitude of wave V (see Stamper and Johnson 2015; Mehraei et al. 2016). In other words, the gain of the auditory pathway between the ANFs and the IC seems to compensate for a weaker than normal ANF response when there is cochlear neuropathy. Such compensation likely helps to ensure that detection thresholds are normal and helps to keep the overall average firing rate of brainstem neurons at the proper level. However, an increase in gain cannot fully restore coding of temporal cues, which rely on the convergence of the noisy, stochastic responses of many ANFs (e.g., Oertel et al. 2000).

Consistent with this hypothesis, following profound cochlear denervation, central compensatory processes restore responses in both cortex and midbrain; however, this compensation cannot overcome deficits in "features encoded by precise spike timing" (Chambers et al. 2016). This kind of thinking helps to resolve the counterintuitive idea that even though experience influences the strength of the EFR, some of the differences in the strength of the response reflect irreversible differences in the number of ANFs encoding sound. Furthermore, the deficits that cannot be overcome by compensatory gain changes in the midbrain and above are those in temporal processing, which explains the pattern of deficits seen in human listeners who are suspected to have cochlear neuropathy.

7.4 Why Hidden Hearing Loss Affects Daily Function

Roughly 5–10% of listeners seeking treatment at audiological clinics have normal hearing thresholds (Kumar et al. 2007; Hind et al. 2011). Typically, these patients are driven to seek help because of difficulty communicating in situations requiring them to focus selective attention. Historically, such listeners were said to have "central auditory processing disorder" (Rosen et al. 2010), a catchall diagnosis that testifies to the fact that underlying causes were not well understood; however, some of these listeners likely are suffering from cochlear neuropathy.

The fact that listeners first notice the effects of cochlear neuropathy when trying to communicate in social settings makes sense, given how neuropathy degrades auditory temporal coding. Spectrotemporal details in a sound mixture are important for grouping of acoustic elements into perceptual objects (Shamma et al. 2011; Christiansen and Oxenham 2014), discrimination of perceptual features like pitch (Smith et al. 2002) and source location (Blauert 1997; Smith et al. 2002), as well as speech perception itself (Zeng et al. 2005). Importantly, subtle hearing deficits may not disrupt speech perception in quiet, yet they still have a debilitating effect on selective auditory attention.

7.4.1 Source Segregation

In order to selectively attend, listeners must be able to segregate sounds making up the acoustic mixture entering the ears. Source segregation depends on harmonic structure, interaural time differences, and other cues computed from acoustic features that are degraded when temporal coding is poor (Bregman 1990; Carlyon 2004). If temporal features are degraded and the target source cannot properly be segregated from the scene, selective attention will fail (Shinn-Cunningham 2008; Shinn-Cunningham and Best 2008). This idea is illustrated by visual analogy in the cartoons shown in Fig. 7.5. In people with good coding fidelity, fine details in the scene ensure that each source is distinct. In the visual analogy, features of each word in the scene are clear: words differ in their color, so are easy to perceive as distinct and separate objects (Fig. 7.5A). In contrast, even if the representation of the scene is weak, an observer may have no difficulty detecting that there are elements present in the scene: they may have normal detection thresholds (Fig. 7.5B). Yet observers may have problems understanding the supra-threshold information in the scene. Elements making up the scene are fuzzy, letter edges are blurry, and colors of different words in the scene are similar, so that the words seem to run together perceptually.

In the auditory domain, when listening to a complex scene, spectrotemporal details (e.g., periodicity, ITD, and amplitude and frequency modulation) are analogous to the letter edges and colors of a visual scene. These features are less clearly represented when a listener suffers from hidden hearing loss, so that the structural elements critical for parsing the acoustic scene are perceptually indistinct.

7.4.2 Source Selection

Successfully listening in a complex setting depends on more than simply segregating the sources from one another; it also requires selecting the desired source from the mixture by focusing selective attention. Selective auditory attention enhances the representation of the auditory object with a desired perceptual feature or attribute (Lee et al. 2012; Maddox and Shinn-Cunningham 2012). The low-level acoustic spectrotemporal structure is what enables a listener to compute perceptual

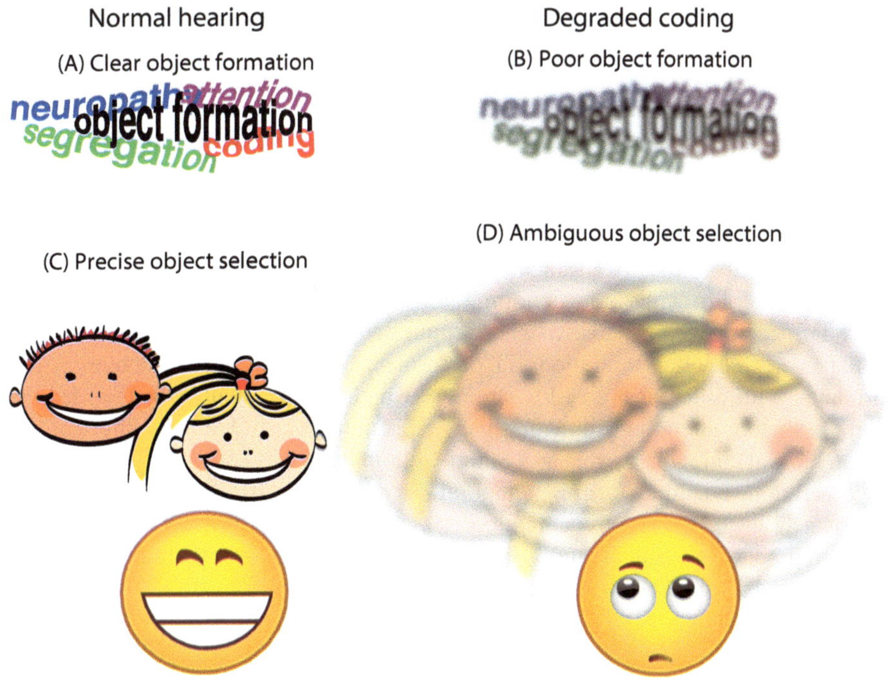

Fig. 7.5 Visual analogy illustrating the effects of poor brainstem coding fidelity on segregating and selecting a target object from a complex scene for a "good" listener (*left*) and a "bad" listener (*right*). (**A**) For a listener with a good sensory representation, each edge of each letter in the scene is represented clearly, and the similarity of the color of the letters making up each word (as well as the dissimilarity of the colors across words) allows each word to be perceived automatically as a distinct unit. In this representation, words are automatically segregated, based on the clear features of the letters and words. (**B**) With a poor representation, individual letters blur together, making the structure of each letter difficult to perceive. The poor quality of the representation also degrades the features that distinguish words from each other, further blurring together the elements of the scene and making it more challenging to separate the words in the scene. (**C**) When the peripheral representation is clear, each object is both distinct and has a clear perceived location, making it easy to deploy spatial selective attention and focus on a target from a particular direction. (**D**) When the peripheral representation is weak, spatial cues are blurry and ambiguous and can be inconsistent over time, making it difficult to focus attention and select out the target object. (Original figure by Shinn-Cunningham)

features of objects in a scene that can be used to focus attention. Specifically, low-level features such as periodicity, ITD, and amplitude and frequency modulation support computation of higher-level perceptual quantities such as pitch, location, and timbre. These attributes can be used to listen to "the high-pitched source," or "the source on the left," or "Sally, not Jim."

One clear example of a high-level feature that is degraded when temporal cues are weak (e.g., due to hidden hearing loss) is spatial location. When temporal cues are weak, the perceived location of a sound source can be broad and diffuse. Listeners with a weak temporal code can fail to select the correct source in the scene based on its less perceptually precise location. For instance, one study found large individual differences in performance on a spatial selective-attention task (Ruggles and Shinn-Cunningham 2011). In this study, when listeners failed, they did not fail to understand speech present in the sound mixture. Instead, they reported the wrong word, coming from the wrong location; that is, perceptual deficits were not severe enough to interfere with understanding the speech that was present in the mixture. The failures happened because listeners could not select the correct talker based on spatial cues. Individual variations in performance on the selective attention task correlated with differences in EFR strength, which is consistent with the idea that spatial-selective attention fails when listeners suffer from hidden hearing loss and poor temporal coding (Ruggles and Shinn-Cunningham 2011; Ruggles et al. 2012). Reverberation, a natural form of temporal degradation in the signals reaching the ears, exacerbated the selective attention errors. In other words, both external noise in the temporal acoustic features important for conveying location (from reverberation) and internal noise in the computation of ITDs (from differences in temporal coding fidelity in the brainstem) had similar, additive effects in disrupting selective auditory attention.

Figure 7.5C–D shows a visual analogy to this kind of problem. In this scene, a listener with a good peripheral representation can focus attention unambiguously to a talker to the left (Fig. 7.5C). If the spatial cues are weakly represented, however, the perceived talker locations overlap and smear into each other (Fig. 7.5D). Even if a listener can parse the scene into a male and a female talker, they may focus on the wrong talker when trying to focus on "the talker on the left" because of the spatial ambiguity in the scene. Such problems can produce communication difficulties in settings where there are multiple sources competing for attention that would not show up on a test of speech perception in quiet or even if there were nonspeech sounds present (i.e., in conditions where competing sound objects are so perceptually dissimilar that failures of selection will not occur).

7.4.3 Understanding Speech in Noisy Settings Reveals Subtle Deficits

The previous examples demonstrate why even modest degradations in temporal processing may lead to communication dysfunction in everyday settings (Shinn-Cunningham and Best 2008). Temporal coding problems interfere with the sound features that support both segregation and selection of the desired source from the mixture. In other words, listening to a talker amid similar, competing talkers reveals deficits that may be too subtle to be observed in other listening situations.

7.5 Interpreting Individual Differences in the Envelope-Following Response

Interpreting EFRs is complicated. The responses that are measured on the scalp are a sum of the electrical activity from different populations of neurons across different stages of processing. While evidence suggests that the EFR strength reflects true differences in the strength of temporal coding in subcortical portions of the auditory pathway, there are a number of issues that arise when interpreting EFRs. This section highlights a few of the issues of which one should be aware when trying to interpret individual differences in EFRs.

7.5.1 Encoding of Modulation in the Auditory Nerve Responses

Any modulation information that drives the EFR must be encoded in the firing patterns of the population of ANFs ascending from the cochlea. Therefore, in order to understand how the EFR is generated, one important thing to understand is how modulation information is first coded in this ANF population.

As described in Sect. 7.2.1, ANF neural spikes are phase locked to a mixture of both TFS and envelope modulation. The degree of phase locking to the TFS rolls off with frequency, with a knee point (in humans) near 2000 Hz. Many EFRs are measured using broadband, periodic inputs, such as a broadband complex tone (Zhu et al. 2013) or a speech syllable that has a fixed fundamental frequency (Russo et al. 2004). With such broadband stimuli, even though most of the cochlea is excited, the EFR is dominated by responses from middle to high frequency regions of the cochlea (Zhu et al. 2013). This can be explained by considering what acoustic energy drives the low-frequency and high-frequency ANFs.

In the lowest frequencies, individual harmonics excite different ANFs, leading to "resolved harmonics" in the excitation pattern. For a steady-state input, a single, resolved sinusoidal harmonic will cause a constant drive, with no envelope fluctuations (recall Fig. 7.1A). In contrast, in high-frequency channels, multiple harmonics fall within a single ANF critical band (unresolved harmonics). For channels responding to unresolved harmonics, the driving signal is periodic with a period corresponding to the fundamental frequency of the input (recall Figs. 7.1B, C). For these channels, envelope fluctuations at the fundamental frequency of the input signal dominate the periodicity in the neural firing pattern, strongly contributing to the EFR. This suggests that for a broadband, periodic input, EFRs come primarily from responses in frequency channels responding to unresolved harmonics.

For narrowband, low-level sounds, phase-locked ANF activity (to both TFS and envelope) is limited to a small region at the tonotopic place tuned to the input (Ananthanarayan and Durrant 1992; Herdman et al. 2002). This specificity led some researchers to propose using ASSRs for objective audiometry (Gardi et al. 1979;

Lins et al. 1996). However, narrowband supra-threshold sounds that are at a comfortable listening level (or louder) cause activity that spreads out from the best place on the cochlea; this spread of excitation can be quite pronounced, especially toward the basal (higher frequency) end of the cochlea. When EFRs are measured for supra-threshold acoustic inputs, the ANFs that drive higher auditory centers are typically spread over a large swath of the cochlea, even if the acoustic input is band limited (e.g., John et al. 1998; Herdman et al. 2002). Therefore it is difficult to deduce how activity from a specific place along the cochlea contributes to EFRs. One approach to minimizing the spread of excitation is to use notched noise maskers, so that contributions from off-frequency channels are attenuated (e.g., Bharadwaj et al. 2015).

The measured EFR is a sum of all neural activity; it therefore depends on the phase alignment of the responses in different frequency channels. If responses in two distinct neural populations are both large, but out of phase with each other, they can cancel one another. While models can predict phase disparities in the responses across the population of ANFs (e.g., Shinn-Cunningham et al. 2013; Verhulst et al. 2015), it is more difficult to predict what phase differences are present in envelope-modulation driven responses in the neural population at the level of the IC or how this impacts the final EFR.

7.5.2 Metrics to Quantify the Envelope-Following Response

Time domain methods make sense for analyzing transient events, such as the waves composing the traditional click-evoked ABR. In contrast, frequency-domain analyses efficiently characterize periodic neural activity such as EFRs. Typically, to measure the EFR, a periodic signal is presented on multiple trials. The response to each trial includes multiple cycles of the underlying periodic signal. Frequency-domain analysis focuses on the degree to which the voltages on the scalp align to the periodic input signal by analyzing the response at the fundamental frequency of the input and possibly at its harmonics by combining the responses across trials. Alternatively, the EFR can also be measured by presenting a long-duration input and then breaking the measured output into equal epochs (ensuring that each epoch has the same starting phase with respect to the periodic input signal; e.g., Schoof 2014). This approach effectively treats each epoch as a separate trial.

Conceptually, the signal that one is trying to measure, $s(t)$, is identical across different trials. The variation in responses from one trial to another is due to noise, $n(t)$, which can be thought of as a random process. This noise comes from various sources, including physiological activity unrelated to the input (electrical activity associated with nonauditory sensory activity, cognitive functions, or electromyographic activity) and/or from the environment (noise from the recording environment or devices, including harmonics of 60 Hz line noise). The noise $n(t)$ is typically assumed to be zero mean and uncorrelated with $s(t)$. Under these assumptions, $n(t)$ adds to the variability in measurements, but does not change the expected mean across trials, which equals the signal $s(t)$. However, the metrics used to quantify the EFR are inherently affected by $n(t)$. This means that when comparing EFRs, one must be aware of the effects of noise on the EFR metric.

A number of frequency-domain measures have been utilized to quantify the EFR. Two that have been applied often are the power spectral density (PSD) and the phase-locking value (PLV), each of which is a function of frequency. To calculate the PSD, the waveforms from each of M trials or epochs are averaged together. $PSD(f)$ is then computed as the square of the absolute value of the Fourier transform of this average. The expected value of the PSD equals the sum of the expected signal power and the expected noise power after averaging:

$$PSD(f) = \left|S^2(f)\right| + \frac{1}{M}\left|N^2(f)\right| \tag{7.1}$$

where $S(f)$ and $N(f)$ are the Fourier transforms of $s(t)$ and $n(t)$, respectively.

The PSD is easy to interpret when the noise floor is the same across conditions and/or subjects. If the noise characteristics differ, however, interpretation of the PSD can be problematic: if the PSD at a particular frequency varies significantly across conditions or subjects, it could either be due to differences in the signal or differences in the noise. Subtracting off an estimate of the noise at each frequency can normalize the PSD and mitigate this problem. For instance, for EFRs, the signal is assumed to be zero for all frequencies except the fundamental frequency and its harmonics. The PSD at these nonsignal frequencies provides a direct estimate of the noise. Typically, the noise floor varies relatively smoothly with frequency (often proportional to the reciprocal of the frequency), allowing the noise to be estimated from neighboring frequency bins.

The PLV measures the phase consistency of the response across individual trials (or epochs), ignoring the magnitude of each trial (see Dobie and Wilson 1993). At low SNRs, the PLV is better able to detect the presence of the signal than is the PSD (Dobie and Wilson 1993; Lachaux et al. 1999). Because it ignores the magnitude of the response, chance performance depends only on the number of trials (epochs) being combined to form the estimate of the PLV, making it easy to determine whether or not there is a significant signal in the measurement (Zhu et al. 2013). The magnitude of the PLV depends on the SNR; thus, just as with the PSD, comparisons of PLVs across conditions or across listeners depends upon appropriately characterizing the noise in the measurements (e.g., using resampling methods).

The choice of what kind of metric to use to quantify the EFR should take into account both the SNR of the measurements and the goal of the study. Because the PLV takes on values between zero (response phases are randomly distributed from $-\pi$ to π) and one (response phases are equal on each trial), it is a compressive function of signal level when the SNR is high. That is, the same amount of change in SNR produces increasingly smaller changes in the PLV as SNR increases. As a result, conditions that differ in the strength of the signal in measurements with a high SNR may be difficult to distinguish using the PLV. Compared to the PLV, the

PSD is more sensitive to changes in signal power when SNR is high; across the range of SNRs, the PSD increases linearly with signal strength (see Eq. 7.1). At low SNRs, the PLV scales approximately linearly with SNR. In this SNR regime, the PLV is more likely to differentiate differences in signal level that might be unobservable using the PSD. Taking these issues together, if the goal is simply to detect the presence of a significant signal rather than to estimate differences in the strength of the EFR, the PLV either equals or outperforms the PSD across all SNRs. However, if the goal is to quantify the magnitude of differences in signal strength across individuals, groups, or conditions, either the PSD or the PLV may be better, depending on the SNR.

This effect of noise on the metrics quantifying the EFR can be especially problematic when comparing different behavioral conditions and trying to conclude whether or not the listener state has an influence on the brainstem response. Cortical activity is one of the main sources of noise in EFR recordings. Moreover, cortical activity depends strongly on task demands. Imagine an experiment exploring the question of whether the EFR strength differs when a subject is attending to an auditory source versus attending to a visual source. The different tasks of listening versus watching will change the distribution of cortical activity on the scalp and thus change the amount of noise in the EFR measurement. Alternatively, imagine a "blocked" experimental design where different listening conditions are presented without sufficient randomization. If a subject's focus varies slowly through time (e.g., due to fatigue or inattentiveness), cortical activity will reflect this shift, and different noise levels will bias EFR measures differently in different conditions. Care should be taken to tease apart changes in noise levels from changes in the signal to avoid misinterpreting differences in the estimates of the EFR strength.

7.5.3 *Effects of Stimulus Characteristics*

Responses in the auditory system exhibit a host of nonlinear effects, including forward masking, adaptation, and the like. EFR measurements often implicitly assume that the response that is being measured is constant across trials and, within each trial, the response has settled into a constant, steady-state response. This is not a fair assumption. Indeed, the one study that explored adaptation effects demonstrated that the ASSR to a periodic stimulus is stronger at stimulus onset compared to the later portion of the stimulus (Gockel et al. 2015).

Adaptation effects will be weaker and EFRs will be stronger when each trial consists of a short stimulus and when inter-trial intervals are long. However, a shorter duration stimulus contains fewer cycles of the periodic input, so that the neural response to the input is more affected by onset and offset transients and is in its pseudo-steady-state for a proportionately briefer portion of time. An alternative approach is to present an ongoing stimulus and to analyze epochs of the output response. With this kind of approach, adaptation effects will be maximal, but the neural response should be more stable (asymptoting toward a true steady-state response).

In considering how to design EFR stimuli and the effects of adaptation, it is worth mentioning that lower-threshold and higher-threshold ANFs differ in their adaptation time constants. Specifically, high-threshold ANFs (i.e., the fibers that may be most susceptible to cochlear neuropathy) have a longer recovery time than do low-threshold fibers (Relkin and Doucet 1991; Furman et al. 2013). Thus, differences in the proportion of high-threshold ANFs versus low-threshold ANFs are likely to affect how adaptation influences EFRs.

7.5.4 Electrode Configuration

The placement of recording electrodes on the scalp and the choice of reference site influence EFR measurements strongly (Stillman et al. 1978; Galbraith 1994). Most EFR studies use a vertical one-channel montage, which emphasizes sustained phase-locked neural activity from the rostral generators in the brainstem (Smith et al. 1975; Stillman et al. 1978). This configuration requires an active lead (usually the vertex channel CZ), reference electrode(s) (usually the earlobes or mastoids), and a ground electrode. Often, the earlobe is the preferred reference (rather than the mastoid) for auditory subcortical recordings because it is a noncephalic site and results in smaller bone vibration artifacts (Hall 2007).

Multiple electrode recordings can be combined to estimate brainstem responses; however, in estimating the EFR, simple time-domain averaging or application of principal component analyses can decrease, rather than increase, the effective SNR of the recorded signal. Specifically, small phase differences in the total signal reaching different recording channels can lead to cancellation of responses. Benefits of multi-channel recording can be realized by averaging frequency-domain amplitudes at the modulation frequency of interest or by using a complex principal components analysis (Bharadwaj and Shinn-Cunningham 2014).

7.6 What the Envelope-Following Response May not Reveal

Although this chapter focuses on why the EFR reflects differences in the precision of subcortical temporal coding, it is also important to mention what kinds of neural processing the EFR response does not index strongly. In particular, there are aspects of auditory processing that are unassailably present in the brainstem (e.g., from electrophysiological animal recordings, neuroimaging techniques, or other approaches) but that do not cause robust effects on the EFR. Two examples are mentioned here as a reminder that the EFR may be insensitive to a manipulation that has a clear effect on subcortical neural processing.

7.6.1 Binaural Processing

The ITD, computed by comparing the timing of inputs reaching the left and right ears through a coincidence detector, is arguably the perceptual feature that relies most heavily on precise temporal coding in the brainstem (for a review, see Joris et al. 1998). Indeed, ITD sensitivity correlates with EFR strength and both correlate with individual differences in the strength of temporal coding in the brainstem. Many neurons in the IC are sensitive to ITDs, responding preferentially to some ITD values over others (e.g., see Kuwada and Yin 1987). Click ABRs reflect binaural processing in the *binaural interaction component* (BIC), which is defined as the difference between the ABR when sound is presented to both ears and the sum of the monaural ABR responses for clicks presented separately to the left and right ears (Wrege and Starr 1981). The ABR BIC has a relatively low SNR but is generally consistent with a response generated at the level of the lateral lemniscus or IC (e.g., Brantberg et al. 1999). In addition to click responses, other transient brainstem responses are sensitive to binaural cues (Parbery-Clark et al. 2013). Since the EFR itself is dominated by responses coming from the level of IC and is a good index of temporal acuity in individual listeners, it stands to reason that binaural processing might also be reflected in EFRs.

A few studies have reported statistically significant differences between the sum of FFRs from left and right monaural inputs and FFRs to binaural stimuli (e.g., Clark et al. 1997; Ballachanda and Moushegian 2000). However, these studies used pure tones as the acoustic stimuli and thus only assessed the subcortical responses that are phase locked to TFS. Moreover, other studies failed to find any signature of binaural processing in FFRs, reporting that the binaural response was roughly equal to the sum of the two monaural responses (Gerken et al. 1975; Zhang and Boettcher 2008).

Conflicting results are seen for other binaural phenomena, such as evidence of physiological correlates of the binaural masking level difference (BMLD: the difference in the detection threshold for a tone in noise when the tone is presented with an ITD that differs from the ITD of the noise, compared to when both are diotic). One study measured the ASSR to diotic 500 Hz tones in the presence of simultaneous noise and found larger ASSR amplitudes when the noises at the two ears were in phase than when either the tone or the noise was 180° out of phase (Wilson and Krishnan 2005). However, another study concluded that the only correlate of the BMLD was in cortical responses, for slow modulations (7 or 13 Hz), with no significant response from brainstem sources for 80 Hz modulations (Wong and Stapells 2004).

Taken together, the results of these various studies suggest that subcortical FFR signatures of binaural processing are weak. Data from an example experiment lends support to the idea that the binaural EFR does not reflect spatial-dependent processing. Figure 7.6 shows EFR responses (quantified by the PLV) to broadband click trains presented at a repetition rate of 100 Hz. Figure 7.6A shows the PLV as a function of frequency for one typical subject for a binaural, diotic input, and for the sum of the left and right ear monaural presentations. Figure 7.5B shows the mean PLV at the 100 Hz fundamental frequency of the input (averaged across subjects and plotted as a function of the binaural stimulus ITD) for binaural responses, the sum of the left and right monaural responses, and "corrected" binaural responses (described below).

For both the individual example subject (Fig. 7.6A) and the mean PLV (Fig. 7.6B), the EFR in the binaural condition is greater than the summed monaural

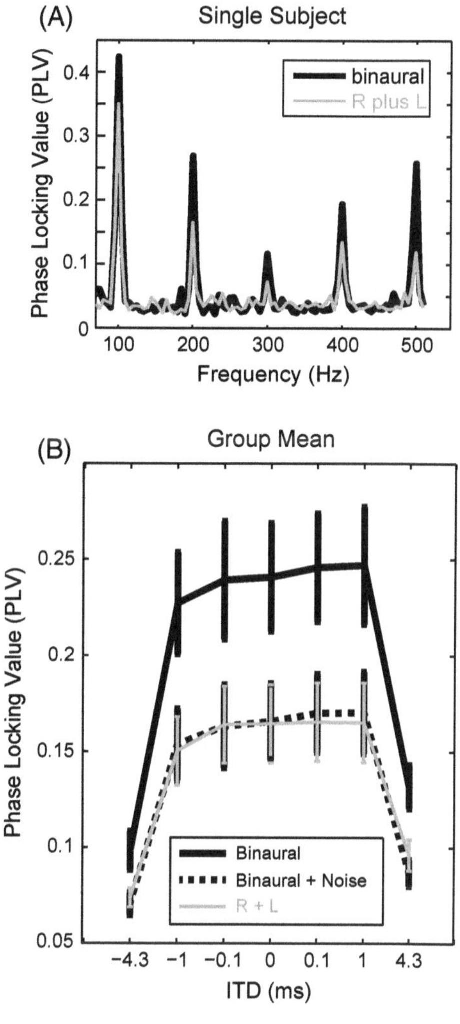

Fig. 7.6 Binaural and monaural envelope-following responses (EFRs) in response to 100 Hz click trains. (**A**) The phase-locking value, plotted as a function of frequency, for one sample subject. The *solid black line* shows the response to a binaural (diotic) presentation, and the *gray line* shows the response to the sum of the monaural responses. (**B**) The across-listener average phase-locking value at 100 Hz, plotted as a function of interaural time difference. *Error bars* show the across-subject standard deviation. The *solid black line* shows the response to binaural stimuli; the *gray line* shows the response to the sum of the monaural responses; the *dashed black line* shows the response to binaural stimuli after compensating for the difference in the noise levels of the summed response and the binaural response. (See text for additional explanation)

responses. This result seems to hint that some component of the response reflects binaural processing. However, this comparison does not take into account the noise in the conditions being compared. Since the presentation order was randomized, the noise floor should be identical across measurement conditions. This means that when the two monaural signals are added, the total noise in the sum has twice the noise power (3 dB more) than the binaural recording. This difference in noise floor actually accounts for the apparent difference between the binaural EFR and the sum of the monaural responses. Once a compensatory level of noise is added to the binaural condition, the binaural response is essentially identical to the sum of the monaural responses (Fig. 7.6B).

Figure 7.6B also shows that the ITD influences the binaural FFR, but only when the ITD is extremely large or extremely small. Importantly, this effect of ITD is also explained by the monaural responses. This reduction at large magnitude ITDs occurs because the left and right ear responses cancel each other; for artificially large ITDs with a magnitude of 4.3 ms, the monaural responses are delayed relative to one another by roughly one-half of the repetition period.

There is no question that binaural cues affect responses in the brainstem, and specifically modulate the synaptic inputs driving the responses of individual neurons in IC. Despite this, there is not a robust, consistent signature of binaural processing in the EFR. This could be due to any number of reasons. For instance, the binaural-specific electrical response may be small compared to responses to monaural stimuli. Alternatively, depending on the ITD, there may be differences in what subpopulation of neurons responds, yet the sum of the responses across the IC population may be roughly constant, independent of ITD. Regardless, the fact that a fundamental feature such as ITD does not have a robust effect on the EFR highlights the limitations of this kind of measure.

7.6.2 Modulation of Subcortical Responses Due to Selective Attention

IC receives many descending projections originating in cortex. These efferent projections create a dynamic feedback loop spanning cortical and subcortical auditory processing stages (for a review, see Kraus and White-Schwoch 2015). Such feedback likely guides long-term learning and plasticity and allows cortical feedback to alter the subcortical sound based on task goals (e.g., Chandrasekaran et al. 2012, 2014).

Experience clearly tunes responses in IC. Direct electrical stimulation of auditory cortex shifts the frequency tuning of IC neurons with changes persisting for hours or longer (Suga and Ma 2003). Long-term learning shapes responses in the midbrain to enhance sound features important for perception and behavior (e.g., Chandrasekaran et al. 2007; Chambers et al. 2016). Moreover, a number of studies show that experience has an impact on EFRs (e.g., Carcagno and Plack 2011; Strait and Kraus 2014). Such effects are considered in other chapters of this volume, including language experience (Krishnan and Gandour, Chap. 3), perceptual learning (Carcagno and Plack, Chap. 4), and musical training (White-Schwoch and Kraus, Chap. 6).

While long-term effects may influence steady-state brainstem responses, the immediate effects of task demands do not show consistent EFR effects. Despite this, it is well established that task demands change physiological responses measured in other ways. In ferrets, spectrotemporal receptive fields of IC neurons change depending on whether the ferrets are actively attending to sounds and performing a listening task compared to when they are passively hearing the same sounds (Slee and David 2015). In humans, selectively attending to a sound in one ear gives rise to higher fMRI activation in the contralateral IC compared to when attention is directed to the opposing ear (Rinne et al. 2008).

Projections from auditory cortex may modulate CN responses (Luo et al. 2008) and underlie changes in CN responses during periods of visual attention (Oatman 1976; Oatman and Anderson 1980). Visually directed attention can even alter responses at the level of the auditory nerve (Oatman 1976). Yet, despite the vast evidence for online modulatory changes in subcortical responses based on subject goals, efforts to demonstrate changes in EFRs due to selective focus of attention have produced mixed results. Although a few studies argued that EFR strength is influenced by exactly which of multiple competing sounds a listener attends (Galbraith et al. 2003; Lehmann and Schonwiesner 2014), the effect sizes are small, the effect directions are inconsistent, and efforts to replicate the effects have failed (see the discussion in Varghese et al. 2015).

7.7 Summary and Conclusions

The EFR provides a window into individual differences in the fidelity of temporal coding in subcortical portions of the auditory pathway. A portion of this variation across listeners reflects compensatory changes and experience-dependent plasticity in brainstem processing. However, a significant portion derives from sensory differences that likely reflect differences in the number of ANFs encoding sound.

Together, noise exposure and aging cause cochlear neuropathy, or death of ANFs. Such cochlear neuropathy reduces the fidelity with which temporal modulation in supra-threshold sounds, such as speech, are encoded in the auditory nerve. This subtle "hidden hearing loss" manifests primarily as perceptual deficits in temporal processing and is especially noticeable when listeners are trying to communicate in noisy social settings (conditions in which listeners must selectively attend in order to understand speech). Individual differences in EFRs quantify these differences in sensory coding and correlate with the resulting differences in perceptual ability. Still, the EFR can be difficult to interpret. When considering how the EFR varies across listeners, across groups, or across experimental conditions, it is important to understand how the EFR is generated and measured and how measurement noise influences EFR measures. Moreover, there are many subcortical aspects of sound processing that do not influence the EFR.

Compliance with Ethics Barbara Shinn-Cunningham, Leonard Varghese, Le Wang, and Hari Bharadwaj declared that they had no conflicts of interest.

References

Aiken, S. J., & Picton, T. W. (2008). Envelope and spectral frequency-following responses to vowel sounds. *Hearing Research, 245*(1–2), 35–47. Doi:10.1016/j.heares.2008.08.004

Ananthanarayan, A. K., & Durrant, J. D. (1992). The frequency-following response and the onset response: Evaluation of frequency specificity using a forward-masking paradigm. *Ear and Hearing, 13*(4), 228–232.

Anderson, S., Parbery-Clark, A., White-Schwoch, T., & Kraus, N. (2012). Aging affects neural precision of speech encoding. *The Journal of Neuroscience, 32*(41), 14156–14164. Doi:10.1523/JNEUROSCI.2176-12.2012

Anderson, S., White-Schwoch, T., Parbery-Clark, A., & Kraus, N. (2013). Reversal of age-related neural timing delays with training. *Proceedings of the National Academy of Sciences of the USA, 110*(11), 4357–4362. Doi:10.1073/pnas.1213555110

Ballachanda, B. B., & Moushegian, G. (2000). Frequency-following response: Effects of interaural time and intensity differences. *Journal of the American Academy of Audiology, 11*(1), 1–11.

Bharadwaj, H. M., Masud, S., Mehraei, G., Verhulst, S., & Shinn-Cunningham, B. G. (2015). Individual differences reveal correlates of hidden hearing deficits. *The Journal of Neuroscience, 35*, 2161–2172.

Bharadwaj, H. M., & Shinn-Cunningham, B. G. (2014). Rapid acquisition of auditory subcortical steady state responses using multichannel recordings. *Clinical Neurophysiology, 125*(9), 1878–1888. Doi:10.1016/j.clinph.2014.01.011

Bharadwaj, H. M., Verhulst, S., Shaheen, L., Liberman, M. C., & Shinn-Cunningham, B. G. (2014). Cochlear neuropathy and the coding of supra-threshold sound. *Frontiers in Systems Neuroscience*. Doi:10.3389/fnsys.2014.00026

Blauert, J. (1997). *Spatial hearing* (2nd ed.). Cambridge, MA: MIT Press.

Boashash, B. (1992). Estimating and interpreting the instantaneous frequency of a signal I. Fundamentals. *Proceedings of the IEEE, 80*(4), 520–538.

Bohne, B. A., & Harding, G. W. (2000). Degeneration in the cochlea after noise damage: Primary versus secondary events. *American Journal of Otology, 21*(4), 505–509.

Bourien, J., Tang, Y., Batrel, C., Huet, A., et al. (2014). Contribution of auditory nerve fibers to compound action potential of the auditory nerve. *Journal of Neurophysiology, 112*(5), 1025–1039. Doi:10.1152/jn.00738.2013

Brantberg, K., Fransson, P. A., Hansson, H., & Rosenhall, U. (1999). Measures of the binaural interaction component in human auditory brainstem response using objective detection criteria. *Scandinavian Audiology, 28*(1), 15–26.

Bregman, A. S. (1990). *Auditory scene analysis: The perceptual organization of sound*. Cambridge, MA: MIT Press.

Carcagno, S., & Plack, C. J. (2011). Subcortical plasticity following perceptual learning in a pitch discrimination task. *Journal of the Association for Research in Otolaryngology, 12*(1), 89–100. Doi:10.1007/s10162-010-0236-1

Carlyon, R. P. (2004). How the brain separates sounds. *Trends in Cognitive Sciences, 8*(10), 465–471.

Chambers, A. R., Resnik, J., Yuan, Y., Whitton, J. P., et al. (2016). Central gain restores auditory processing following near-complete cochlear denervation. *Neuron, 89*(4), 867–879. Doi:10.1016/j.neuron.2015.12.041

Chandrasekaran, B., Kraus, N., & Wong, P. C. (2012). Human inferior colliculus activity relates to individual differences in spoken language learning. *Journal of Neurophysiology, 107*(5), 1325–1336. Doi:10.1152/jn.00923.2011

Chandrasekaran, B., Krishnan, A., & Gandour, J. T. (2007). Experience-dependent neural plasticity is sensitive to shape of pitch contours. *NeuroReport, 18*(18), 1963–1967. Doi:10.1097/WNR.0b013e3282f213c5

Chandrasekaran, B., Skoe, E., & Kraus, N. (2014). An integrative model of subcortical auditory plasticity. *Brain Topography, 27*(4), 539–552. Doi:10.1007/s10548-013-0323-9

Christiansen, S. K., & Oxenham, A. J. (2014). Assessing the effects of temporal coherence on auditory stream formation through comodulation masking release. *The Journal of the Acoustical Society of America, 135*(6), 3520–3529. Doi:10.1121/1.4872300

Clark, J. L., Moushegian, G., & Rupert, A. L. (1997). Interaural time effects on the frequency-following response. *Journal of the American Academy of Audiology, 8*(5), 308–313.

Cohen, L. T., Rickards, F. W., & Clark, G. M. (1991). A comparison of steady-state evoked potentials to modulated tones in awake and sleeping humans. *Journal of the Acoustical Society of America, 90*(5), 2467–2479.

Dobie, R. A., & Wilson, M. J. (1993). Objective response detection in the frequency domain. *Electroencephalography and Clinical Neurophysiology, 88*(6), 516–524.

Dolphin, W. F., & Mountain, D. C. (1992). The envelope-following response: Scalp potentials elicited in the Mongolian gerbil using sinusoidally AM acoustic signals. *Hearing Research, 58* (1), 70–78.

Escabi, M. A., & Read, H. L. (2003). Representation of spectrotemporal sound information in the ascending auditory pathway. *Biological Cybernetics, 89*(5), 350–362.

Fitzgibbons, P. J., & Gordon-Salant, S. (2010). Age-related differences in discrimination of temporal intervals in accented tone sequences. *Hearing Research, 264*(1–2), 41–47. Doi:10.1016/j.heares.2009.11.008

Fullgrabe, C., Moore, B. C., & Stone, M. A. (2014). Age-group differences in speech identification despite matched audiometrically normal hearing: Contributions from auditory temporal processing and cognition. *Frontiers in Aging Neuroscience, 6*, 347. Doi:10.3389/fnagi.2014.00347

Furman, A. C., Kujawa, S. G., & Liberman, M. C. (2013). Noise-induced cochlear neuropathy is selective for fibers with low spontaneous rates. *Journal of Neurophysiology*. Doi:10.1152/jn.00164.2013

Galambos, R., Makeig, S., & Talmachoff, P. J. (1981). A 40-Hz auditory potential recorded from the human scalp. *Proceedings of the National Academy of Sciences of the USA, 78*(4), 2643–2647.

Galbraith, G. C. (1994). Two-channel brainstem frequency-following responses to pure tone and missing fundamental stimuli. *Electroencephalography and Clinical Neurophysiology, 92*(4), 321–330.

Galbraith, G. C., Olfman, D. M., & Huffman, T. M. (2003). Selective attention affects human brainstem frequency-following response. *NeuroReport, 14*(5), 735–738. Doi:10.1097/01.wnr.0000064983.96259.49

Gardi, J., Merzenich, M., & McKean, C. (1979). Origins of the scalp recorded frequency-following response in the cat. *Audiology, 18*(5), 358–381.

Gerken, G. M., Moushegian, G., Stillman, R. D., & Rupert, A. L. (1975). Human frequency-following responses to monaural and binaural stimuli. *Electroencephalography and Clinical Neurophysiology, 38*(4), 379–386.

Goblick, T. J., Jr., & Pfeiffer, R. R. (1969). Time-domain measurements of cochlear nonlinearities using combination click stimuli. *The Journal of the Acoustical Society of America, 46*(4), 924–938.

Gockel, H. E., Krugliak, A., Plack, C. J., & Carlyon, R. P. (2015). Specificity of the human frequency-following response for carrier and modulation frequency assessed using adaptation. *Journal of the Association for Research in Otolaryngology, 16*(6), 747–762. Doi:10.1007/s10162-015-0533-9

Grose, J. H., & Mamo, S. K. (2010). Processing of temporal fine structure as a function of age. *Ear and Hearing, 31*, 755–760. Doi:10.1097/AUD.0b013e3181e627e7

Grose, J. H., & Mamo, S. K. (2012). Frequency modulation detection as a measure of temporal processing: Age-related monaural and binaural effects. *Hearing Research, 294*(1–2), 49–54. Doi:10.1016/j.heares.2012.09.007

Grose, J. H., Mamo, S. K., Buss, E., & Hall, J. W., III. (2015). Temporal processing deficits in middle age. *American Journal of Audiology, 24*(2), 91–93. Doi:10.1044/2015_AJA-14-0053

Grose, J. H., Mamo, S. K., & Hall, J. W., III. (2009). Age effects in temporal envelope processing: Speech unmasking and auditory steady state responses. *Ear and Hearing, 30*(5), 568–575. Doi:10.1097/AUD.0b013e3181ac128f

Hall, J. (2007). *New handbook of auditory evoked responses*. Boston: Pearson.

Hamalainen, M., Hari, R., Ilmoniemi, R. J., Knuutila, J., & Lounasmaa, O. V. (1993). Magnetoencepahlography: Theory, instrumentation, and applications to noninvasive studies of the working human brain. *Reviews of Modern Physics, 65*(2), 413–497. Doi:10.1103/RevModPhys.65.413

He, N. J., Mills, J. H., Ahlstrom, J. B., & Dubno, J. R. (2008). Age-related differences in the temporal modulation transfer function with pure-tone carriers. *Journal of the Acoustical Society of America, 124*(6), 3841–3849. Doi:10.1121/1.2998779

Helfer, K. S. (2015). Competing speech perception in middle age. *American Journal of Audiology, 24*(2), 80–83. Doi:10.1044/2015_AJA-14-0056

Helfer, K. S., & Vargo, M. (2009). Speech recognition and temporal processing in middle-aged women. *Journal of the American Academy of Audiology, 20*(4), 264–271.

Herdman, A. T., Picton, T. W., & Stapells, D. R. (2002). Place specificity of multiple auditory steady-state responses. *The Journal of the Acoustical Society of America, 112*(4), 1569–1582.

Hickox, A. E., & Liberman, M. C. (2014). Is noise-induced cochlear neuropathy key to the generation of hyperacusis or tinnitus? *Journal of Neurophysiology, 111*(3), 552–564. Doi:10.1152/jn.00184.2013

Hind, S. E., Haines-Bazrafshan, R., Benton, C. L., Brassington, W., et al. (2011). Prevalence of clinical referrals having hearing thresholds within normal limits. *International Journal of Audiology, 50*(10), 708–716. Doi:10.3109/14992027.2011.582049

Hornickel, J., Chandrasekaran, B., Zecker, S., & Kraus, N. (2011). Auditory brainstem measures predict reading and speech-in-noise perception in school-aged children. *Behavioral and Brain Research, 216*(2), 597–605. Doi:10.1016/j.bbr.2010.08.051

Jin, S. H., Liu, C., & Sladen, D. P. (2014). The effects of aging on speech perception in noise: Comparison between normal-hearing and cochlear-implant listeners. *Journal of the American Academy of Audiology, 25*(7), 656–665. Doi:10.3766/jaaa.25.7.4

John, M. S., Lins, O. G., Boucher, B. L., & Picton, T. W. (1998). Multiple auditory steady-state responses (MASTER): Stimulus and recording parameters. *Audiology, 37*(2), 59–82.

Joris, P. X., Schreiner, C. E., & Rees, A. (2004). Neural processing of amplitude-modulated sounds. *Physiology Review, 84*(2), 541–577. Doi:10.1152/physrev.00029.2003

Joris, P. X., Smith, P. H., & Yin, T. C. (1998). Coincidence detection in the auditory system: 50 years after Jeffress. *Neuron, 21*(6), 1235–1238. S0896-6273(00)80643-1 [pii].

Joris, P. X., & Yin, T. C. (1992). Responses to amplitude-modulated tones in the auditory nerve of the cat. *The Journal of the Acoustical Society of America, 91*(1), 215–232.

Kiren, T., Aoyagi, M., Furuse, H., & Koike, Y. (1994). An experimental study on the generator of amplitude-modulation following response. *Acta Otolaryngolica Supplement, 511*, 28–33.

Kraus, N., & White-Schwoch, T. (2015). Unraveling the biology of auditory learning: A cognitive-sensorimotor-reward framework. *Trends in Cognitive Sciences, 19*(11), 642–654.

Kujawa, S. G., & Liberman, M. C. (2006). Acceleration of age-related hearing loss by early noise exposure: Evidence of a misspent youth. *The Journal of Neuroscience, 26*(7), 2115–2123. Doi:10.1523/JNEUROSCI.4985-05.2006

Kujawa, S. G., & Liberman, M. C. (2009). Adding insult to injury: Cochlear nerve degeneration after "temporary" noise-induced hearing loss. *The Journal of Neuroscience, 29*(45), 14077–14085. Doi:10.1523/JNEUROSCI.2845-09.2009

Kujawa, S. G., & Liberman, M. C. (2015). Synaptopathy in the noise-exposed and aging cochlea: Primary neural degeneration in acquired sensorineural hearing loss. *Hearing Research, 330*(Pt B), 191–199. Doi:10.1016/j.heares.2015.02.009

Kumar, G., Amen, F., & Roy, D. (2007). Normal hearing tests: Is a further appointment really necessary? *Journal of the Royal Society of Medicine, 100*(2), 66. Doi:10.1258/jrsm.100.2.66-a

Kuwada, S., Anderson, J. S., Batra, R., Fitzpatrick, D. C., et al. (2002). Sources of the scalp-recorded amplitude-modulation following response. *Journal of the American Academy of Audiology, 13*(4), 188–204.

Kuwada, S., & Yin, T. C. T. (1987). Physiological studies of directional hearing. In W. A. Yost & G. Gourevitch (Eds.), *Directional hearing* (pp. 146–176). New York: Springer.

Lachaux, J. P., Rodriguez, E., Martinerie, J., & Varela, F. J. (1999). Measuring phase synchrony in brain signals. *Human Brain Mapping, 8*(4), 194–208. Doi:10.1002/(SICI)1097-0193(1999)8:4<194:AID-HBM4>3.0.CO;2-C [pii]

Lee, A. K. C., Larson, E., & Maddox, R. K. (2012). Mapping cortical dynamics using simultaneous MEG/EEG and anatomically-constrained minimum-norm estimates: An auditory attention example. *Journal of Visual Experiments, 68*, e4262. Doi:10.3791/4262

Lehmann, A., & Schonwiesner, M. (2014). Selective attention modulates human auditory brainstem responses: Relative contributions of frequency and spatial cues. *PLoS ONE, 9*(1), e85442. Doi:10.1371/journal.pone.0085442

Liberman, M. C. (1980). Morphological differences among radial afferent fibers in the cat cochlea: An electron-microscopic study of serial sections. *Hearing Research, 3*(1), 45–63.

Liberman, M. C. (2015). Hidden hearing loss. *Scientific American, 313*(2), 48–53.

Liberman, M. C., Chesney, C., & Kujawa, S. (1997). Effects of selective inner hair cell loss on DPOAE and CAP in carboplatin-treated chinchillas. *Auditory Neuroscience, 3*(3), 255–268.

Lin, H. W., Furman, A. C., Kujawa, S. G., & Liberman, M. C. (2011). Primary neural degeneration in the guinea pig cochlea after reversible noise-induced threshold shift. *Journal of the Association for Research in Otolaryngology, 12*(5), 605–616. Doi:10.1007/s10162-011-0277-0

Lins, O. G., Picton, T. W., Boucher, B. L., Durieux-Smith, A., et al. (1996). Frequency-specific audiometry using steady-state responses. *Ear and Hearing, 17*(2), 81–96.

Lobarinas, E., Salvi, R., & Ding, D. (2013). Insensitivity of the audiogram to carboplatin induced inner hair cell loss in chinchillas. *Hearing Research, 302*, 113–120. Doi:10.1016/j.heares.2013.03.012

Luo, F., Wang, Q., Kashani, A., & Yan, J. (2008). Corticofugal modulation of initial sound processing in the brain. *The Journal of Neuroscience, 28*(45), 11615–11621. Doi:10.1523/JNEUROSCI.3972-08.2008

Maddox, R. K., & Shinn-Cunningham, B. G. (2012). Influence of task-relevant and task-irrelevant feature continuity on selective auditory attention. *Journal of the Association for Research in Otolaryngology, 13*(1), 119–129. Doi:10.1007/s10162-011-0299-7

Makary, C. A., Shin, J., Kujawa, S. G., Liberman, M. C., & Merchant, S. N. (2011). Age-related primary cochlear neuronal degeneration in human temporal bones. *Journal of the Association for Research in Otolaryngology, 12*(6), 711–717. Doi:10.1007/s10162-011-0283-2

Marsh, J. T., Brown, W. S., & Smith, J. C. (1975). Far-field recorded frequency-following responses: Correlates of low pitch auditory perception in humans. *Electroencephalography and Clinical Neurophysiology, 38*(2), 113–119.

Mehraei, G., Hickox, A. E., Bharadwaj, H. M., Goldberg, H., et al. (2016). Auditory brainstem response latency in noise as a marker of cochlear synaptopathy. *The Journal of Neuroscience, 36*(13), 3755–3764. Doi:10.1523/JNEUROSCI.4460-15.2016

Mehta, A., Prabhakar, M., Kumar, P., Deshmukh, R., & Sharma, P. L. (2013). Excitotoxicity: Bridge to various triggers in neurodegenerative disorders. *European Journal of Pharmacology, 698*(1–3), 6–18. Doi:10.1016/j.ejphar.2012.10.032

Milstein, J. N., & Koch, C. (2008). Dynamic moment analysis of the extracellular electric field of a biologically realistic spiking neuron. *Neural Computation, 20*(8), 2070–2084. Doi:10.1162/neco.2008.06-07-537

Moore, B. C. J. (2003). *An introduction to the psychology of hearing* (5th ed.). San Diego, CA: Academic Press.

Oatman, L. C. (1976). Effects of visual attention on the intensity of auditory evoked potentials. *Experimental Neurology, 51*(1), 41–53.

Oatman, L. C., & Anderson, B. W. (1980). Suppression of the auditory frequency-following response during visual attention. *Electroencephalography and Clinical Neurophysiology, 49* (3–4), 314–322.

Oertel, D., Bal, R., Gardner, S. M., Smith, P. H., & Joris, P. X. (2000). Detection of synchrony in the activity of auditory nerve fibers by octopus cells of the mammalian cochlear nucleus. *Proceedings of the National Academy of Sciences of the USA, 97*(22), 11773–11779. Doi:10.1073/pnas.97.22.11773

Okada, Y. C., Wu, J., & Kyuhou, S. (1997). Genesis of MEG signals in a mammalian CNS structure. *Electroencephalography and Clinical Neurophysiology, 103*(4), 474–485.

Parbery-Clark, A., Strait, D. L., Hittner, E., & Kraus, N. (2013). Musical training enhances neural processing of binaural sounds. *The Journal of Neuroscience, 33*(42), 16741–16747. Doi:10.1523/JNEUROSCI.5700-12.2013

Parbery-Clark, A., Strait, D. L., & Kraus, N. (2011). Context-dependent encoding in the auditory brainstem subserves enhanced speech-in-noise perception in musicians. *Neuropsychologia, 49* (12), 3338–3345. Doi:10.1016/j.neuropsychologia.2011.08.007

Pauli-Magnus, D., Hoch, G., Strenzke, N., Anderson, S., et al. (2007). Detection and differentiation of sensorineural hearing loss in mice using auditory steady-state responses and transient auditory brainstem responses. *Neuroscience, 149*(3), 673–684. Doi:10.1016/j.neuroscience.2007.08.010

Plack, C. J., Barker, D., & Prendergast, G. (2014). Perceptual consequences of "hidden" hearing loss. *Trends in Hearing,* 18. Doi:10.1177/2331216514550621

Pujol, R., Puel, J. L., Gervais d'Aldin, C., & Eybalin, M. (1993). Pathophysiology of the glutamatergic synapses in the cochlea. *Acta Otolaryngolica, 113*(3), 330–334.

Purcell, D. W., John, S. M., Schneider, B. A., & Picton, T. W. (2004). Human temporal auditory acuity as assessed by envelope-following responses. *Journal of the Acoustical Society of America, 116*(6), 3581–3593.

Quaranta, A., Portalatini, P., & Henderson, D. (1998). Temporary and permanent threshold shift: An overview. *Scandinavian Audiology Supplement, 48*, 75–86.

Relkin, E. M., & Doucet, J. R. (1991). Recovery from prior stimulation. I: Relationship to spontaneous firing rates of primary auditory neurons. *Hearing Research, 55*(2), 215–222.

Rinne, T., Balk, M. H., Koistinen, S., Autti, T., et al. (2008). Auditory selective attention modulates activation of human inferior colliculus. *Journal of Neurophysiology, 100*(6), 3323–3327. Doi:10.1152/jn.90607.2008

Rosen, S., Cohen, M., & Vanniasegaram, I. (2010). Auditory and cognitive abilities of children suspected of auditory processing disorder (APD). *International Journal of Pediatric Otorhinolaryngology, 74*(6), 594–600. Doi:10.1016/j.ijporl.2010.02.021

Ruggles, D., Bharadwaj, H., & Shinn-Cunningham, B. G. (2011). Normal hearing is not enough to guarantee robust encoding of suprathreshold features important in everyday communication. *Proceedings of the National Academy of Sciences of the USA, 108*(37), 15516–15521. Doi:10.1073/pnas.1108912108

Ruggles, D., Bharadwaj, H., & Shinn-Cunningham, B. G. (2012). Why middle-aged listeners have trouble hearing in everyday settings. *Current Biology, 22*(15), 1417–1422. Doi:10.1016/j.cub.2012.05.025

Ruggles, D., & Shinn-Cunningham, B. (2011). Spatial selective auditory attention in the presence of reverberant energy: Individual differences in normal-hearing listeners. *Journal of the Association for Research in Otolaryngology, 12*(3), 395–405. Doi:10.1007/s10162-010-0254-z

Russo, N., Nicol, T., Musacchia, G., & Kraus, N. (2004). Brainstem responses to speech syllables. *Clinical Neurophysiology, 115*(9), 2021–2030. Doi:10.1016/j.clinph.2004.04.003

Schaette, R., & McAlpine, D. (2011). Tinnitus with a normal audiogram: Physiological evidence for hidden hearing loss and computational model. *The Journal of Neuroscience, 31*(38), 13452–13457. Doi:10.1523/JNEUROSCI.2156-11.2011

Schmiedt, R. A., Mills, J. H., & Boettcher, F. A. (1996). Age-related loss of activity of auditory-nerve fibers. *Journal of Neurophysiology, 76*(4), 2799–2803.

Schoof, T. (2014). *The effects of ageing on the perception of speech in noise.* Dissertation, University College London, London, UK.

Sergeyenko, Y., Lall, K., Liberman, M. C., & Kujawa, S. G. (2013). Age-related cochlear synaptopathy: An early-onset contributor to auditory functional decline. *Journal of Neuroscience, 33*(34), 13686–13694. Doi:10.1523/JNEUROSCI.1783-13.2013

Shaheen, L. A., Valero, M. D., & Liberman, M. C. (2015). Towards a diagnosis of cochlear neuropathy with envelope-following responses. *Journal of the Association for Research in Otolaryngology, 16*(6), 727–745. Doi:10.1007/s10162-015-0539-3

Shamma, S. A., Elhilali, M., & Micheyl, C. (2011). Temporal coherence and attention in auditory scene analysis. *Trends in Neurosciences, 34*(3), 114–123. Doi:10.1016/j.tins.2010.11.002. S0166-2236(10)00167-0 [pii]

Shinn-Cunningham, B. G. (2008). Object-based auditory and visual attention. *Trends in Cognitive Sciences, 12*(5), 182–186. Doi:10.1016/j.tics.2008.02.003

Shinn-Cunningham, B. G., & Best, V. (2008). Selective attention in normal and impaired hearing. *Trends in Amplification, 12*(4), 283–299. Doi:10.1177/1084713808325306

Shinn-Cunningham, B., Ruggles, D. R., & Bharadwaj, H. (2013). How early aging and environment interact in everyday listening: From brainstem to behavior through modeling. *Basic Aspects of Hearing: Physiology and Perception, 787*, 501–510. Doi:10.1007/978-1-4614-1590-9_55

Skoe, E., Chandrasekaran, B., Spitzer, E. R., Wong, P. C., & Kraus, N. (2014). Human brainstem plasticity: The interaction of stimulus probability and auditory learning. *Neurobiological Learning and Memory, 109*, 82–93. Doi:10.1016/j.nlm.2013.11.011

Skoe, E., & Kraus, N. (2010). Auditory brainstem response to complex sounds: A tutorial. *Ear and Hearing, 31*(3), 302–324. Doi:10.1097/AUD.0b013e3181cdb272

Slater, J., Skoe, E., Strait, D. L., O'Connell, S., et al. (2015). Music training improves speech-in-noise perception: Longitudinal evidence from a community-based music program. *Behavioral and Brain Research, 291*, 244–252. Doi:10.1016/j.bbr.2015.05.026

Slee, S. J., & David, S. V. (2015). Rapid task-related plasticity of spectrotemporal receptive fields in the auditory midbrain. *The Journal of Neuroscience, 35*(38), 13090–13102. Doi:10.1523/JNEUROSCI.1671-15.2015

Smith, Z. M., Delgutte, B., & Oxenham, A. J. (2002). Chimaeric sounds reveal dichotomies in auditory perception. *Nature, 416*(6876), 87–90.

Smith, J. C., Marsh, J. T., & Brown, W. S. (1975). Far-field recorded frequency-following responses: Evidence for the locus of brainstem sources. *Electroencephalography and Clinical Neurophysiology, 39*(5), 465–472.

Snell, K., & Frisina, D. R. (2000). Relationship among age-related differences in gap detection and word recognition. *The Journal of the Acoustical Society of America, 107*(3), 1615–1626.

Snell, K. B., Mapes, F. M., Hickman, E. D., & Frisina, D. R. (2002). Word recognition in competing babble and the effects of age, temporal processing, and absolute sensitivity. *The Journal of the Acoustical Society of America, 112*(2), 720–727.

Sohmer, H., Pratt, H., & Kinarti, R. (1977). Sources of frequency-following responses (FFR) in man. *Electroencephalography and Clinical Neurophysiology, 42*(5), 656–664.

Stamper, G. C., & Johnson, T. A. (2015). Auditory function in normal-hearing, noise-exposed human ears. *Ear and Hearing, 36*(2), 172–184. Doi:10.1097/AUD.0000000000000107

Stapells, D. R., Linden, D., Suffield, J. B., Hamel, G., & Picton, T. W. (1984). Human auditory steady state potentials. *Ear and Hearing, 5*(2), 105–113.

Starr, A., Picton, T. W., Sininger, Y., Hood, L. J., & Berlin, C. I. (1996). Auditory neuropathy. *Brain, 119*(Pt 3), 741–753.

Stillman, R. D., Crow, G., & Moushegian, G. (1978). Components of the frequency-following potential in man. *Electroencephalography and Clinical Neurophysiology, 44*(4), 438–446.

Strait, D. L., Hornickel, J., & Kraus, N. (2011). Subcortical processing of speech regularities underlies reading and music aptitude in children. *Behavioral Brain Function, 7*(1), 44. Doi:10.1186/17449081-7-44

Strait, D. L., & Kraus, N. (2014). Biological impact of auditory expertise across the life span: Musicians as a model of auditory learning. *Hearing Research, 308*, 109–121. Doi:10.1016/j.heares.2013.08.004

Strelcyk, O., & Dau, T. (2009). Relations between frequency selectivity, temporal fine-structure processing, and speech reception in impaired hearing. *Journal of the Acoustical Society of America, 125*(5), 3328–3345. Doi:10.1121/1.3097469

Suga, N., & Ma, X. (2003). Multiparametric corticofugal modulation and plasticity in the auditory system. *Nature Review Neuroscience, 4*(10), 783–794. Doi:10.1038/nrn1222

Szydlowska, K., & Tymianski, M. (2010). Calcium, ischemia and excitotoxicity. *Cell Calcium, 47*(2), 122–129. Doi:10.1016/j.ceca.2010.01.003

Valero, M. D., Hancock, K. E., & Liberman, M. C. (2016). The middle ear muscle reflex in the diagnosis of cochlear neuropathy. *Hearing Research, 332*, 29–38. Doi:10.1016/j.heares.2015.11.005

Varghese, L., Bharadwaj, H. M., & Shinn-Cunningham, B. G. (2015). Evidence against attentional state modulating scalp-recorded auditory brainstem steady-state responses. *Brain Research, 1626*, 146–164. Doi:10.1016/j.brainres.2015.06.038

Verhulst, S., Bharadwaj, H. M., Mehraei, G., Shera, C. A., & Shinn-Cunningham, B. G. (2015). Functional modeling of the human auditory brainstem response to broadband stimulation. *The Journal of the Acoustical Society of America, 138*(3), 1637–1659. Doi:10.1121/1.4928305

Whitton, J. P., Hancock, K. E., & Polley, D. B. (2014). Immersive audiomotor game play enhances neural and perceptual salience of weak signals in noise. *Proceedings of the National Academy of Sciences of the USA, 111*(25), E2606–E2615. Doi:10.1073/pnas.1322184111

Wible, B., Nicol, T., & Kraus, N. (2005). Correlation between brainstem and cortical auditory processes in normal and language-impaired children. *Brain, 128*(Pt 2), 417–423. Doi:10.1093/brain/awh367

Wilson, J. R., & Krishnan, A. (2005). Human frequency-following responses to binaural masking level difference stimuli. *Journal of the American Academy of Audiology, 16*(3), 184–195.

Wong, P. C., Skoe, E., Russo, N. M., Dees, T., & Kraus, N. (2007). Musical experience shapes human brainstem encoding of linguistic pitch patterns. *Nature Neuroscience, 10*(4), 420–422. Doi:10.1038/nn1872

Wong, W. Y., & Stapells, D. R. (2004). Brainstem and cortical mechanisms underlying the binaural masking level difference in humans: An auditory steady-state response study. *Ear and Hearing, 25*(1), 57–67. Doi:10.1097/01.AUD.0000111257.11898.64

Wrege, K. S., & Starr, A. (1981). Binaural interaction in human auditory brainstem evoked potentials. *Archives of Neurology, 38*(9), 572–580.

Zeng, F. G., Nie, K., Stickney, G. S., Kong, Y. Y., et al. (2005). Speech recognition with amplitude and frequency modulations. *Proceedings of the National Academy of Sciences of the USA, 102*(7), 2293–2298. Doi:10.1073/pnas.0406460102

Zhang, F., & Boettcher, F. A. (2008). Effects of interaural time and level differences on the binaural interaction component of the 80 Hz auditory steady-state response. *Journal of the American Academy of Audiology, 19*(1), 82–94.

Zhu, L., Bharadwaj, H., Xia, J., & Shinn-Cunningham, B. (2013). A comparison of spectral magnitude and phase-locking value analyses of the frequency-following response to complex tones. *The Journal of the Acoustical Society of America, 134*(1), 384–395. Doi:10.1121/1.4807498

Chapter 8
Communicating in Challenging Environments: Noise and Reverberation

Gavin M. Bidelman

Abstract In everyday listening situations, speech perception is challenged by interfering noise and other adverse room acoustics (e.g., reverberation). These intrusions hinder verbal communication and prevent audible access to salient cues by masking (noise) and smearing ("reverb") spectrotemporal features of the speech signal. The brainstem frequency-following response (FFR) provides a detailed window into the early neural transcription of complex sounds and how normal and degraded speech signals are coded by the human auditory nervous system. This chapter provides an overview of noise-related and reverb-related changes in brainstem representations for speech as reflected in the scalp-recorded FFR. Although noise and reverberation affect behavior to a similar extent, they have a differential effect on neural speech representations, noise being a larger detriment to the speech code than reverberation. Acoustic interferences also produce distinct effects within the speech signal: the neural encoding of "timbre" is more affected than voice "pitch" cues. Applications for the FFR as a "biomarker" for understanding the neural basis and individual differences in degraded speech perception skills are also discussed.

Keywords Auditory scene analysis · Autocorrelogram · Degraded speech processing · FFR · Figure-ground analysis · Frequency-following response · SIN · Speech-in-noise perception · Voice pitch · Voice timbre

8.1 Introduction

In nearly all real-world listening environments, acoustic interferences hinder the successful extraction of speech information. This chapter concerns the neural basis of human communication in adverse listening conditions and focuses on the effects

G.M. Bidelman (✉)
School of Communication Sciences and Disorders, University of Memphis, 4055 North Park Loop, Memphis, TN 38152, USA
e-mail: g.bidelman@memphis.edu

N. Kraus et al. (eds.), *The Frequency-Following Response*,
Springer Handbook of Auditory Research 61,
DOI 10.1007/978-3-319-47944-6_8

of "energetic masking" on neurophysiological speech processing. This problem, classically known as the "cocktail party scenario," represents a fundamental challenge for the auditory system and a barrier to effective communication. Speech-in-noise (SIN) understanding is often exacerbated in cases of hearing impairment, but the issues typically persist even after restoring audiometric thresholds via hearing aids (for review, see Popelka et al. 2016). Moreover, SIN comprehension is problematic even for individuals without substantial hearing loss (Middelweerd et al. 1990; Song et al. 2011). These findings have led to the increasing notion that speech intelligibility and SIN listening skills are determined by more than simple audibility (i.e., peripheral hearing status) (Humes and Roberts 1990). In particular, recent interest in the physiological basis of speech processing has focused on the role of central auditory brain mechanisms in SIN listening and how robust neural coding supports successful listening skills. In this regard, the human frequency-following response (FFR) has provided considerable insight into human communication and central auditory processing in adverse listening environments.

8.2 Listening at the "Cocktail Party"

8.2.1 Acoustical Consequences of Noise and Reverberation

Listeners face two primary challenges when extracting speech from the auditory scene: *noise* and *reverberation*. Each has a distinct effect on the speech signal, yet both hinder intelligibility (Nabelek and Dagenais 1986; Helfer and Wilber 1990). Noise is caused by the addition of external competing sound(s) to target speech and acts as a simultaneous masker, obscuring less intense portions of the speech signal and reducing its *signal-to-noise ratio* (SNR). In contrast, reverberation (reverb) is an interference caused by the internal room acoustics of an enclosed space (Kinsler et al. 2000). Formally defined, reverberation is the persistence of acoustic energy in the sound field after it is produced. Reflected sound waves are exaggerated in reverberant settings (e.g., a concrete stairwell), resulting in a slow decay of energy and a temporal overlap of incident and reflected wave fronts. The overlap between direct and indirect sounds results in a "smearing" of the signal's spectrum.

There are qualitative differences in the way in which noise and reverberation obstruct signals of interest. Reverberation is based on reflection and absorption characteristics of materials within an enclosed space (Sabine 1962). Hence, the effectiveness of reverberation to occlude a signal is largely determined by the acoustic properties of the room itself. However, in the case of additive noise, signal occlusion is mainly determined by the similarity between the masker and the signal spectra.

The systematic effects of noise and reverberation on speech perception can be studied by parametrically changing the amount of interference added to a "clean" speech signal (containing no interference). Conveniently, the degree of noise and reverberation superimposed onto a target signal can be quantified by similar metrics. For additive noise, the relative contribution of "noise" and "signal" are quantified via the SNR. Specified in decibels (dB), positive SNRs reflect more favorable noise conditions (i.e., signal > noise), whereas negative SNRs reflect listening conditions where the noise dominates and masks the signal (i.e., noise > signal). Similarly, the proportion of acoustic energy attributable to signal and reverberant energy can be characterized (in dB) by a metric called the *direct-to-reverberant energy ratio* (D/R) (von Békésy 1938; Zahorik 2002). Other metrics can be used to characterize reverberation including *reverberation time* (RT_{60}), a measure describing how long it takes for reverberant sound energy to decay before attenuating by 60 dB. However, D/R is most comparable to SNR and allows the most direct comparison between the two forms of interference. Sometimes D/R is referred to as "wet-to-dry" ratio. Behavioral studies in human listeners have shown that the *just noticeable difference* (JND) for D/R sensitivity is on the order of 5–6 dB (Zahorik 2002; Larsen et al. 2008), slightly higher than the 3 dB JND for noise SNR (McShefferty et al. 2015).

In a reverberant space, D/R decreases with increasing source-to-receiver distance as direct ("dry") energy becomes swamped by sound energy from indirect ("wet") specular (mirror-like) reflections. In signal processing, a system is fully described by its impulse response (i.e., response to a broadband transient). Similarly, the reverberant characteristics of a particular room are described by its acoustic impulse response, which can be recorded with a microphone at different source–receiver distances to the presentation of the impulsive sound (e.g., a balloon pop). By convolving a room's impulse response with a signal (e.g., speech), the resulting output is heard as if the target sound were recorded in the reverberant space. By employing impulse responses measured at different source–receiver distances, D/R can be parametrically manipulated akin to varying the SNR for noise.

Figure 8.1 shows the effects of additive noise and reverberation to the vowel token /i/ (Bidelman and Krishnan 2010) at comparable signal-to-interference levels (i.e., SNR $\approx$ D/R of ± 5 dB). While the relative intensity of the signal to noise/reverberation is identical in this example, it is clear that the two forms of interference have different acoustic effects on the speech signal. With increasing reverberation, the dynamic (i.e., time-varying) change in voice fundamental frequency (F_0) and its harmonics (integer-related frequencies) show a smearing effect; portions of the signal persist, distorting the sequencing of spectral cues in speech as it unfolds in time. This overlap results in a spectrotemporal smearing that distorts ongoing and subsequent speech information. In reverberant settings, target cues essentially act as their own forward maskers (Nabelek et al. 1989; Wang and Brown 2006); yet, the strong harmonic structure of speech and its F_0 contour (cues that convey voice pitch) are largely preserved. Static (i.e., steady-state) signal features would be even less affected by reverberation. Contrastively, increasing levels of broadband noise are seen to "fill in" the peaks and troughs of the spectrum, reducing the spectral contrast of important speech cues necessary for proper identification (e.g., formants). Thus, despite having comparable relative intensity

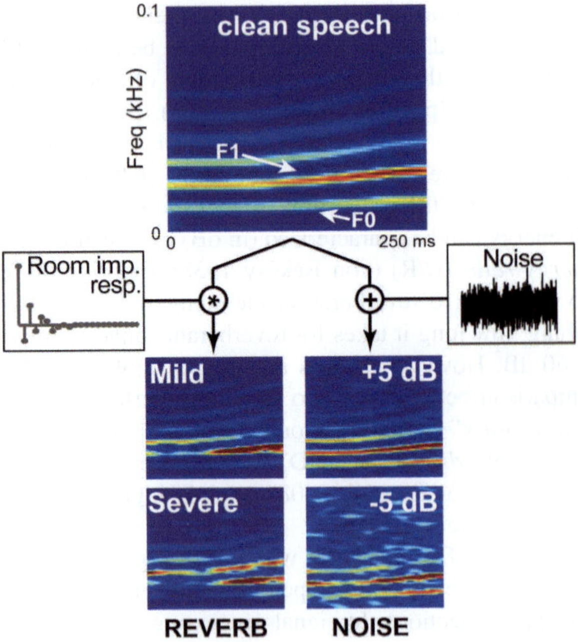

Fig. 8.1 Comparison of noise and reverb-related changes in speech acoustics. Spectrogram of the clean speech vowel token /i/ (*top*) containing a time-varying F_0 (~100 Hz) and fixed formant frequencies of F1 = 300; F2 = 2500, F3 = 3500 and F4 = 4530 Hz (Bidelman and Krishnan 2010). The parametric effects of reverberation on speech processing can be studied by convolving the clean speech signal with various room impulse responses recorded in reverberant spaces at different source-receiver distances (e.g., Watkins 2005). Larger source-receiver distances decrease the direct-to-reverberant energy ratio (D/R), a metric akin to signal-to-noise ratio (SNR). Similarly, the parametric effect of noise is studied by varying the SNR of additive noise superimposed on clean speech. Acoustically, reverberation has the effect of "smearing" the speech spectrum, distorting the spectrotemporal timing of acoustic landmarks. In contrast, noise "fills in" the signal's spectrum with decreasing SNR, reducing spectral contrast between signal and noise. *FO*, F_0; *imp resp*, impulse response

between the signal and interference, it is clear that noise and reverberation act very differently and even differentially across the various cues important to speech perception (e.g., voice pitch versus timbre; see Sect. 8.3.2).

Lastly, it should be noted that unlike noise, which is nearly always considered to be a negative interference, reverberation can sometimes provide positive benefits to auditory perception. In fact, reverberation is tolerable (and often desirable) in concert music halls (Lifshitz 1925; Backus 1977), where pitch dominates the signal, but not in classrooms (Yang and Bradley 2009), where target acoustics are geared toward speech intelligibility (see Sect. 8.3.2). Beyond aesthetic considerations, the D/R of a reverberant signal can also facilitate the perceptual judgments of source distance when interaural level and intensity cues are ambiguous (Zahorik 2002; Larsen et al. 2008).

8.2.2 *Behavioral Basis and Individual Variability in Degraded Speech Perception*

8.2.2.1 Speech Perception in Noise

The effect of noise on speech perception has enjoyed a long history in the hearing sciences. Early work on speech audiometry recognized two components of hearing impairment: (1) loss of acuity (*audibility*) and (2) loss of clarity (*distortion*) (Carhart 1951; Plomp 1978). Audibility (i.e., signal attenuation) is a linear process and easily predicted from the pure-tone audiogram or articulation index (French and Steinberg 1947; ANSI 1969). In contrast, the distortion is a non-linear component of hearing loss and, more problematically, is poorly predicted from the pure-tone audiogram or word recognition scores (in quiet). Degraded speech perception tests therefore became routine in the late 1960s as a means to quantify the distortion component of hearing (Carhart and Tillman 1970) and to address a common complaint of hearing impaired listeners: poor speech recognition despite restored audibility through hearing aids (for review, see Wilson and McArdle 2005).

SIN perception is now measured using a number of standardized audiological tests, for example: Hearing-in-Noise test (HINT), Nilsson et al. (1994; QuickSIN^{TM}, Killion et al. (2004). SIN tests vary in the semantic and contextual cues they offer the listener (e.g., sentences versus words; high versus low predictability). Nonetheless, at their core the basic premise of these tests is similar: listeners are presented with speech stimuli and are asked to detect certain key words. The SNR is varied at fixed intervals or adaptively to obtain the individual's speech reception threshold in noise.

While the acoustic effects of additive noise on speech perception are somewhat predictable, the behavioral consequences are all but trivial. Speech intelligibility in adverse conditions is influenced by a number of factors including the spectrotemporal characteristics of the noise, its semantic content, whether or not the noise is stationary, modulated or continuous, presented monaurally or binaurally, and the spatial proximity of the noise to the target signal (reviewed by Assmann and Summerfield 2004). Irrespective of lexical-semantic or contextual cues, noise reduces SNR, obscuring the perceptually salient cues of speech by masking contrastive portions of the signal's spectrum. However, in addition to spectral masking effects, noise can have detrimental effects on temporal aspects of the speech signal. For instance, one prominent finding of perceptual studies is that listeners exploit the temporal envelope of speech (i.e., slow amplitude fluctuations) for robust comprehension. This is true even when the speech signal's "fine-structure" (carrier) is noise containing no spectral cues or, in cases of cochlear implant signal processing, where only envelope cues are delivered to the stimulating electrodes (Shannon et al. 1995; Swaminathan and Heinz 2012). This has led to the prevailing view that the reduction in speech intelligibility in noise results in degradations to the speech envelope. In noisy conditions, speech fine structure is often exploited to help aid spoken word recognition (e.g., Lorenzi et al. 2006; Swaminathan and Heinz 2012).

8.2.2.2 Speech Perception in Reverberation

The deleterious effects of reverberation on speech intelligibility can be ascribed to consequences of both "overlap-masking" (i.e., forward) and "self-masking" (Nabelek et al. 1989). As segments of the speech signal reflect in a reverberant space they act as forward maskers, overlapping subsequent syllables and inhibiting their discrimination. In addition, reflections concurrent with the incident (i.e., direct) sound dramatically change the dynamics of speech by blurring the waveform's fine-structure. When acting on a time-varying signal, this "temporal smearing" tends to transfer spectral features of the signal from one time epoch into later ones, inducing a smearing effect in the spectrogram (Wang and Brown 2006). As a consequence, this internal temporal smearing distorts the energy within each phoneme such that a signal can effectively act as its own masker (i.e., self-masking). With such distortions, normal hearing listeners have difficulty identifying and discriminating consonantal features (Gelfand and Silman 1979; Nabelek et al. 1989), vowels (Nabelek and Letowski 1988; Drgas and Blaszak 2009), and time-varying formant cues (Nabelek and Dagenais 1986) in reverberant listening conditions. It should be noted that speech confusions in reverberation are further exacerbated with hearing impairment (Nabelek and Letowski 1985; Nabelek and Dagenais 1986; Nabelek 1988).

Despite its difference from noise SNR (see Sect. 8.2.1), decreased D/R for reverberation has a similar negative effect on speech intelligibility. Figure 8.2A shows closed set vowel identification performance in noise (0 dB SNR) and reverberation (RT_{60} = 1.2 s) for listeners with binaural sensorineural hearing loss (Nabelek and Dagenais 1986). Vowels are typically highly identifiable in noise and reverberation for normal-hearing listeners. As seen in the figure, both noise and reverberation interference reduce speech identification in hearing impaired listeners by a similar magnitude (~15–20%). However, the pattern of specific vowel confusion errors is typically not the same in noise as it is in reverberation. Nabelek and Dagenais (1986) suggest that in noise, misidentifications are related to the spectral proximity of formant frequencies for confused pairs. In contrast, confusions in reverberation are probably attributable to changes in the relative weighting of the formants as they are smeared in time and frequency by the prolonged reverberant energy.

8.2.2.3 Comparisons Between Noise and Reverberation on Speech Perception

While behavioral studies have mainly focused on the independent consequences of noise and reverberation on speech perception, it should be noted that real-world listening environments typically contain both multiple noise sources and reverberant acoustics. Thus, in most cases, noise and reverberation occur simultaneously

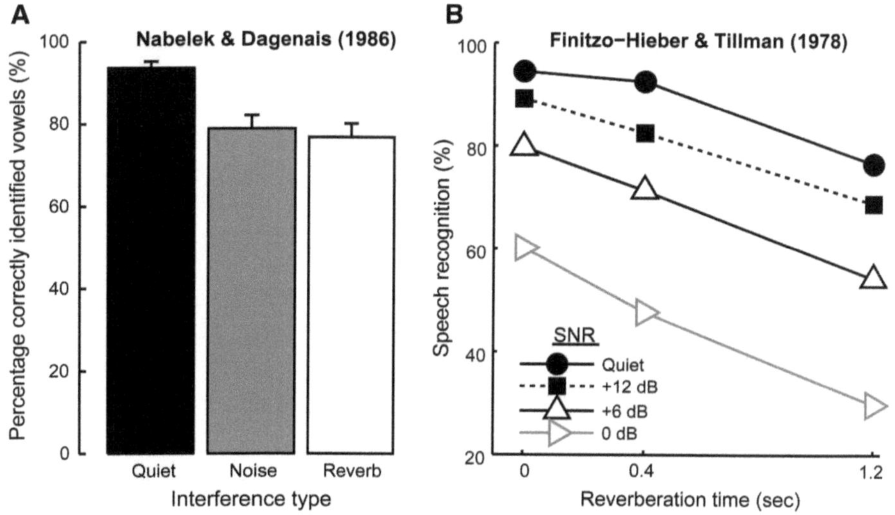

Fig. 8.2 Perceptual speech comprehension in noise and reverberation. (**A**) Closed-set vowel identification performance in noise (0 dB SNR) and comparable degrees of reverberation (RT = 1.2 s) reported by Nabelek and Dagenais (1986). In the absence of contextual speech cues, both noise and reverberation reduce speech intelligibility. (**B**) Interaction of noise and reverberation on speech recognition (Finitzo-Hieber and Tillman 1978). While both noise and reverberation impair speech understanding independently, their combined effect yields poorer speech intelligibility than either factor alone. *SNR*, signal-to-noise ratio

and can act synergistically to further impair understanding (George et al. 2008). Figure 8.2B illustrates the combined effects of noise and reverberation on speech recognition scores reported by Finitzo-Hieber and Tillman (1978) for normal hearing children. Both the SNR of additive noise and reverberation time (*RT*) were manipulated in the experiment. Although RT was the independent variable for reverberation in this experiment, higher RTs correspond with decreased D/R, and hence, less favorable reverberation levels. Both main effects of noise and reverberation were observed on speech recognition scores when each type of interference was considered alone. However, Finitzo-Hieber and Tillman (1978) also noted an interaction between reverberation and noise. That is, the influence of noise or reverberation in isolation was further increased by the introduction of the other interference.

8.2.2.4 Individual Differences and Normal Variability in SIN

Current hearing aids provide little benefit for SIN understanding despite restoring audiometric thresholds (Chmiel and Jerger 1996). Consequently, it is now well accepted that SIN perception cannot be reliably predicted from the audiogram (Killion and Niquette 2000). This might not be entirely surprising in light of the

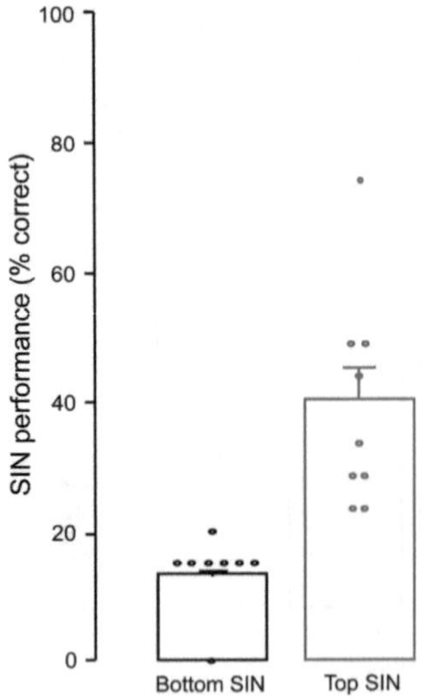

Fig. 8.3 Individual differences in speech-in-noise (SIN) perception. Grand average (*bars*) and individual participant responses (*points*) on the QuickSIN^{TM} for audiometrically normal-hearing young adults (Song et al. 2011). Despite having normal hearing thresholds, there is considerable variability in SIN performance, which can be categorized into top (≥25%, $n = 9$) and bottom (<25%, $n = 8$) SIN perceiving groups. In the (presumed) absence of differences in peripheral hearing function, individual differences in degraded speech perception might be related to differences in the neural encoding of speech (also see Fig. 8.8). (Data from Song et al. 2011)

"distortion" that often accompanies reduced audibility in cases of hearing impairment (see Sect. 8.2.2.1). Furthermore, SIN perception is problematic and performance varies considerably among individuals without substantial hearing impairment (Middelweerd et al. 1990; Frisina and Frisina 1997). Even normal-hearing young adults show individual variability in SIN in the absence of any known audiological or peripheral hearing deficit (Song et al. 2011, 2012). In particular, Song et al. (2011) measured behavioral performance on the QuickSIN^{TM} in normal-hearing young adults with normal pure tone hearing thresholds (≤20 dB HL from 125 to 8000 Hz) (Fig. 8.3). Despite normal hearing, no listener performed the task at ceiling and in fact, there was considerable variability between listeners (i.e., individual differences), ranging from 0 to 75% speech recognition.

These findings challenge conventional and longstanding views that speech intelligibility is determined solely by audibility, i.e., peripheral hearing status (Plomp 1986; Humes and Christopherson 1991). Rather, hearing sensitivity alone seems to be inadequate to account for SIN perception issues (Humes and Christopherson 1991; Parbery-Clark et al. 2011). Consequently, a growing body of evidence suggests that central auditory processing—as early as the brainstem—plays a critical role in mediating robust perceptual SIN abilities.

8.3 Effects of Acoustic Interference on FFR Representations of Speech

The brainstem FFR has provided critical insight toward understanding the neurobiological encoding of clean and degraded speech from a subcortical perspective (Parbery-Clark et al. 2009; Bidelman and Krishnan 2010; Song et al. 2011). Although there are multiple sources of FFRs throughout the hearing pathway (e.g., cochlear microphonic, Sohmer and Pratt 1977; auditory nerve, Bidelman 2015b; and brainstem, Sohmer et al. 1977; Bidelman 2015b), the inferior colliculus of the midbrain is considered the primary generator of the scalp-recorded FFR (Sohmer et al. 1977; Bidelman 2015b). The brainstem FFR is also distinct in its response characteristics from the more conventional click-evoked auditory brainstem response (ABR) familiar to audiologists, differing in rate susceptibility (Krizman et al. 2010), frequency specificity (Picton et al. 1977), spectral content (Bidelman 2015b), susceptibility to noise masking (Cunningham et al. 2002; Russo et al. 2004), and latency-intensity changes (Akhoun et al. 2008). These response properties make the FFR a unique window into auditory neurophysiological function that is distinct from the traditional brainstem ABR.

Also different from the transient ABR, FFRs code dynamic, spectrotemporal features of periodic acoustic stimuli. This unique feature makes FFRs a quasi "neural fingerprint" of the acoustic signal within the human EEG. Indeed, the remarkable fidelity of FFRs is evident in listening experiments in which the neural responses are replayed to human listeners as audio signals and can be reliably identified as intelligible speech (Galbraith et al. 1995; Weiss and Bidelman 2015). Moreover, although debated, brainstem responses are largely unaffected by attention (Woods and Hillyard 1978; Galbraith et al. 2003; also see Shinn-Cunningham, Varghese, Wang, and Bharadwaj, Chap. 7). Thus, unlike their cortical event-related potential (ERP) counterparts that are highly malleable to subject state, habituation, and overlap with endogenous neural activity, FFRs provide a stable window into the neural transcription of speech signals that can be obtained under passive listening paradigms. Consequently, FFRs have provided important insight into individual differences in SIN listening skills and the neural encoding of speech in normal and clinical populations who cannot participate in traditional, behavioral (i.e., subjective) auditory assessments (e.g., Cunningham et al. 2001).

8.3.1 Noise-Related Changes in Brainstem Speech Processing

Noise-related changes in the FFR elicited by complex sounds are evident in both the time and frequency domain. Figure 8.4A shows brainstem FFRs recorded in response to a complex tone containing the 12th–17th harmonics of a 90 Hz F_0 (unresolved harmonics) (Smalt et al. 2012). In the Smalt et al. (2012) study,

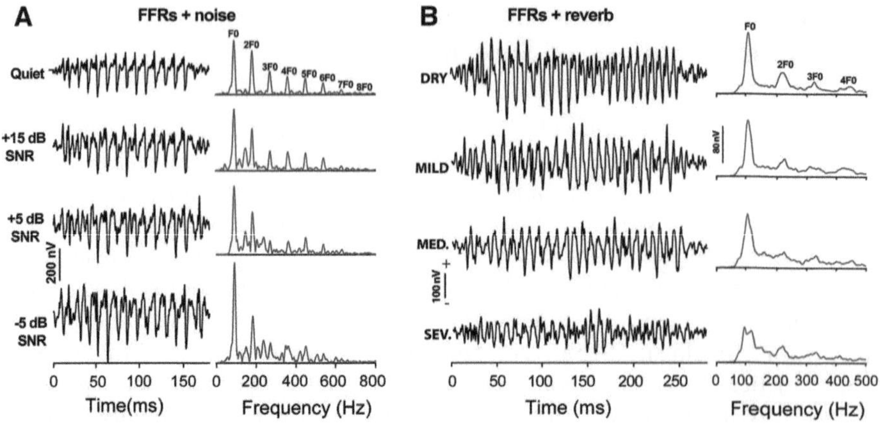

Fig. 8.4 Brainstem FFRs as a function of signal-to-noise ratio (SNR) (Smalt et al. 2012) and reverberation level (Bidelman and Krishnan 2010). (**A**) With increasing noise, phase-locking to the stimulus diminished. Most notable is the reduction in periodicity at higher harmonics of the F_0 frequency (i.e., $2F_0$–$8F_0$), components useful for the calculation of the common pitch. In contrast, the F_0 component remains largely intact or is enhanced in noise. (**B**) Similarly, speech FFRs evoked by the token /i/ (see Fig. 8.1) show a reduction in overall response energy with notable declines in neural synchronization (reduced periodicity) with increasing reverberation, particularly at higher frequencies. Neural coding at F_0 is largely maintained until the most severe levels of reverberation. *FO*, F_0; *MED*, medium; *SEV*, severe. (Reprinted from Smalt et al. 2012; Bidelman and Krishnan 2010, with permissions from Elsevier)

lowpass filtered noise was used to mask audible distortion products. The first observation apparent from these FFRs is the prominent neural energy at the F_0 and its lower-order harmonics (F_0, $3F_0$, etc.), despite the fact that these components did not occur in the stimulus. The presence of response energy at the F_0 indicates that the FFR phase locks at the common periodicity of the stimulus, providing a neural correlate of the missing fundamental (Greenberg et al. 1987). Secondly, with the addition of noise, it is apparent that the neural encoding of the sustained F_0 periodicity (i.e., stimulus envelope) is well-maintained at decreasing (poorer) SNRs; little degradation in FFR F_0 is observed even at higher noise levels. In contrast, broadband white noise often delays, attenuates, or even eradicates the onset components of the transient ABR (Burkard and Hecox 1983; Russo et al. 2004). The resilience of the brainstem FFR at F_0 (but not its higher harmonics or onset) in the presence of noise has been noted by a number of investigators (e.g., Russo et al. 2004; Li and Jeng 2011; Prevost et al. 2013) and suggests that neural synchronization at the fundamental F_0 periodicity is relatively robust to acoustic interference. In speech perception, F_0 provides a correlate of voice pitch and a robust cue for stream segregation and identification of the number of sources in complex auditory scenes (Assmann and Summerfield 1990). Thus, the low susceptibility of the F_0 steady-state portion of the FFR is consistent with the notion that pitch remains a robust cue for segregating target speech from a sound mixture (Assmann and Summerfield 1990). In contrast to the F_0 of speech, higher spectral components captured by the FFR (e.g., formant-related harmonics) are systematically degraded with noise, paralleling their rapid deterioration behaviorally (Liu and Kewley-Port 2004).

Some studies have even reported a facilitation of F_0 with additive noise (Smalt et al. 2012; Prevost et al. 2013). Two possible mechanisms have been proposed to account for the enhancement of F_0 in noise. It has been suggested that the maintenance and/or strengthening of the response at the F_0 in noise occurs as the result of *stochastic resonance* in the auditory system (Henry 1999; Cunningham et al. 2002). As stated by Prevost et al. (2013), this phenomenon is described as "improved detection and physiological representation of a weak periodic signal by the addition of noise." In other words, degraded F_0 representations can actually be counteracted (enhanced) by further neural entrainment to the signal in the presence of a noise masker. The specific neuronal mechanisms responsible for this phenomenon are unknown. However, stochastic resonance has been reported in the envelope following responses of auditory nerve fibers in the Mongolian gerbil (*Meriones unguiculatis*) (Henry 1999), so it is conceivable that resonance occurs as a result of cochlear nonlinearities (Jaramillo and Wiesenfeld 1998). Alternatively, the higher stimulus presentation levels used in typical brainstem experiments (>70–80 dB SPL) mean that FFRs reported in most studies reflect contributions from a wide range of the cochlear partition due to upward spread of excitation with increasing level (Dau 2003). Thus, the robustness of F_0 within the FFR amid noise may not reflect stochastic resonance per se, but instead reflects the additional engagement of low-frequency "tails" of basal, high-frequency neurons as they begin to phase lock to the common, high intensity F_0 across cochlear channels (Kiang and Moxon 1973). Multiple points of phase locking at F_0 across the cochlear array would tend to reinforce one another and, consequently, offer some resilience or redundancy in a pitch cue in the presence of noise interference.

8.3.2 Reverberation-Related Changes in Brainstem Speech Processing

To date, only a single study has investigated changes in the FFR under reverberation. Bidelman and Krishnan (2010) measured speech FFRs in response to the speech token /i/ (250 ms; time-varying F_0) presented in dry, mild, medium, and severe levels of reverberation (see Fig. 8.1). FFRs showed a systematic degradation in neural periodicity with increasing levels of reverberation (Fig. 8.4B). Thus, as with additive noise, reverberation degrades the normal phase-locking capacity of the FFR and reduces the response's ability to "tag" acoustic features of the speech signal.

Spectral analyses are typically more informative than time-domain analysis for sustained evoked potentials and can reveal specific stimulus-related changes in the FFR not apparent in raw waveforms. In the time domain, response autocorrelograms (ACGs) can be computed to index variation in neural periodicities over the

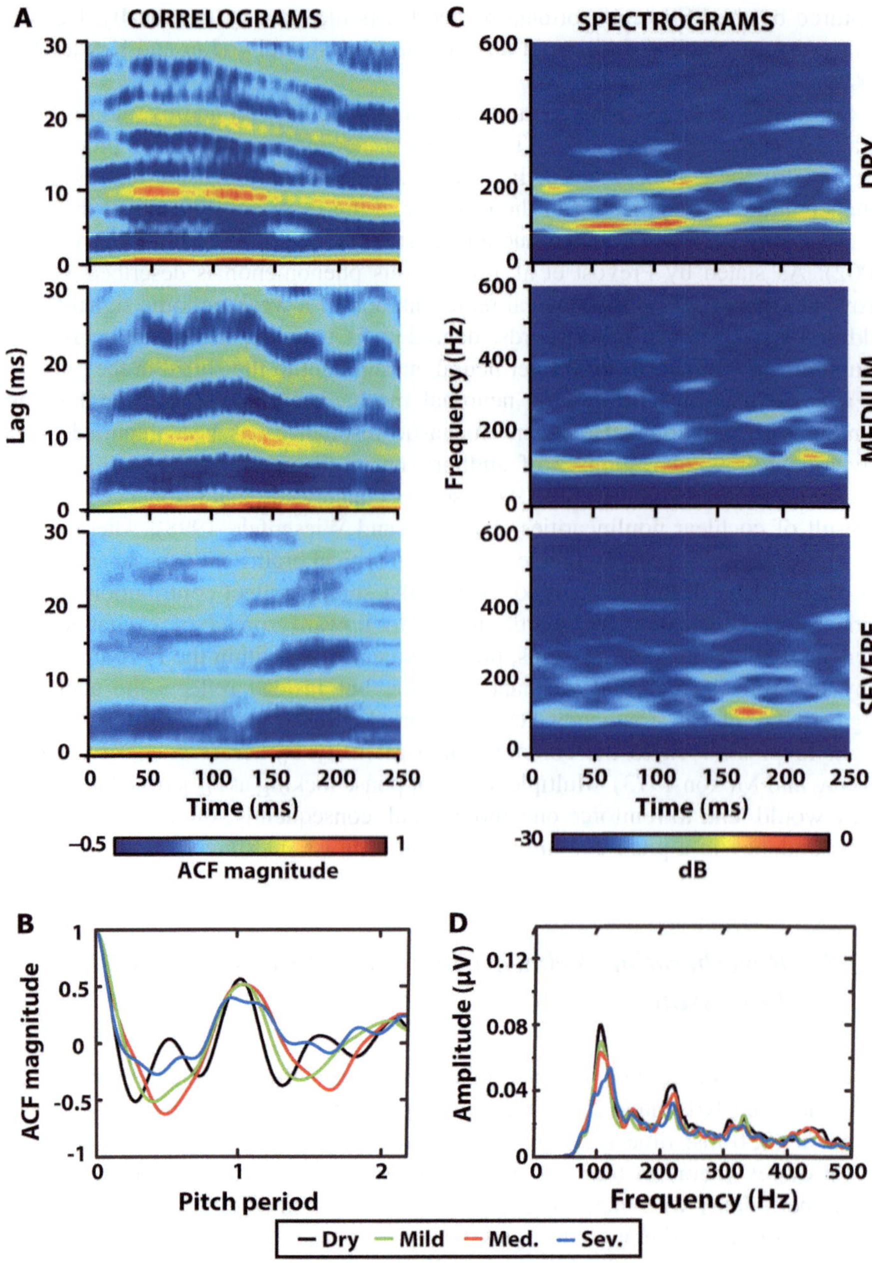

duration of the response (Fig. 8.5A). ACGs represent the short term (i.e., running) autocorrelation function of windowed frames of a compound signal:

◀ **Fig. 8.5** Temporal and spectral changes in speech-FFR with increasing reverberation. (**A**) Autocorrelograms, (**B**) time-averaged autocorrelation functions, (**C**) spectrograms, and (**D**) time-averaged FFTs derived from FFR waveforms in response to the vowel /i/ in various amounts of reverberation. As indexed by the invariance of the autocorrelation function (*ACF*) magnitude at the fundamental period (i.e., F_0 pitch period = 1), increasing levels of reverberation have little effect on the neural encoding of pitch-relevant information (**A–B**). The effect of reverberation on FFR encoding of the formant-related harmonics is much more pronounced (**C–D**). As with temporal measures, the representation of F_0 (100–130 Hz) remains more intact across conditions than higher harmonics (>200 Hz), which are smeared and intermittently lost in more severe amounts of reverberation (e.g., compare strength of F_0 to strength of harmonics across conditions). *Med.*, medium; *Sev.*, severe. (Reprinted from Bidelman and Krishnan 2010, with permission from Elsevier)

$$ACG(\tau, t) = X(t) \times X(t - \tau)$$

for each time t and time-lag τ. It is a three-dimensional plot quantifying the variations in periodicity and "neural pitch strength" (i.e., degree of phase locking) as a function of time. The horizontal axis represents the time at which single autocorrelation function (ACF) "slices" are computed while the vertical axis represents their corresponding time-lags (i.e., periods). The intensity of each point in the image represents the instantaneous ACF magnitude computed at a given time within the FFR response. Mathematically, the running autocorrelogram is the time-domain analog to the frequency-domain spectrogram (Fig. 8.5C). In terms of neurophysiology, it represents the running distribution of all-order interspike intervals present in the population neural activity (Cariani and Delgutte 1996; Sayles and Winter 2008), which may underlie the farfield FFR. From the time-varying ACG, global neural periodicity strength can be obtained by pooling the running ACG. This results in a summary ACF (Fig. 8.5B), which is a time-domain analog to a spectrum in the frequency domain (Fig. 8.5D). The magnitude of the ACF at the lag of the fundamental pitch period (i.e., $\tau = 1/T_0$) has been used as a unitary measure of FFR pitch strength (Krishnan et al. 2005; Bidelman and Krishnan 2010).

Temporal and spectral analyses of FFR recordings to reverberant speech (Bidelman and Krishnan 2010) are shown in Fig. 8.5. Apparent from the FFR's ACF is the strong phase-locked response at the fundamental pitch period ($T_0 \approx 10$ ms; $F_0 = 100$ Hz) and its subharmonic periodicities both with and without reverberation. Time-averaged ACFs more clearly show that the FFR representation of F_0 remains largely intact with the addition of reverberant energy. That is, the magnitude of summary ACF energy at fundamental period (i.e., voice pitch) is invariant to increasing reverberation (Fig. 8.5B). Only in the most severe reverberation tested (D/R = −12 dB; $RT_{60} = 900$ ms) was there a noticeable change in response magnitude at F_0. In other words, pitch cues remain largely intact in the presence of reverberation consistent with single unit responses in the cochlear nucleus (Sayles and Winter 2008). In contrast, spectral analysis of the FFR (Fig. 8.5C–D) reveals a smearing of the response spectrum with increasing reverberation, particularly in higher harmonics of the signal's spectrum. The weaker, more diffuse encoding of higher spectral components in the brainstem FFR is particularly evident near harmonics proximal to the first formant ($\sim$300 Hz).

These examples demonstrate that the brainstem's ability to encode speech cues is not an all or nothing phenomenon. Rather, pitch (F_0) and timbre (F1) cues that are differentially affected by reverberation at the acoustic level are similarly differentially encoded in subcortical FFRs. As observed for noise-degraded speech (Cunningham et al. 2002; Russo et al. 2004), FFRs largely preserve F_0 cues with increasing reverberation, whereas formant cues are rapidly degraded. Interestingly, these neural effects appear to parallel listeners' behavioral responses. Perceptual discrimination thresholds (difference limens) for speech F_0 are largely invariant to increasing reverberation, whereas F1 discrimination thresholds worsen in even minimal amounts of reverberation (Bidelman and Krishnan 2010). A larger impact on lower compared to higher spectral components of speech may at least partly be due to the effects of harmonic resolvability. Lower-order harmonics that are said to be "resolved" by the peripheral auditory filters dominate the FFR and are also more resilient to noise than higher, "unresolved" harmonics (Laroche et al. 2013). Conceivably, the differential perception of voice pitch and timbre cues in the presence of reverberation (and resilience of the former) might be driven by the differential encoding of pitch and timbre cues and/or resolved versus unresolved harmonics at subcortical, sensory levels of auditory processing.

The dissociable effects of reverberation (and noise) on the neural encoding of F_0 versus higher speech harmonics suggest that acoustic interferences have a differential effect on the 'source-filter' components of speech (Fant 1960). The *source-filter theory* of speech postulates that speech acoustics result from the glottal source being filtered by the vocal tract's articulators. The fundamental frequency at which the vocal folds vibrate determines the pitch of a talker independently from the configuration of the vocal tract and oral cavity, which determine formant structure (i.e., voice quality) (Fant 1960). Together, voice pitch and timbre cues provide adequate information for identifying who is speaking (e.g., male versus female talker) and what is being said (e.g., distinguishing vowel sounds) (Assmann and Summerfield 1989, 1990). Cast in terms of the source-filter model, it appears that source-related response components (i.e., F_0) coded in the FFR are relatively immune to additive background interferences, whereas filter-related components (i.e., formant structure/upper harmonics) are easily degraded (Russo et al. 2004; Bidelman and Krishnan 2010). This differential effect may be one reason, for instance, why reverberation is desirable in concert music halls (Lifshitz 1925; Backus 1977) but not in classrooms designed for speech intelligibility (Yang and Bradley 2009).

8.3.3 Direct Comparisons Between Noise-Related and Reverberation-Related Changes in FFR

To date, there has not been a direct comparison in a single study between speech-evoked FFRs recorded in noise and reverberation. Nevertheless, one can compare across studies to appreciate potential differences in how these two forms of interference have both similar and unique effects on the subcortical encoding of speech. A direct comparison between noise-related and reverb-related changes in FFR would also be useful in light of the differential acoustic and perceptual effects that these two forms of interference have on the speech signal (see Sect. 8.2).

Figure 8.6 illustrates speech-evoked FFRs recorded in clean, reverberant, and noise-degraded listening conditions. The noise and reverberation levels reflect an SNR and D/R of roughly +5dB and thus represent a comparable degree of masking between the two classes of interference. Despite a similar relative level between target speech and the masker, it is clear that noise and reverberation produce unique changes in the FFR. As observed acoustically, noise largely "fills in" the spectral peaks and troughs of the speech signal, masking the spectral contrast between perceptually salient cues (e.g., formant peaks) and the noise floor. In contrast, comparable levels of reverberation blur the speech spectrogram, although harmonics are still clearly visible in the response. These differences are more apparent in time-averaged spectra, which show (at least qualitatively) clearer neural representation of the speech spectrum in reverberation compared to speech presented in noise.

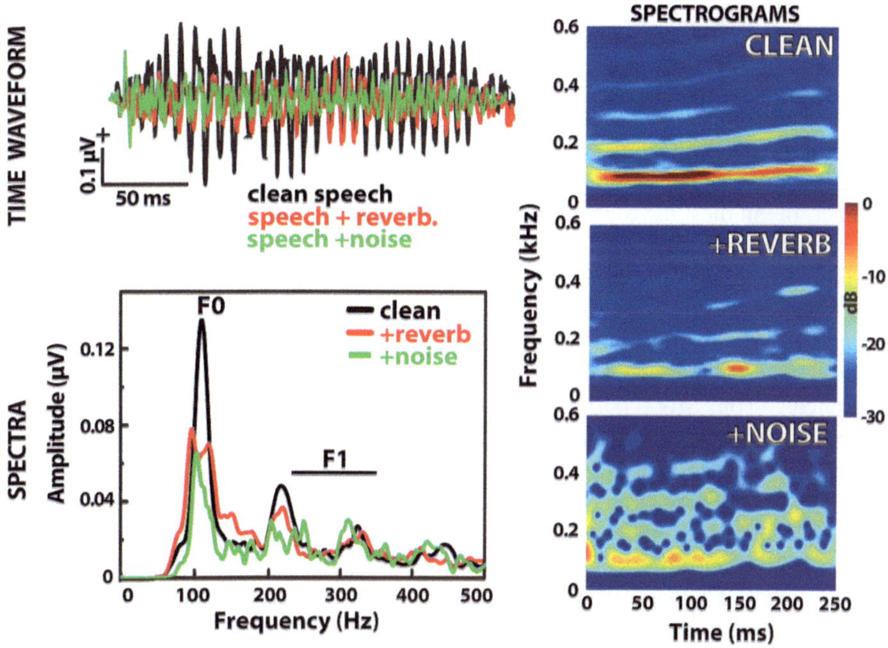

Fig. 8.6 Direct comparison between noise-related and reverb-related changes in speech-FFRs. Response waveforms (*upper left*) elicited by clean, noisy, and reverberant speech (both SNRs ≈ +5 dB). Response spectra (*bottom left*) show reduced encoding of voice pitch (F_0 ≈ 100 Hz) and higher spectral information in noise/reverb. FFR spectrograms illustrate that voice "pitch" (F_0) and "timbre" (F1) are hindered more by noise than reverberation. [Based on data from Bidelman and Krishnan (2010) and unpublished data]

Though a more comprehensive characterization is needed, these findings point to a qualitative difference (and even uniqueness) in how noise and reverberation interference affect the subcortical transcription of speech. A differential neural coding in noise compared to reverberation may help account for the challenges observed by hearing impaired listeners in certain acoustic environments (but not others) as well as the unique types of perceptual confusions listeners experience in noise compared to reverberation (Nabelek and Dagenais 1986). It is possible that signal processing performed by hearing aids and other listener devices may need to be specifically tailored to restore the neurophysiological representation for speech, dependent on the type of interference in a given listening situation (i.e., noisy versus reverberant setting). Future studies are needed (particularly examining reverberation) to better characterize the potentially nuanced effects of different acoustic interferences on speech coding. Given the limited number of FFR studies examining speech in reverberation (Bidelman and Krishnan 2010), the remainder of this chapter will focus mainly on the psychophysiological processing of speech amidst additive noise.

8.3.4 Brain-Behavior Connections Underlying Perceptual SIN

To satisfy a meaningful neural correlate of SIN perceptual skills, changes in the subcortical encoding of speech, as reflected in the FFR, should parallel listener's perceptual performance in noise-degraded listening tasks. Indeed, a growing number of studies have investigated noise-related changes in the speech FFR concurrent with individuals' behavioral SIN measures and have reported robust predictive relationships. Figure 8.7 illustrates brain-behavior correlations between FFR F_0 amplitude (a proxy measure for pitch encoding) and SIN performance on the *QuickSIN*^{TM} test in young, audiometrically normal listeners (Song et al. 2011). Song et al. (2011) showed that FFR F_0 magnitudes positively predicted SIN performance: "Top SIN" performers (see Fig. 8.3) had more robust subcortical responses than "Bottom SIN" performers, who had both weaker neural representation of the speech F_0 and poorer perceptual scores. Complementary findings were reported by Anderson et al. (2010) who showed that poorer (i.e., lower median) SIN listeners experienced greater ($\sim$0.5–1 ms) noise-related shifts in the timing of their speech FFR from quiet to noise than top performing listeners. Collectively, these results suggest a strong relationship between both the magnitude and timing of the brainstem FFR and perceptual SIN skills, whereby faster and more robust subcortical speech encoding is associated with better behavioral outcomes. Bidelman and Krishnan (2010) similarly showed a relationship between FFR encoding of speech in reverberation and behavioral discrimination for pitch and timbre cues whereby more robust neural responses predicted higher perceptual sensitivity.

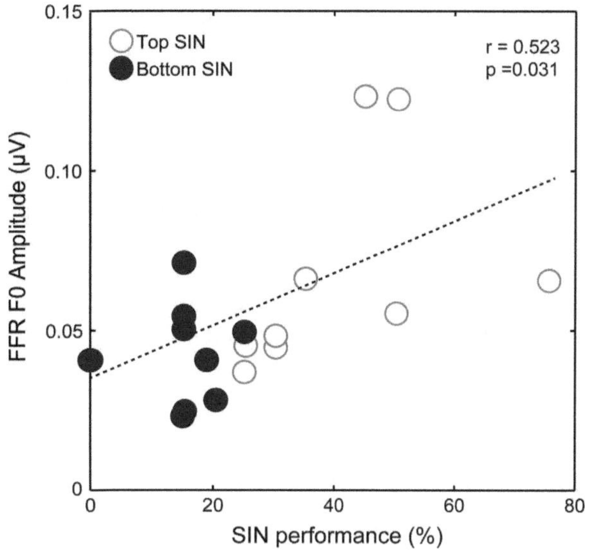

Fig. 8.7 Correlations between speech-in-noise (SIN) listening skills and FFR response properties. FFR F_0 amplitude reflects the brainstem response amplitude during the 40-ms formant transition of the 170-ms syllable/da/, presented in six-talker babble. SIN performance on the QuickSIN^{TM} test (see Fig. 8.3) is associated with larger, more robust subcortical responses to the speech F_0. Top performers on SIN perception behaviorally also show larger brainstem responses to noise-degraded speech. This suggests that individual differences in SIN processing are at least partially accounted for based on how well the auditory nervous system transcribes speech at pre-attentive levels of brain processing. (Data based on Song et al. 2011)

These findings provide evidence for at least a relationship between subcortical auditory processing and SIN perception. However, most studies to date examining the FFR and SIN perception have been correlational in nature. Consequently, it is currently unclear if improved neural encoding of speech (as reflected in the brainstem FFR) causes improved listening skills in noise (but see Sects. 8.4.3, 8.4.4). Moreover, investigators typically manipulate the amount of acoustic information in the stimulus (e.g., SNR) and observe parallel changes in neural responses. In such experimental designs, modulations in the evoked response and human behavior both covary with the acoustic properties of the signal. This confounding of variables further obscures if changes in FFR reflect a true neural correlate of the auditory percept or merely reflect properties of the stimulus itself. This is an important distinction as recent studies employing stimuli that dissociate acoustics from the actual auditory percept suggest that the FFR may not reflect a true neural correlate of the auditory percept but rather reflects more exogenous stimulus properties (Gockel et al. 2011; Bidelman et al. 2013). Regardless of whether FFRs reflect a perceptual correlate of speech phenomena, it is clear that the FFR can be a useful tool in examining individual differences in stimulus coding and auditory function prior to the cerebral cortex.

Although beyond the scope of this chapter, it should also be noted that there are other factors involved in SIN behaviors, including cognitive (post-perceptual) mechanisms. In addition to the encoding (pre-perceptual) factors reviewed here, SIN listening skills are also dependent on certain cognitive abilities, including working memory (Anderson et al. 2010) and selective auditory attention (Ruggles et al. 2011). Nevertheless, the clear connection between FFR response properties and behavioral SIN abilities in young, normal hearing adults implies that at least some of the individual variation in auditory scene analysis is partly determined by how well acoustic information is transcribed at early (and pre-attentive) levels of sensory processing—before the engagement of these cognitive mechanisms (for a review of cognitive and sensory factors, see Anderson et al. 2013a).

8.3.5 Brainstem Versus Cortical Encoding of Degraded Speech

Noise-induced changes in the magnitude and timing of the auditory cortical ERPs have been reported by comparing responses to clean speech sounds relative to noise-degraded speech sounds. The cortical encoding of auditory stimuli amidst noise reflects a complex interaction between the types of signal/noise, as well as the evoking stimulus paradigm (e.g., sequential versus oddball paradigm) (Billings et al. 2010). Component waves of the ERPs (e.g., P1-N1-P2) can be suppressed (i.e., delayed and reduced in amplitude) (Billings et al. 2009, 2010) or facilitated (i.e., enhanced in amplitude) (Alain et al. 2009; Bidelman and Dexter 2015; Bidelman and Howell 2016) depending on the type and effectiveness of a concurrent noise in masking the target signal (e.g., white noise, multi-talker babble). Moreover, like the correlation observed between the FFR and behavior, several studies have shown a relationship between the cortical N1 component latency/amplitude and better SIN perception (Parbery-Clark et al. 2011; Billings et al. 2013; Bidelman and Howell 2016), in the form of earlier/larger responses that are associated with improved behavior. It is conceivable that this type of degraded signal analysis in early auditory cortex is at least partially inherited or influenced by structures much lower in the auditory pathway, including the brainstem (Bidelman and Krishnan 2010; Song et al. 2011) and/or cochlea (Bidelman and Bhagat 2015). This raises the intriguing question of how subcortical FFRs relate to cortical ERPs in the processing of normal and noise-degraded speech.

A handful of studies have directly investigated the correspondence and interactions between brainstem and cortical-evoked activity for clean speech signals (Bidelman et al. 2013, 2014), but few have examined concurrent brainstem-cortical responses under noisy conditions (Parbery-Clark et al. 2011). Parbery-Clark et al. (2011) recorded FFRs and ERPs in normal-hearing listeners in response to the speech syllable /da/presented in quiet and +10 dB SNR of noise (multi-talker babble). Direct comparisons between brainstem and cortical measures showed strong correlations between FFR fidelity (measured via stimulus-to-response correlation) and the cortical N1 magnitude, both of which also related to behavioral SIN perception (i.e., HINT scores). Parbery-Clark et al. (2011) posited that background noise produces a "system-wide" degradation in the neural representation of speech observable across the auditory pathway. Moreover, the strong link between FFR and ERP activity suggested that the neural representations for speech at a cortical level are partially determined (i.e., inherited) from encoding at the level of the brainstem.

Yet, interpreting the relationship between brainstem and cortical speech processing is all but straightforward. While the FFR is generated primarily by deep, brainstem nuclei (Smith et al. 1975; Bidelman 2015b), the cortical ERPs are more diffuse, reflecting overlapping activity generated from multiple sources including the auditory cortices and contributions from the frontal lobes (Knight et al. 1999; Picton et al. 1999). This blurring of sources in the scalp potentials precludes firm interpretations between auditory ERPs and SIN perception, as behaviors cannot be definitively attributed to changes in a single auditory (or nonauditory) brain structure. Indeed, analysis of source-level (as opposed to sensory-level) neural responses reveals that both primary and inferior frontal regions (e.g., proximal to Broca's area) are engaged during noise-degraded speech processing (Bidelman and Dexter 2015; Bidelman and Howell 2016). Moreover, it is often neural activity in these frontal (rather than lemniscal) auditory brain regions that best predicts SIN behaviors (Wong et al. 2008; Bidelman and Dexter 2015). How brainstem activity exactly relates to auditory and perhaps nonauditory cerebral processing remains to be seen.

8.4 Experiential and Clinical Influences on SIN Abilities

8.4.1 Hearing Impairment and Auditory Disorders

Perceiving speech in noise is particularly problematic for older adults and listeners with hearing loss who often experience poorer speech understanding than would be predicted by their hearing thresholds (Glasberg and Moore 1989; Gordon-Salant and Fitzgibbons 1993; also see Anderson, Chap. 11). Disentangling the contributions of aging and hearing loss to auditory function is challenging as these factors typically covary over the lifespan and are often comorbid with other age-related deficits (e.g., cognitive decline). Nevertheless, several studies suggest that age and hearing loss have independent effects on the FFR encoding of complex sounds (Clinard et al. 2010; Marmel et al. 2013; Bidelman et al. 2014). Aging alone tends to weaken and delay the neural encoding of voice pitch and timbre cues coded by the FFR (Anderson et al. 2012; Bidelman et al. 2014), resulting in a less precise neural transcription of the speech waveform. These age-related changes in FFR phase locking (often most prominent at F_0) reflect a reduction in neural synchronization. Age-related changes in the precision of synchronization may be related to declines in neural inhibition (Caspary et al. 2008) and increased deafferentation (Kujawa and Liberman 2006; Makary et al. 2011) that occur along the aging mammalian auditory nervous system.

Hearing impaired listeners' difficulty understanding speech in noise is thought to reflect an inability to use temporal fine structure of speech (Lorenzi et al. 2006), a prominent cue for pitch perception and signal extraction in noise. Given that the FFR reflects phase-locked activity to the spectral fine structure of complex sounds, it provides a useful objective assay to track changes in auditory function that accompany hearing loss. Indeed, FFR studies reveal that even for clean speech, listeners with sensorineural hearing loss show poorer representations for the onset (Anderson et al. 2010; Bidelman et al. 2014), formant transitions (Plyler and Ananthanarayan 2001; Anderson et al. 2010), and upper harmonics of speech sounds (Anderson et al. 2013c; Bidelman et al. 2014). In addition, hearing loss has a differential impact on how the auditory system extracts envelope versus fine-structure cues. Paralleling single-unit responses in animal models (Kale and Heinz 2010), the population FFR response shows an over-exaggeration in the stimulus envelope and imbalance in the normal encoding of envelope and fine-structure cues in hearing impaired humans (Anderson et al. 2013c). Distortions between speech envelope and fine structure in terms of both their neural encoding and perception may partly underlie deficits in SIN perception in hearing-impaired listeners that are not accounted for by purely reduced audibility.

Deficiencies in brainstem encoding for noise-degraded speech have also been observed in individuals with auditory processing disorders (Billiet and Bellis 2011; also see Schochat, Rocha-Muniz, and Filippini, Chap. 9), language-learning impairments (Cunningham et al. 2001; Banai et al. 2009), and autism (Russo et al. 2008). These deficits are not always apparent in neural responses to clean speech. From a clinical standpoint, this suggests that some central auditory processing disorders might be best detected when the auditory system is taxed by challenging listening conditions (e.g., noise or reverberation) or is required to perform difficult figure-ground analysis (see Reetzke, Xie, and Chandrasekaran, Chap. 10). In this regard, the FFR may provide a useful "biomarker" for identifying certain central auditory processing deficits using an objective electrophysiological technique.

8.4.2 Language Experience

Seminal FFR studies have demonstrated that extensive auditory experiences introduce functional reorganization in the human midbrain, as reflected in the FFR. First examined in the context of language, studies have shown that long-term experience with a tonal language—in which changes in pitch alter word meaning—enhances the subcortical representation of pitch-relevant information as indicated by the smoother, more robust voice fundamental-frequency tracking in the FFRs of native Chinese relative to English-speaking listeners (Krishnan et al. 2005; Krishnan and Gandour 2009). Krishnan et al. (2005) provided the first evidence that (1) long-term auditory experience can enhance subcortical auditory processing in an experience-dependent manner, and (2) that neuroplasticity extends below cerebral cortex. The effects of life-long language experience on subcortical FFRs are treated more thoroughly elsewhere in this volume (see Krishnan and Gandour, Chap. 3). However, an interesting question that emerges from those studies is whether these cross-linguistic enhancements in brainstem auditory processing confer any advantage to signal-in-noise extraction and SIN perception.

To address this question, Krishnan et al. (2010) measured FFRs in native Chinese and English-speaking listeners in response to a dynamic pitch pattern modeled after the rising lexical tone in the Mandarin language (i.e., "T2"). The tone was synthesized using *iterated rippled noise* (IRN), which allowed for the parametric control of the tone's degree of periodicity ("noisiness") by varying the number of iterations in the IRN generation circuit (Yost 1996). Both clean and degraded T2 responses were recorded and "pitch tracking accuracy" was used to quantify the degree to which neural FFRs followed the stimulus pitch contour. Results showed that Chinese FFRs show more faithful response tracking of the T2 contour in clean and noise-degraded conditions compared to English listeners (Fig. 8.8). This illustrates that long-term language experience enhances not only the fidelity of brainstem pitch processing for clean signals but also confers advantages to extracting degraded speech information. The fact that noise-related enhancements were not observed in non-native listeners

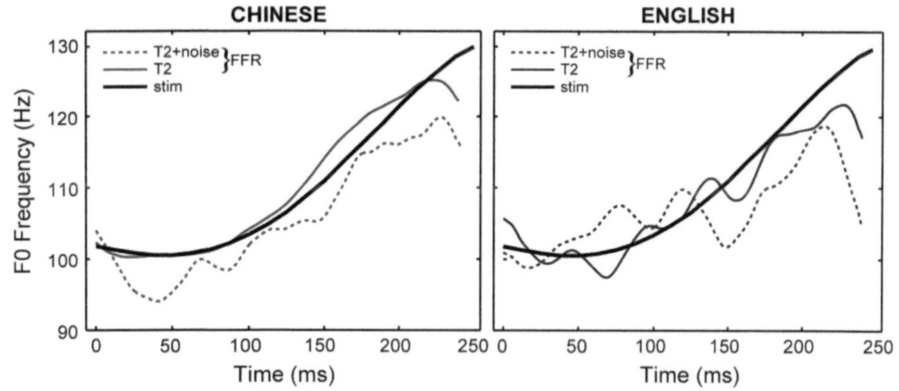

Fig. 8.8 Tone-language experience improves the fidelity of linguistic pitch processing and its extraction from noise. FFRs were measured in response to rising linguistic pitch pattern (T2) found in Mandarin Chinese. Linguistic pitches were created using dynamic iterated rippled noise (IRN) stimuli and parametrically varied in their "noisiness" by varying the number of iterations in the IRN generation. Fewer iterations disrupt the periodicity of the IRN pitch and result in a noisy pitch percept; more iterations result in clearer pitch percepts. Traces represent the pitch track (i.e., F_0 contour) extracted from the stimulus (*thick line*) and corresponding FFRs (*thin and dashed lines*) recorded from native Chinese and nonnative English speaking listeners elicited by clean and noise-degraded versions of the Mandarin tone. Chinese FFRs show more faithful response tracking of the T2 contour in clean and noise-degraded conditions compared to English listeners. This illustrates that long-term language experience can enhance the fidelity with which the brainstem encodes important acoustic features of listeners' native language. *FO*, F_0; *stim*, stimulus. (Data based on Krishnan et al. 2010)

further suggests that brainstem enhancements are restricted to specific features of speech that are most congruent with a listeners' native language (Krishnan et al. 2009; Bidelman et al. 2011).

Somewhat at odds with the aforementioned brainstem findings are behavioral and cortical evoked-potential studies, which often show the opposite effect: a clear bilingual disadvantage for SIN processing. Behavioral studies, for instance, have consistently shown that bilinguals show much poorer (i.e., upwards of ~ 10 dB SNR difference) SIN perception for their second (non-native) language relative to monolinguals (Rogers et al. 2006; Bidelman and Dexter 2015). Moreover, in response to noise-degraded speech contrasts, cortical mismatch negativity responses—reflecting the brain's automatic differentiation of speech—are both prolonged and weaker in bilingual compared to monolingual speakers (Bidelman and Dexter 2015). Thus, bilinguals' more faithful encoding of noise-degraded speech observed at the level of the brainstem (Krishnan et al. 2010; Krizman et al. 2012) does not seem to be paralleled in higher level responses at a cortical level or perception in non-native listeners.

The dissociable effects of SIN processing between brainstem and cortex indicate that bilinguals' sensory gains in auditory processing are, by themselves, likely insufficient to improve perceptual SIN. Differences in brainstem-cortical speech processing could reflect the fact that complex listening skills like SIN require additional engagement of higher-order cortical brain areas more specialized for language processing that override any sensory gains from the brainstem. Alternatively, if FFR representations reflect primarily stimulus properties rather than the behavioral percept (Gockel et al. 2011; see Sect. 8.3.5), dissociations between brainstem and cortical speech processing might be expected. More research is needed using integrative brainstem-cortical recording paradigms (e.g., Bidelman 2015a) to clarify the role of subcortical and cortical activity in SIN processing as well as potential transformations between these two stages of auditory signal processing (Bidelman et al. 2013).

8.4.3 Musical Experience

A detailed treatment of the effects of musical experience on brainstem FFRs, speech processing, and listening skills is reviewed elsewhere in this volume (see White-Schwoch and Kraus, Chap. 6). In relation to degraded speech perception, several studies have noted that the neural enhancements in quiet/clean speech encoding observed for musically trained individuals also extend to the processing of acoustically degraded speech sounds (Parbery-Clark et al. 2009; Bidelman and Krishnan 2010; Zendel and Alain 2012). Parbery-Clark et al. (2009) first reported a "musician advantage" for SIN listening. They found correlations between the speech-FFR onset latency, transition latency (i.e., timing during formant transition period), and fidelity (i.e., stimulus-to-response correlations) and behavioral measures of SIN. Similar neural and behavioral advantages have been observed for musicians

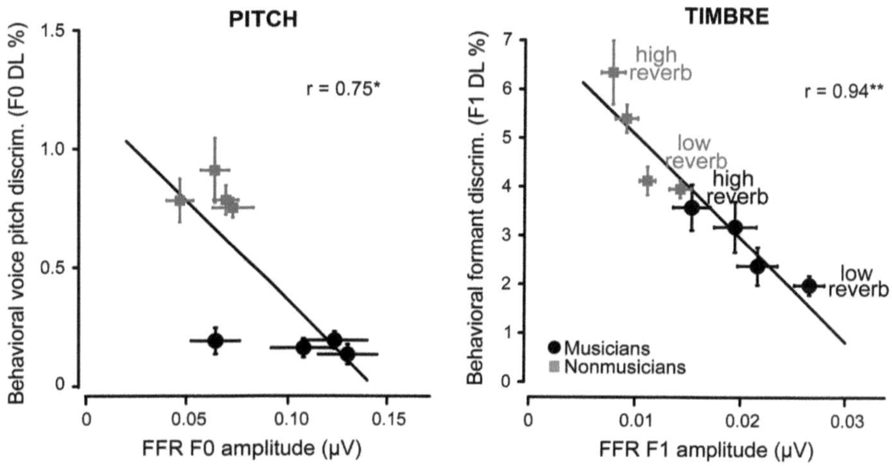

Fig. 8.9 Effects of musical training on behavioral and neural processing of degraded speech. Frequency-following responses (FFRs) were recorded in listeners with (>10 years) and without (<3 years) formal musical training in response to speech sounds varying in the severity of concurrent reverberation (see Fig. 8.4). Additionally, behavioral difference limens (DLs) were obtained for voice fundamental frequency (*FO* = F_0) and first formant (F1) discrimination to index behavior acuity for voice "pitch" and "timbre" elements of speech, respectively. The magnitude of F_0 and F1 encoding in reverberation, as indexed by the FFR, strongly predicts perceptual discrimination for both speech cues; larger, more robust subcortical activity is associated with better (i.e., smaller) DLs. Group comparisons demonstrate that musicians have both stronger neurological coding of these speech cues and improved behavioral discrimination in reverberation. *Points* show the group mean responses across dry, mild, medium, and severe reverberation conditions. *Error bars* reflect ±1 s.e.m. in either the neural or behavioral dimension. (Based on unpublished data from Bidelman and Krishnan 2010)

in reverberation (Bidelman and Krishnan 2010) (Fig. 8.9). Both pitch (F_0) and timbre (F1 formant) magnitudes are stronger in the FFRs of listeners with musical expertise (>10 years formal training) compared to their nonmusician peers (<3 years training). Importantly, neural enhancements in the FFR are well-correlated with behavioral discrimination of these cues for both groups, but they are more highly coupled to behavior in musicians.

Collectively, these studies indicate that musicians might parse and segregate competing signals in complex auditory scenes more effectively (Munte et al. 2001; Zendel and Alain 2009). Musicians are also less influenced by information masking (Oxenham et al. 2003) than nonmusicians. Given the importance of these factors in auditory scene analysis (e.g., "cocktail party" scenarios), it is possible that musical expertise improves important aspects of real-world listening required for robust communication (Alain et al. 2014). While most studies to date have only assessed musicians' SIN advantages via cross-sectional comparisons, newer training studies with randomized group assignments are promising in that they show improvement in FFR and SIN listening skills with 1–2 years of musical training (Kraus et al. 2014; Slater et al. 2015).

8.4.4 Short-Term Auditory Training and Perceptual Learning

The putative benefits of long-term auditory experiences (e.g., bilingualism, musicianship) on the FFR and SIN provoke the question of whether or not short-term training regimens might similarly boost SIN processing skills. In this regard, there have been few investigations of how auditory training impacts SIN perception (Burk and Humes 2007; Yund and Woods 2010). Of the reports showing successful training effects, gains are often modest (improvements of a few dB), fail to generalize to untrained material (lack generalizability outside the laboratory), or quickly fade (no long-term retention) (also see Plack and Carcagno, Chap. 4).

Nevertheless, a handful of recent studies have demonstrated that short-term auditory training can improve FFR encoding of degraded speech as well as SIN perception. In particular, Song et al. (2012) assessed speech FFRs to clean and noise-degraded speech before and after training on the *Listening and Communication Enhancement* (LACE®) program (Sweetow and Sabes 2006). LACE is a commercially available auditory training suite that tests users' open-set speech recognition in various listening conditions and is often provided to hearing aid patients for acclimatization after the first fitting. FFRs and SIN perception were measured before and after LACE training (twenty, 30-min sessions over 4 weeks) as well as 6 months for a long-term follow-up. A group of noncontact, control listeners was measured at similar intervals but did not participant in the training schedule. Results showed improvements in both the neural encoding of voice pitch (F_0) and upper harmonics of speech in the trained group (but not the control group) coupled with improvements in SIN perception (*QuickSIN™* and HINT scores). Interestingly, training-related enhancements were correlated with changes in the FFR and persisted at six months after the cessation of the program. The findings of Song et al. (2012) provide provocative evidence that even short-term auditory training can improve SIN listening skills in young adults and might be mediated, at least partially, by experience-dependent mechanisms of the auditory brainstem.

Whether similar training-related benefits are observed for older adults—for whom SIN issues are more problematic—remains somewhat equivocal. Anderson et al. (2013b) trained older adults on an adaptive, computer-based auditory training program for eight weeks. Results showed that the normal timing deficits (i.e., prolonged responses) observed in older adults' speech-FFR with noise (Anderson et al. 2010) is partially ameliorated with training. Trained older adults showed faster FFR timing, less noise-related change, and less trial-to-trial jitter in brainstem activity to speech sounds than age-matched controls. Unfortunately, subsequent follow-up testing at six months suggested that while training-related improvements in FFR responses were maintained, perceptual SIN gains did not persist (Anderson et al. 2014). Thus, while SIN training might be successful for younger adults, it is currently unclear what dosage of training is required to achieve the same longevity (i.e., "sticking power") of SIN benefits in older listeners. It is possible that the aged auditory nervous system is less plastic than earlier in life (Stiles 2000); hence, auditory training may not yield the same degree of neurological and listening benefits in older individuals compared to younger individuals.

8.5 Summary

This chapter has described several perceptual and neurophysiological consequences of real-world human communication in noise and reverberant environments. Both forms of acoustic interference have similar detriments to behavioral recognition of speech but each challenges the neural coding of speech sounds in different manners. In this regard, the scalp-recorded FFR has provided considerable insight into the neural representations of speech at early, pre-attentive levels of the auditory pathway and how such representations are altered in challenging listening conditions. Noise largely masks the spectral details of speech, reducing the contrast between perceptually salient frequency characteristics (e.g., formants) and the surrounding noise floor. In contrast, the effect of reverberation is to smear spectrotemporal details of a signal, producing a temporal overlap (i.e., self-masking) of time-frequency information in the running speech signal. These acoustic consequences are closely paralleled in human FFRs, which similarly show changes in "neural SNR" and spectral smearing with additive noise and reverberation, respectively. Yet, even within the speech signal, acoustic interferences do not produce uniform impairments. To a certain extent, voice pitch (F_0) cues of speech are largely immune to noise/reverberation interferences, whereas timbral cues (i.e., higher harmonics and spectral envelope) degrade quickly with increasing noise/reverberation.

The subcortical encoding of speech and SIN listening skills are not static. Rather they naturally decline across the lifespan and are impaired with certain auditory disorders. Promisingly, the neural encoding and perception of SIN can be enhanced by aspects of the human experience (e.g., language and musical experience) and short-term auditory training. Neuroplastic changes within training suggest that the brainstem plays an active (but underappreciated) role in molding functional properties of the auditory system. Moreover, perceptual and neurobiological changes in FFR imply that impoverished auditory sensory coding, resulting from aging, hearing impairment, or central auditory processing disorders, might be partially counteracted with certain forms of auditory learning and experience.

Several empirical and theoretical questions of the FFR were raised but remain unanswered. First, while a number of studies have focused on the brainstem encoding of speech in noise (e.g., Song et al. 2011; Prevost et al. 2013), at present there is an unfortunate dearth of studies examining how reverberation changes the FFR coding of speech (Bidelman and Krishnan 2010). Nevertheless, studies reviewed herein reveal a possible differential neural coding of speech in noise compared to reverberation. Conceivably, the unique impact of noise versus reverberation on speech processing may help account for the challenges observed by hearing impaired listeners in certain acoustic environments but not others. Indeed, while both noise and reverberation have a negative impact of similar magnitude on speech perception, their effects are reflected in different types of perceptual errors (e.g., confusions) (Nabelek and Dagenais 1986) and associated differences in neural coding (Bidelman and Krishnan 2010). Future studies are warranted that directly compare noise and reverberation. Such contrasts will surely reveal a more nuanced understanding of the effects of acoustic interference(s) on auditory brainstem processing and might provide new insight into both specific deficits and biomarkers of degraded speech perception abilities.

Secondly, while brainstem response properties often covary with perceptual abilities in SIN tasks, it is often unclear how (and even if) neural enhancements seen in the FFR (e.g., faster timing, more robust magnitudes) are causally related to improvements in behavior. While FFR studies do suggest experience-dependent effects in brainstem, auditory-processing training studies will help clarify cross-sectional findings that dominate the literature. These studies are beginning to reveal causal links between biological changes in degraded speech coding with behavior (e.g., Song et al. 2012; Slater et al. 2015), suggesting that improvements in SIN skills are directly related to enhancements in the brainstem's transcription of speech.

Thirdly, it is possible that perhaps subcortical FFRs do not carry a perceptual code, but rather they reflect mainly stimulus properties (Gockel et al. 2011; Bidelman et al. 2013). Experimental designs that pit stimulus acoustics at least partially orthogonal to the resulting percept (e.g., as in categorical perception: Bidelman et al. 2013; Bidelman and Alain 2015) might be used to clarify exogenous and endogenous contributions to the FFR that are not yet fully clarified.

Finally, while the FFR has provided a rich window into how the human auditory system encodes speech acoustics, it is but a single snapshot of the auditory brain at work. A holistic account of human communication in challenging environments would presumably need to account for both sensory and cognitive mechanisms subserving SIN processing (e.g., Anderson et al. 2013a). Moreover, how spectrotemporal features coded by the FFR are transformed as they arrive in auditory cortex and how brainstem following responses relate to the more diffuse waves of the cortical ERPs remain to be elucidated (Bidelman et al. 2013). Integrative electrophysiological paradigms that include simultaneous recordings of brainstem and cortical-evoked activity are needed (Bidelman 2015a) to help clarify how the auditory brainstem works in concert with auditory (and nonauditory) brain regions to analyze the auditory scene.

Acknowledgments Portions of this work were supported by grants from the American Hearing Research Foundation (AHRF) and American Academy of Audiology Foundation (AAF) awarded to G.M.B. The author thanks Amy Fehrenbach, Jill Lowther, and Sarah Lawson for comments on an earlier version of this manuscript.

Compliance with Ethics Requirements G. Bidelman received grants from the American Hearing Research Foundation (AHRF) and American Academy of Audiology Foundation (AAF) to support portions of this work.

References

Akhoun, I., Gallego, S., Moulin, A., Menard, M., et al. (2008). The temporal relationship between speech auditory brainstem responses and the acoustic pattern of the phoneme /ba/ in normal-hearing adults. *Clinical Neurophysiology, 119*(4), 922–933.

Alain, C., Quan, J., McDonald, K., & Van Roon, P. (2009). Noise-induced increase in human auditory evoked neuromagnetic fields. *European Journal of Neuroscience, 30*(1), 132–142.

Alain, C., Zendel, B. R., Hutka, S., & Bidelman, G. M. (2014). Turning down the noise: The benefit of musical training on the aging auditory brain. *Hearing Research, 308*, 162–173.

Anderson, S., Parbery-Clark, A., White-Schwoch, T., Drehobl, S., & Kraus, N. (2013a). Effects of hearing loss on the subcortical representation of speech cues. *Journal of the Acoustical Society of America, 133*(5), 3030–3038.

Anderson, S., Parbery-Clark, A., White-Schwoch, T., & Kraus, N. (2012). Aging affects neural precision of speech encoding. *The Journal of Neuroscience, 32*(41), 14156–14164.

Anderson, S., Skoe, E., Chandrasekaran, B., & Kraus, N. (2010). Neural timing is linked to speech perception in noise. *The Journal of Neuroscience, 30*(14), 4922–4926.

Anderson, S., White-Schwoch, T., Choi, H. J., & Kraus, N. (2014). Partial maintenance of auditory-based cognitive training benefits in older adults. *Neuropsychologia, 62*, 286–296.

Anderson, S., White-Schwoch, T., Parbery-Clark, A., & Kraus, N. (2013b). A dynamic auditory-cognitive system supports speech-in-noise perception in older adults. *Hearing Research, 300*, 18–32.

Anderson, S., White-Schwoch, T., Parbery-Clark, A., & Kraus, N. (2013c). Reversal of age-related neural timing delays with training. *Proceedings of the National Academy of Sciences of the USA, 110*(11), 4357–4362.

ANSI. (1969). ANSI American national standards methods for the calculation of the articulation index. ANSI S3.5-69-1969. New York.

Assmann, P. F., & Summerfield, Q. (1989). Modeling the perception of concurrent vowels: Vowels with the same fundamental frequency. *The Journal of the Acoustical Society of America, 85*(1), 327–338.

Assmann, P. F., & Summerfield, Q. (1990). Modeling the perception of concurrent vowels: Vowels with different fundamental frequencies. *The Journal of the Acoustical Society of America, 88*(2), 680–697.

Assmann, P. F., & Summerfield, A. Q. (Eds.). (2004). *The perception of speech under adverse conditions*. New York: Springer Science+Business Media.

Backus, J. (1977). *The acoustical foundations of music* (2nd ed.). New York: Norton.

Banai, K., Hornickel, J., Skoe, E., Nicol, T., et al. (2009). Reading and subcortical auditory function. *Cerebral Cortex, 19*(11), 2699–2707.

Bidelman, G. M. (2015a). Towards an optimal paradigm for simultaneously recording cortical and brainstem auditory evoked potentials. *The Journal of Neuroscience Methods, 241*, 94–100.

Bidelman, G. M. (2015b). Multichannel recordings of the human brainstem frequency-following response: Scalp topography, source generators, and distinctions from the transient ABR. *Hearing Research, 323*, 68–80.

Bidelman, G. M., & Alain, C. (2015). Hierarchical neurocomputations underlying concurrent sound segregation: Connecting periphery to percept. *Neuropsychologia, 68*, 38–50.

Bidelman, G. M., & Bhagat, S. P. (2015). Right ear advantage drives the link between olivocochlear efferent "antimasking" and speech-in-noise listening benefits. *NeuroReport, 26*, 483–487.

Bidelman, G. M., & Dexter, L. (2015). Bilinguals at the "cocktail party": Dissociable neural activity in auditory-linguistic brain regions reveals neurobiological basis for nonnative listeners' speech-in-noise recognition deficits. *Brain and Language, 143*, 32–41.

Bidelman, G. M., Gandour, J. T., & Krishnan, A. (2011). Cross-domain effects of music and language experience on the representation of pitch in the human auditory brainstem. *Journal of Cognitive Neuroscience, 23*(2), 425–434.

Bidelman, G. M., & Howell, M. (2016). Functional changes in inter- and intra-hemispheric auditory cortical processing underlying degraded speech perception. *NeuroImage, 124*, 581–590.

Bidelman, G. M., & Krishnan, A. (2010). Effects of reverberation on brainstem representation of speech in musicians and non-musicians. *Brain Research, 1355*, 112–125.

Bidelman, G. M., Moreno, S., & Alain, C. (2013). Tracing the emergence of categorical speech perception in the human auditory system. *NeuroImage, 79*(1), 201–212.

Bidelman, G. M., Villafuerte, J. W., Moreno, S., & Alain, C. (2014). Age-related changes in the subcortical-cortical encoding and categorical perception of speech. *Neurobiology of Aging, 35* (11), 2526–2540.

Billiet, C. R., & Bellis, T. J. (2011). The relationship between brainstem temporal processing and performance on tests of central auditory function in children with reading disorders. *Journal of Speech, Language, and Hearing Research, 54*, 228–242.

Billings, C. J., Bennett, K. O., Molis, M. R., & Leek, M. R. (2010). Cortical encoding of signals in noise: Effects of stimulus type and recording paradigm. *Ear and Hearing, 32*(1), 53–60.

Billings, C. J., McMillan, G. P., Penman, T. M., & Gille, S. M. (2013). Predicting perception in noise using cortical auditory evoked potentials. *Journal of the Association for Research in Otolaryngology, 14*(6), 891–903.

Billings, C. J., Tremblay, K. L., Stecker, G. C., & Tolin, W. M. (2009). Human evoked cortical activity to signal-to-noise ratio and absolute signal level. *Hearing Research, 254*(1–2), 15–24.

Burk, M. H., & Humes, L. E. (2007). Effects of training on speech recognition performance in noise using lexically hard words. *Journal of Speech, Language, and Hearing Research, 50*, 25–40.

Burkard, R., & Hecox, K. (1983). The effect of broadband noise on the human brainstem auditory evoked response. I. Rate and intensity effects. *The Journal of the Acoustical Society of America, 74*(4), 1204–1213.

Carhart, R. (1951). Basic principles of speech audiometry. *Acta Oto-Laryngologica, 40*, 62–71.

Carhart, R., & Tillman, T. W. (1970). Interaction of competing speech signals with hearing losses. *Archives of Otolaryngology, 91*, 274–279.

Cariani, P. A., & Delgutte, B. (1996). Neural correlates of the pitch of complex tones. I. Pitch and pitch salience. *Journal of Neurophysiology, 76*(3), 1698–1716.

Caspary, D. M., Ling, L., Turner, J. G., & Hughes, L. F. (2008). Inhibitory neurotransmission, plasticity and aging in the mammalian central auditory system. *Journal of Experimental Biology and Medicine, 211*, 1781–1791.

Chmiel, R., & Jerger, J. (1996). Hearing aid use, central auditory disorder, and hearing handicap in elderly persons. *Journal of the American Academy of Audiology, 7*(3), 190–202.

Clinard, C. G., Tremblay, K. L., & Krishnan, A. R. (2010). Aging alters the perception and physiological representation of frequency: Evidence from human frequency-following response recordings. *Hearing Research, 264*(1–2), 48–55.

Cunningham, J., Nicol, T., King, C., Zecker, S. G., & Kraus, N. (2002). Effects of noise and cue enhancement on neural responses to speech in auditory midbrain, thalamus and cortex. *Hearing Research, 169*, 97–111.

Cunningham, J., Nicol, T., Zecker, S. G., Bradlow, A., & Kraus, N. (2001). Neurobiologic responses to speech in noise in children with learning problems: Deficits and strategies for improvement. *Clinical Neurophysiology, 112*(5), 758–767.

Dau, T. (2003). The importance of cochlear processing for the formation of auditory brainstem and frequency following responses. *The Journal of the Acoustical Society of America, 113*(2), 936–950.

Drgas, S., & Blaszak, M. A. (2009). Perceptual consequences of changes in vocoded speech parameters in various reverberation conditions. *Journal of Speech, Language, and Hearing Research, 52*(4), 945–955.

Fant, G. (1960). *Acoustic theory of speech production*. The Hague: Mouton.

Finitzo-Hieber, T., & Tillman, T. (1978). Room acoustics effects on monosyllabic word discrimination ability for normal and hearing-impaired children. *Journal of Speech and Hearing Research, 21*, 440–458.

French, N., & Steinberg, J. (1947). Factors governing the intelligibility of speech sounds. *The Journal of the Acoustical Society of America, 19*, 90–119.

Frisina, D. R., & Frisina, R. D. (1997). Speech recognition in noise and presbycusis: Relations to possible neural mechanisms. *Hearing Research, 106*(1–2), 95–104.

Galbraith, G., Arbagey, P. W., Branski, R., Comerci, N., & Rector, P. M. (1995). Intelligible speech encoded in the human brain stem frequency-following response. *NeuroReport, 6*(17), 2363–2367.

Galbraith, G., Olfman, D. M., & Huffman, T. M. (2003). Selective attention affects human brain stem frequency-following response. *NeuroReport, 14*(5), 735–738.

Gelfand, S. A., & Silman, S. (1979). Effects of small room reverberation upon the recognition of some consonant features. *The Journal of the Acoustical Society of America, 66*(1), 22–29.

George, E. L., Festen, J. M., & Houtgast, T. (2008). The combined effects of reverberation and nonstationary noise on sentence intelligibility. *The Journal of the Acoustical Society of America, 124*(2), 1269–1277.

Glasberg, B. R., & Moore, B. C. J. (1989). Psychoacoustic abilities of subjects with unilateral and bilateral cochlear hearing impairments and their relationship to the ability to understand speech. *Scandinavian Audiology, 32*, 1–25.

Gockel, H. E., Carlyon, R. P., Mehta, A., & Plack, C. J. (2011). The frequency following response (FFR) may reflect pitch-bearing information but is not a direct representation of pitch. *Journal of the Association for Research in Otolaryngology, 12*(6), 767–782.

Gordon-Salant, S., & Fitzgibbons, P. J. (1993). Temporal factors and speech recognition performance in young and elderly listeners. *Journal of Speech and Hearing Research, 36*, 1276–1285.

Greenberg, S., Marsh, J. T., Brown, W. S., & Smith, J. C. (1987). Neural temporal coding of low pitch. I. Human frequency-following responses to complex tones. *Hearing Research, 25*(2–3), 91–114.

Helfer, K., & Wilber, L. (1990). Hearing loss, aging, and speech perception in reverberation and in noise. *Journal of Speech and Hearing Research, 33*, 149–155.

Henry, K. R. (1999). Noise improves transfer of near-threshold, phase-locked activity of the cochlear nerve: Evidence for stochastic resonance? *Journal of Comparative Physiology A: Neuroethology, Sensory, Neural, and Behavioral Physiology, 184*(6), 577–584.

Humes, L. E., & Christopherson, L. (1991). Speech identification difficulties of hearing-impaired elderly persons: The contributions of auditory processing deficits. *Journal of Speech and Hearing Research, 34*(3), 686–693.

Humes, L. E., & Roberts, L. (1990). Speech-recognition difficulties of the hearing-impaired elderly: The contributions of audibility. *Journal of Speech and Hearing Research, 33*(4), 726–735.

Jaramillo, F., & Wiesenfeld, K. (1998). Mechanoelectrical transduction assisted by Brownian motion: A role for noise in the auditory system. *Nature Neuroscience, 1*, 384–388.

Kale, S., & Heinz, M. G. (2010). Envelope coding in auditory nerve fibers following noise-induced hearing loss. *Journal of the Association for Research in Otolaryngology, 11* (4), 657–673.

Kiang, N. Y., & Moxon, E. C. (1973). Tails of tuning curves of auditory-nerve fibers. *The Journal of the Acoustical Society of America, 55*(2), 620–630.

Killion, M., & Niquette, P. (2000). What can the pure-tone audiogram tell us about a patient's SNR loss? *The Hearing Journal, 53*(3), 46–53.

Killion, M. C., Niquette, P. A., Gudmundsen, G. I., Revit, L. J., & Banerjee, S. (2004). Development of a quick speech-in-noise test for measuring signal-to-noise ratio loss in normal-hearing and hearing-impaired listeners. *The Journal of the Acoustical Society of America, 116*(4 Pt 1), 2395–2405.

Kinsler, L. E., Frey, A. R., Coppens, A. B., & Sanders, J. V. (2000). *Fundamentals of acoustics*. New York: Wiley.

Knight, R. T., Staines, W. R., Swick, D., & Chao, L. L. (1999). Prefrontal cortex regulates inhibition and excitation in distributed neural networks. *Acta Psychologica, 101*(2–3), 159–178.

Kraus, N., Slater, J., Thompson, E. C., Hornickel, J., et al. (2014). Music enrichment programs improve the neural encoding of speech in at-risk children. *The Journal of Neuroscience, 34*(36), 11913–11918.

Krishnan, A., & Gandour, J. T. (2009). The role of the auditory brainstem in processing linguistically-relevant pitch patterns. *Brain and Language, 110*(3), 135–148.

Krishnan, A., Gandour, J. T., & Bidelman, G. M. (2010). Brainstem pitch representation in native speakers of Mandarin is less susceptible to degradation of stimulus temporal regularity. *Brain Research, 1313*, 124–133.

Krishnan, A., Gandour, J. T., Bidelman, G. M., & Swaminathan, J. (2009). Experience-dependent neural representation of dynamic pitch in the brainstem. *NeuroReport, 20*(4), 408–413.

Krishnan, A., Xu, Y., Gandour, J. T., & Cariani, P. (2005). Encoding of pitch in the human brainstem is sensitive to language experience. *Cognitive Brain Research, 25*(1), 161–168.

Krizman, J., Marian, V., Shook, A., Skoe, E., & Kraus, N. (2012). Subcortical encoding of sound is enhanced in bilinguals and relates to executive function advantages. *Proceedings of the National Academy of Sciences of the USA, 109*(20), 7877–7881.

Krizman, J., Skoe, E., & Kraus, N. (2010). Stimulus rate and subcortical auditory processing of speech. *Audiology and Neuro-Otology, 15*, 332–342.

Kujawa, S. G., & Liberman, M. C. (2006). Acceleration of age-related hearing loss by early noise exposure: Evidence of a misspent youth. *The Journal of Neuroscience, 26*(7), 2115–2123.

Laroche, M., Dajani, H. R., Prevost, F., & Marcoux, A. M. (2013). Brainstem auditory responses to resolved and unresolved harmonics of a synthetic vowel in quiet and noise. *Ear and Hearing, 34*(1), 63–74.

Larsen, E., Iyer, N., Lansing, C. R., & Feng, A. S. (2008). On the minimum audible difference in direct-to-reverberant energy ratio. *The Journal of the Acoustical Society of America, 124*(1), 450–461.

Li, X., & Jeng, F.-C. (2011). Noise tolerance in human frequency-following responses to voice pitch. *The Journal of the Acoustical Society of America, 129*, 21–26.

Lifshitz, S. (1925). Optimum reverberation for an auditorium. *Physical Review, 25*(3), 391–394.

Liu, C., & Kewley-Port, D. (2004). Formant discrimination in noise for isolated vowels. *The Journal of the Acoustical Society of America, 116*(5), 3119–3129.

Lorenzi, C., Gilbert, G., Carn, H., Garnier, S., & Moore, B. C. (2006). Speech perception problems of the hearing impaired reflect inability to use temporal fine structure. *Proceedings of the National Academy of Sciences of the USA, 103*(49), 18866–18869.

Makary, C. A., Shin, J., Kujawa, S. G., Liberman, M. C., & Merchant, S. N. (2011). Age-related primary cochlear neuronal degeneration in human temporal bones. *Journal of the Association for Research in Otolaryngology, 12*(6), 711–717.

Marmel, F., Linley, D., Carlyon, R. P., Gockel, H. E., et al. (2013). Subcortical neural synchrony and absolute thresholds predict frequency discrimination independently. *Journal of the Association for Research in Otolaryngology, 14*, 755–766.

McShefferty, D., Whitmer, W. M., & Akeroyd, M. A. (2015). The just-noticeable difference in speech-to-noise ratio. *Trends in Hearing, 19*, 1–9.

Middelweerd, M. J., Festen, J. M., & Plomp, R. (1990). Difficulties with speech intelligibility in noise in spite of a normal pure-tone audiogram. *Audiology, 29*(1), 1–7.

Munte, T. F., Kohlmetz, C., Nager, W., & Altenmuller, E. (2001). Superior auditory spatial tuning in conductors. *Nature, 409*(6820), 580.

Nabelek, A. K. (1988). Identification of vowels in quiet, noise, and reverberation: Relationships with age and hearing loss. *The Journal of the Acoustical Society of America, 84*(2), 476–484.

Nabelek, A. K., & Dagenais, P. A. (1986). Vowel errors in noise and in reverberation by hearing-impaired listeners. *The Journal of the Acoustical Society of America, 80*(3), 741–748.

Nabelek, A. K., & Letowski, T. R. (1985). Vowel confusions of hearing-impaired listeners under reverberant and nonreverberant conditions. *Journal of Speech and Hearing Disorders, 50*(2), 126–131.
Nabelek, A. K., & Letowski, T. R. (1988). Similarities of vowels in nonreverberant and reverberant fields. *The Journal of the Acoustical Society of America, 83*(5), 1891–1899.
Nabelek, A. K., Letowski, T. R., & Tucker, F. M. (1989). Reverberant overlap- and self-masking in consonant identification. *The Journal of the Acoustical Society of America, 86*(4), 1259–1265.
Nilsson, M., Soli, S. D., & Sullivan, J. A. (1994). Development of the hearing in noise test for the measurement of speech reception thresholds in quiet and in noise. *The Journal of the Acoustical Society of America, 95*(2), 1085–1099.
Oxenham, A. J., Fligor, B. J., Mason, C. R., & Kidd, G., Jr. (2003). Informational masking and musical training. *The Journal of the Acoustical Society of America, 114*(3), 1543–1549.
Parbery-Clark, A., Marmel, F., Bair, J., & Kraus, N. (2011). What subcortical-cortical relationships tell us about processing speech in noise. *European Journal of Neuroscience, 33*(3), 549–557.
Parbery-Clark, A., Skoe, E., & Kraus, N. (2009). Musical experience limits the degradative effects of background noise on the neural processing of sound. *The Journal of Neuroscience, 29*(45), 14100–14107.
Picton, T. W., Alain, C., Woods, D. L., John, M. S., et al. (1999). Intracerebral sources of human auditory-evoked potentials. *Audiology and Neuro-Otology, 4*(2), 64–79.
Picton, T. W., Woods, D. L., Baribaeu-Braun, J., & Healy, T. M. G. (1977). Evoked potential audiometry. *Journal of Otolaryngology, 6*(2), 90–119.
Plomp, R. (1978). Auditory handicap of hearing impairment and the limited benefit of hearing aids. *The Journal of the Acoustical Society of America, 63*, 533–549.
Plomp, R. (1986). A signal-to-noise ratio model for the speech-reception threshold of the hearing impaired. *Journal of Speech and Hearing Research, 29*(2), 146–154.
Plyler, P. N., & Ananthanarayan, A. K. (2001). Human frequency-following responses: Representation of second formant transitions in normal-hearing and hearing-impaired listeners. *Journal of the American Academy of Audiology, 12*(10), 523–533.
Popelka, G. R., Moore, B. C. J., Popper, A. N., & Fay, R. R. (2016). *Hearing Aids*. New York: Springer Science+Business Media.
Prevost, F., Laroche, M., Marcoux, A. M., & Dajani, H. R. (2013). Objective measurement of physiological signal-to-noise gain in the brainstem response to a synthetic vowel. *Clinical Neurophysiology, 124*(1), 52–60.
Rogers, C. L., Lister, J. J., Febo, D. M., Besing, J. M., & Abrams, H. B. (2006). Effects of bilingualism, noise, and reverberation on speech perception by listeners with normal hearing. *Applied Psycholinguistics, 27*(03), 465–485.
Ruggles, D. R., Bharadwaj, H., & Shinn-Cunningham, B. (2011). Normal hearing is not enough to guarantee robust encoding of suprathreshold features important in everyday communication. *Proceedings of the National Academy of Sciences of the USA, 108*(37), 15516–15521.
Russo, N. M., Bradlow, A. R., Skoe, E., Trommer, B. L., et al. (2008). Deficient brainstem encoding of pitch in children with autism spectrum disorders. *Clinical Neurophysiology, 119*(8), 1720–1731.
Russo, N., Nicol, T., Musacchia, G., & Kraus, N. (2004). Brainstem responses to speech syllables. *Clinical Neurophysiology, 115*(9), 2021–2030.
Sabine, W. (1962). *Collected Papers on Acoustics*. Los Altos Hills, CA: Dover.
Sayles, M., & Winter, I. M. (2008). Reverberation challenges the temporal representation of the pitch of complex sounds. *Neuron, 58*(5), 789–801.
Shannon, R. V., Zeng, F. G., Kamath, V., Wygonski, J., & Ekelid, M. (1995). Speech recognition with primarily temporal cues. *Science, 270*(5234), 303–304.
Slater, J., Skoe, E., Strait, D., O'Connell, S., et al. (2015). Music training improves speech-in-noise perception: Longitudinal evidence from a community-based music program. *Behavioural Brain Research, 291*, 244–252.

Smalt, C. J., Krishnan, A., Bidelman, G. M., Ananthakrishnan, S., & Gandour, J. T. (2012). Distortion products and their influence on representation of pitch-relevant information in the human brainstem for unresolved harmonic complex tones. *Hearing Research, 292*(1–2), 26–34.

Smith, J. C., Marsh, J. T., & Brown, W. S. (1975). Far-field recorded frequency-following responses: Evidence for the locus of brainstem sources. *Electroencephalography and Clinical Neurophysiology, 39*(5), 465–472.

Sohmer, H., & Pratt, H. (1977). Identification and separation of acoustic frequency following responses (FFRs) in man. *Electroencephalography and Clinical Neurophysiology, 42*(4), 493–500.

Sohmer, H., Pratt, H., & Kinarti, R. (1977). Sources of frequency-following responses (FFR) in man. *Electroencephalography and Clinical Neurophysiology, 42*, 656–664.

Song, J. H., Skoe, E., Banai, K., & Kraus, N. (2011). Perception of speech in noise: Neural correlates. *Journal of Cognitive Neuroscience, 23*(9), 2268–2279.

Song, J. H., Skoe, E., Banai, K., & Kraus, N. (2012). Training to improve hearing speech in noise: Biological mechanisms. *Cerebral Cortex, 22*(5), 1180–1190.

Stiles, J. (2000). Neural plasticity and cognitive development. *Developmental Neuropsychology, 18*(2), 237–272.

Swaminathan, J., & Heinz, M. G. (2012). Psychophysiological analyses demonstrate the importance of neural envelope coding for speech perception in noise. *The Journal of Neuroscience, 32*(5), 1747–1756.

Sweetow, R. W., & Sabes, J. H. (2006). The need for and development of an adaptive listening and communication enhancement (LACE) program. *Journal of the American Academy of Audiology, 17*(8), 538–558.

von Békésy, G. (1938). Über die Enstehung der Entfernungsempfindung beim Hören (On the origin of distance perception in hearing). *Akustische Zeitschrift, 3*, 21–31.

Wang, D. L., & Brown, G. J. (2006). *Computational auditory scene analysis*. Hoboken, NJ: Wiley.

Watkins, A. J. (2005). Perceptual compensation for effects of reverberation in speech identification. *The Journal of the Acoustical Society of America, 118*(1), 249–262.

Weiss, M. W., & Bidelman, G. M. (2015). Listening to the brainstem: Musicianship enhances intelligibility of subcortical representations for speech. *The Journal of Neuroscience, 35*(4), 1687–1691.

Wilson, R. H., & McArdle, R. (2005). Speech signals used to evaluate functional status of the auditory system. *Journal of Rehabilitation Research and Development, 42*(4 Suppl 2), 79–94.

Wong, P. C. M., Uppunda, A. K., Parrish, T. B., & Dhar, S. (2008). Cortical mechanisms of speech perception in noise. *Journal of Speech, Language, and Hearing Research, 51*, 1026–1041.

Woods, D. L., & Hillyard, S. A. (1978). Attention at the cocktail party: Brainstem evoked responses reveal no peripheral gating. In D. A. Otto (Ed.), *Multidisciplinary perspectives in event-related brain potential research* (pp. 230–233). Washington, DC: U.S. Government Printing Office. (EPA 600/9-77-043).

Yang, W., & Bradley, J. S. (2009). Effects of room acoustics on the intelligibility of speech in classrooms for young children. *The Journal of the Acoustical Society of America, 125*(2), 922–933.

Yost, W. A. (1996). Pitch of iterated rippled noise. *The Journal of the Acoustical Society of America, 100*(1), 511–518.

Yund, E. W., & Woods, D. L. (2010). Content and procedural learning in repeated sentence tests of speech perception. *Ear and Hearing, 31*, 769–778.

Zahorik, P. (2002). Direct-to-reverberant energy ratio sensitivity. *The Journal of the Acoustical Society of America, 112*(5 Pt 1), 2110–2117.

Zendel, B. R., & Alain, C. (2009). Concurrent sound segregation is enhanced in musicians. *Journal of Cognitive Neuroscience, 21*(8), 1488–1498.

Zendel, B. R., & Alain, C. (2012). Musicians experience less age-related decline in central auditory processing. *Psychology and Aging, 27*(2), 410–417.

Chapter 9
Understanding Auditory Processing Disorder Through the FFR

Eliane Schochat, Caroline Nunes Rocha-Muniz, and Renata Filippini

Abstract This chapter gives an overview of the importance of auditory processing for successful language learning and describes assessment measures that are not influenced by factors such as alertness and fatigue. The frequency-following response (FFR) to speech is similar to the evoking stimulus both acoustically and visually, that is, it has good accuracy in encoding specific speech features. Thus, FFR can assess sound processing to an extent that is not possible with slower, cortical potentials such as the middle latency response (MLR) and late latency responses. The fidelity of the FFR to the stimulus enables the evaluation of the strength of subcortical encoding of multiple acoustic aspects of complex sounds, including timing, pitch, and harmonics. Taken as a whole, FFR to speech shows patterns in subcomponents of the FFR that are associated with clinical populations. These distinct patterns of neural processing are described and possible mechanisms underlying abnormalities of the FFR associated with auditory processing disorders are discussed.

Keywords Auditory-based learning impairments · Auditory processing disorders · Central auditory function · Electrophysiology · FFR · Objective assessment · Temporal processing

---

E. Schochat (✉) · C.N. Rocha-Muniz · R. Filippini
Department of Physiotherapy, Audiology and Speech Therapy and Occupational Therapy, LIM 34, Neuroaudiology Lab, Universidade de São Paulo, Rua Cipotânea, 51, Cidade Universitária, São Paulo SP 05360-160, Brazil
e-mail: eschocha@usp.br

C.N. Rocha-Muniz
e-mail: carolrocha@usp.br

R. Filippini
e-mail: refilippini@usp.br

N. Kraus et al. (eds.), *The Frequency-Following Response*, Springer Handbook of Auditory Research 61, DOI 10.1007/978-3-319-47944-6_9

9.1 Introduction

Auditory processing disorder (APD) has been the subject of much debate and dissension. The controversies surrounding APD are related to its definition, its heterogeneous nature, the lack of a "gold standard" for its assessment and study, and the consequent lack of an appropriate test battery for clinical evaluation. Initially, the study of the underlying physiological mechanisms of auditory processing was based on psychophysical procedures and behavioral responses. However, such measures require the involvement of cognitive processes, attention, intelligence, motivation, fatigue, motor skills, language experience, and language impairments (Jirsa and Clontz 1990; Jerger and Musiek 2000), thus hindering the definition of the sensory deficit as the sole causative factor in APD. By contrast, electrophysiological procedures are independent of the subject's ability to provide behavioral responses (Musiek et al. 2002); thus, electrophysiological procedures have been useful in establishing basic structure-function relationships in the human auditory system. Such relationships have provided researchers and clinicians with valuable information regarding the sequence, timing, and neural location of auditory processes.

Although electrophysiological measures in the form of auditory evoked potentials (AEPs) have important roles in auditory processing studies, the reliability of these responses can be disputed. Some AEPs, such as the N1P2N2 complex, P300, and mismatch negativity (MMN), are widely used but their responses present great variability, and their slow voltage fluctuations, occurring hundreds of milliseconds after the evoking sound, are poor renderings of the acoustics of the stimulus (Kraus and Nicol 2014).

The frequency-following response (FFR) or auditory brainstem response to complex stimuli (cABR), assesses auditory neural functions, especially those believed to be involved in the neural coding of auditory processes. Studies using cABR have provided a more faithful measure of acoustic processing compared with other AEPs because of the physical similarity between the stimulus and response waveforms. Furthermore, cABR also reflects that auditory processing is profoundly affected by external factors, such as communication skills and training (Kraus and Nicol 2014).

The proper identification and quantification of auditory dysfunction is key to providing researchers and clinicians with a clear and irrefutable description of APD. In this regard, this chapter presents research demonstrating the valuable role of electrophysiology in general and of cABR/FFR in the comprehensive study and assessment of APD.

9.2 Auditory Processing Disorder: Definitions and Mechanisms

Although central auditory processing is a relatively recent field of research for audiologists, the first studies of APD date back to the 1950s. These classic studies were performed in subjects with temporal lobe lesions and auditory complaints in the presence of normal peripheral hearing (Bocca et al. 1954; Sanchez-Longo et al. 1957; Kimura 1961). It was not until the 1970s and 1980s that researchers began to effectively study APD in children with language and learning difficulties despite normal hearing thresholds (Tallal 1976; Sanger et al. 1987). The significant increase in interest in APD provided not only a substantial amount of data but also many controversies and a lack of consensus on the nomenclature, concepts, and diagnostic criteria used in research and the clinic.

Currently, the American Speech-Language-Hearing Association (ASHA), corroborated by the American Academy of Audiology (AAA 2010), defines APD as a deficit in the "perceptual processing of auditory information in the central nervous system (CNS) and the neurobiological activity that underlies that processing and gives rise to the electrophysiological auditory potentials" (p. 2, ASHA 2005). The ASHA definition is based on the concept that APD is a primary dysfunction of the auditory substrate (Musiek et al. 2005) that might be influenced by, but not determined by, cognitive deficits. In contrast, the British Society of Audiology (BSA) places increased weight on cognitive deficits, stating: "the mechanisms underlying APD can include both afferent and efferent pathways in the auditory system, as well as *higher level processing that provides 'top-down' modulation of such pathways*" (p. 3, BSA 2011).

In essence, both ASHA and BSA agree on the underpinnings of APD as the primary consequence of a known lesion to the central auditory nervous system (CANS) or secondary to peripheral hearing impairment. However, when discussing APD as a developmental disorder, these groups disagree to some extent, and this disagreement is the origin of many of the controversies surrounding APD.

In children, APD may be related to auditory deprivation linked to long-term otitis media (Moore 2007), delayed maturation of the auditory central pathway, neurological conditions arising from tumors of the CANS, prematurity and low weight at birth, cerebrovascular and metabolic disorders, epilepsy, and extrinsic damage to the brain caused by bacterial meningitis, head trauma, or heavy metal exposure (Chermak and Musiek 2014; Bellis and Bellis 2015); however, there are no clear etiological factors strongly related to APD diagnosis (Dawes et al. 2008). The prevalence of APD is also unclear, but it is predicted to be 2–5% in school-aged children and approximately 50% in children with learning problems (Chermak and Musiek 1997).

The difficulty in establishing definitions, etiology, and prevalence data is because APD is heterogeneous (Jerger and Musiek 2000; Banai and Kraus 2014), lacks diagnostic "gold standards" (BSA 2011; Moore et al. 2013), and frequently occurs with concurrent difficulties in reading, spelling, and language (Sharma et al. 2009; Witton 2010). APD can be seen as a constellation of different behavioral manifestations from not being able to localize sounds or understand speech in noise to an inability in dealing with competing sounds, following verbal instructions, and, the more radical, central deafness. Such symptoms may overlap with other developmental disorders or cognitive deficits. However, the extent to which these overlaps represent causal relationships (with one primary disorder being responsible for the other deficits) or are simply different disorders sharing the same underlying deficit is unclear (Bellis 2007).

The central auditory pathway consists of afferent, efferent, and parallel pathways arranged as an intricate circuitry connecting different auditory and nonauditory structures that conduct, analyze, compare, code, and decode acoustic information. Such complex interactions of neuronal fibers are the basis for the perspective that the auditory system is actually a non modular system (Musiek et al. 2005). Based on this perspective, neurons in the auditory areas, which are mainly activated by acoustic signals, might also be activated by other signals (e.g., visual). Thus, a lesion around the auditory areas, even if well-limited within the area, may give rise to nonauditory symptoms. In addition, the auditory cortex also presents a large number of connections with regions related to nonsensory processes that influence how people respond to sounds (e.g., pre-frontal cortex and decision making) (Moore 2012). In summary, with such intricate multisensory and cognitive interactions, it seems natural that lesions and dysfunctions related to the auditory pathway generate such a heterogeneous disorder (for a more detailed description of the interactivity of the auditory system refer to Kraus and White-Schwoch 2015).

If APD is considered a neurobiological alteration, children's deficits could be related to acquired neurological disorders (lesions), neuromaturational delays, or neuroanatomical abnormalities (Chermak and Musiek 2014). Neural maturation follows a caudal-rostral sequence, with peripheral functions reaching maturity around birth and axonal, dendritic, synaptic maturation, and myelination continuing to develop in the brainstem into early childhood and, in the cerebral cortex, into late childhood (Moore and Linthicum 2007; Moore 2012). Auditory development and speech perception are guided by relevant acoustic and linguistic information experienced early in life to ensure cortical maturation (Kuhl 2000). Consequently, children with delayed development, either related to sound deprivation or delayed myelination, might not present with the expected auditory performance of their age, possibly delaying language acquisition and learning (see also Jeng, Chap. 2; Reetzke, Xie, and Chandrasekaran, Chap. 10).

Sometimes, children might have variations in the development of their brain structures, thus presenting both anatomical and functional abnormalities (Chermak and Musiek 2014). For instance, Boscariol et al. (2011) observed that children with language learning problems who had perisylvian polymicrogyria (i.e., an abnormal number of gyri along the Sylvian fissure on the cortex) had poorer performance on auditory processing tests than their peers who also had language problems but normal magnetic resonance imaging (MRI) scans.

It is thought that poor synchrony of neural firing can result in poor representation of sound at all system levels, deficits in conduction velocities of such a signal, abnormalities in inter-hemispheric transmission, desynchronized activation of interneurons affecting the inhibition/activation of specific neurons, etc. Such alterations at the neuronal level could lead to the deficits typically observed in APD, such as the following: difficulties in sound localization, which depend on simultaneous activation of coincidence cells and binaural hearing; pattern recognition, which depends on frequency and duration cues; hearing performance for competitive and degraded sounds, which depends on interhemispheric analysis and intrinsic redundancies, respectively; and analysis of temporal aspects (Musiek and Chermak 2014).

Electrophysiological measures can provide an objective biological means of investigating the auditory processing of sound and the underlying physiological mechanisms (He et al. 2012; Schochat and Musiek 2006). Integrity of the auditory pathways is a necessary condition for the normal development of auditory abilities, and obtaining direct information on the function of the auditory pathway assists in intervention programming and contributes to monitoring treatment outcomes. Given the frequent comorbidity of APD with language, learning, and attention deficits, the inclusion of electrophysiological measures in clinical assessments is gaining traction in clinical practice.

9.3 Electrophysiological Evaluation of Auditory Processing Disorder

Davis (1939) was the first to report an auditory-evoked response. Knowledge of AEPs has grown as a result of increases in computing power, enhanced signal processing strategies, and the urgent need to develop an objective measure of hearing. Audiologists first focused on the auditory brainstem response (ABR). By contrast, psychologists, psychiatrists, and neurologists focused on late potentials because these evoked response potentials (ERPs) are associated with both perceptual and cognitive processes (McPherson 1996). The AEPs are briefly reviewed to provide a broader perspective on how cABR/FFR contributes to the study, diagnosis, and management of APD.

9.3.1 Click Auditory Brainstem Response (ABR)

Jewett and Williston (1971) explained the clinical advantages of ABR over the already existing late-evoked-potential because of its reliability and independence from the patient′s state of arousal and its value in neurological applications. The conventional ABR is generated by neurobiological activity within the auditory nerve and the central auditory pathways in response to click, tone, or chirp stimuli, and it has a strong track record of being sensitive (able to detect a true positive) and specific (able to detect a true negative) to confirmed lesions of the auditory brainstem (see Starr and Achor 1975; Musiek and Lee 1995; and for a review, Hall 2015). However, click-evoked ABRs are generally normal in children with auditory based learning problems, such as dyslexia and specific language impairment (Billiet and Bellis 2011; Kumar and Singh 2015), and learning disorders (Song et al. 2006; Reetzke, Xie, and Chandrasekaran, Chap. 10).

Since in individuals with APD the proportion of abnormalities in click-evoked ABR is much lower than in middle and long latency potentials (Hall and Johnson 2007), it was suggested that individuals with APD either do not have brainstem lesions/dysfunctions, or do not have difficulties in processing simple acoustic information such as the click (Filippini and Schochat 2009). Recently, there has been an effort to use more complex stimuli (e.g., speech segments) in an attempt to measure subtle processes in the brainstem related to more complex sound processing. This potential will be described in depth in Sect. 9.4.

9.3.2 Auditory Middle Latency Response (MLR)

The MLR is typically evoked by a click or tone burst for neurodiagnosis or audiometry (Hall 2015). The *wave Pa* is the most reliable of the MLR waves (i.e., it is the most visible and robust), but amplitude and latency of the waves are widely variable and, as a consequence, this potential is not yet being used in clinical settings. Furthermore, it is highly affected by sleep as Kraus et al. (1989) found out in their study. Wave Pa was present during wakefulness, stage 1, and REM sleep but poorly detected during sleep stage 4. During sleep stages 2 and 3, Pa detectability was variable. This inconsistency of MLRs in children is another reason that limits its use for a clinical purpose.

Since the early 1980s, the MLR has attracted interest as an electrophysiological measure of central auditory function. There is considerable evidence to suggest that the MLR is especially sensitive to lesions of the auditory cortex and thalamo-cortical connections (Özdamar and Kraus 1983; Schochat et al. 2004). Hit rates in the mid-70% range or better have been reported (Musiek et al. 1999). The amplitudes of the MLR are reduced when the auditory cortex has been damaged; conversely, the MLR amplitude is preserved in the presence of lesions in other areas of the cortex (Shehata-Dieler et al. 1991).

The goal of MLR waveform analysis is to ascertain the symmetry of Pa wave amplitude among the two-electrode or three-electrode arrays used. Although the MLR typically shows considerable inter-subject variability, there is reasonable intra-subject consistency with electrodes located over the auditory cortical regions (e.g., regions C3 and C4) and from a frontal midline electrode site (e.g., Fz or Cz). There is general agreement that latency is less clinically useful than amplitude in the detection of auditory system dysfunction (Hall 2015).

In a study of children with APD, Schochat et al. (2010) observed smaller amplitudes for C3-A1 and C3-A2 waveforms in comparison to the control group. After auditory training, the amplitudes of MLR's waveforms for both electrode positions were increased in the APD group, commensurate with control group values. Figure 9.1 shows an abnormal MLR (electrode effect) from a patient before auditory training and the normal MLR waveform after training.

Weihing et al. (2012) assessed children with normal audiograms and children with APD and observed that the relative difference measurements of the MLR were

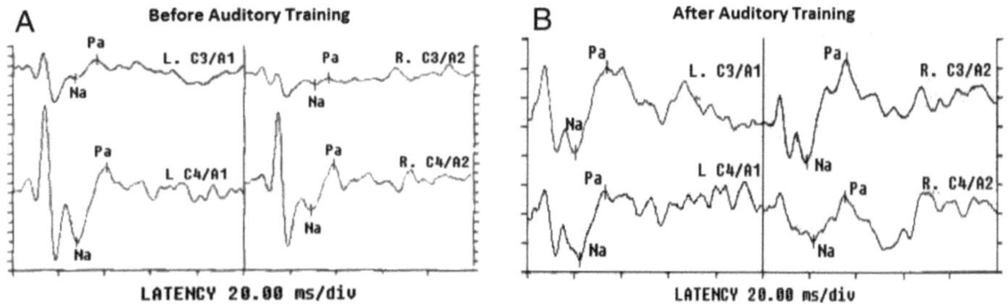

Fig. 9.1 **(A)** Abnormal electrode effect (left hemisphere) of middle latency response (MLR) waveform obtained from C3 and C4 electrode placements for a neurologically involved patient (Pilocytic Astrocytoma) *before* auditory training and **(B)** normal MLR waveform *after* training. *L*, left; *R*, right

less variable than the raw measurements when examined across subjects. The authors concluded that the analysis of the relative differences may provide better utility in the clinical diagnosis of central auditory deficits in children than absolute amplitude measures.

9.3.3 Auditory Late Potentials

Auditory late potentials (ALPs, also known as late-latency responses or LLRs) are determined more by the state of the nervous system when stimuli are presented rather than by the physical properties of the stimuli themselves. In this discussion, the late potentials include the N1P2N2 complex, P3 (P300), and MMN.

9.3.3.1 N1P2N2 Complex

The N1P2N2 complex is believed to originate from the supratemporal plane in or near the primary auditory cortex (Tremblay et al. 2001). This potential is believed to reflect the pre-attentive sensory encoding of the auditory stimulus attributes, including spectral and temporal cues critical to speech perception (McPherson 1996). Temporal cues are crucial to the differential sensitivity of intensity, frequency, and duration, and N1P2N2 components may provide not only a neurophysiological means of identifying dysfunction in this area but also objective monitoring of therapeutic progress in an intervention program (Musiek and Berge 1998). Studies in individuals with neurological lesions have shown latency and or amplitude effects on the N1P2N2 complex related to central auditory involvement in the auditory cortex and associated areas (Knight et al. 1988). In fact, these late potentials are compromised when auditory regions are involved, but this is not the case for nonauditory areas of the brain (e.g., frontal lobe) (Knight et al. 1980).

Studies in young adults with normal hearing have demonstrated that the objective cortical AEPs, in particular the N1 peak latency, can be used as a sensitive measure of temporal changes in sound stimuli at the level of the auditory cortex (Tremblay et al. 2001). However, the N1 component is either inconsistent or undeveloped from 3 to 8 years. The presence of the N1 becomes more reliable at approximately 9–11 years of age (Sharma and Mitchell 2013). In general, the dominant features of the cortical response in young children are the P1, which varies in latency as a function of age, and the N2, a negativity following the P1 at approximately 200 or 250 ms.

9.3.3.2 Acoustic Change Complex (ACC)

The ACC is a large response that can be elicited by changes in an ongoing sound, such as intensity or frequency modulations of a sustained tone, or in response to acoustic changes in more complex ongoing sounds such as speech (Small and Werker 2012). The ACC appears essentially as a second ALR waveform that follows the conventional onset-evoked N1P2N2 complex (Hall 2015). Despite the ACC's potential clinical applications, research on the relationship between the ACC response and auditory discrimination abilities is sparse, and the studies conducted so far have used different measures and procedures, which can affect the interpretation of results (He et al. 2012).

9.3.3.3 P300

The P300 is an endogenous potential that is triggered by the use of the "oddball paradigm" (i.e., an experimental method in which an infrequent or odd stimulus is presented among more frequent standard stimuli), which requires the listener to detect the infrequent stimulus. The P300 is highly dependent on attention as well as sensory processing. Generators of the P300 arise within the temporal lobe, limbic system, thalamus, and frontoparietal cortex. Although nonauditory regions contribute to the P300, there is evidence that lesions in auditory cortical regions compromise the latency and amplitude of the P300.

The P300 complex is affected by central auditory disorders in the auditory areas of the cerebrum. Knight et al. (1988) reported that individuals with lesions of the temporoparietal junction exhibited reduced P300 amplitudes compared with individuals with lesions limited to the parietal lobe. Notably, the P300 successfully identified those with acute and long-standing auditory effects due to traumatic brain injury (Segalowitz et al. 2001; Musiek et al. 2012), which is associated with auditory processing difficulties.

Jirsa and Clontz (1990) observed that the latency of the P300 was later for children with APD compared with a control group. Interestingly, Krishnamurti (2001) showed P300 latencies were delayed for an adult group with APD (diagnosed using behavioral central auditory tests) compared with a control group for binaural (tone) stimulation and for a competing noise condition. The latencies were later in the noise condition than the binaural condition for the APD group but not for the control group.

9.3.3.4 Mismatch Negativity (MMN)

The MMN is an event-related response that reflects the detection of acoustic change. The MMN is elicited using an oddball paradigm, which is based on the premise that a neural trace or template is formed to represent the standard stimulus and is held in short-term memory (Näätänen 1995), and, in contrast to the P300, can be elicited even when subjects ignore the sounds presented to them (Martin et al. 2007).

The advantage of the MMN is that it can be reliably recorded (Escera et al. 2000) even in the absence of a behavioral response or the individual's attention, for example, in sleeping infants (Ruusuvirta et al. 2009), stroke patients, and in comatose (Fischer and Luauté 2005) and persistent-vegetative-state patients (Wijnen et al. 2007). Two major theories have been proposed within the past years: one model that posits the formation of an auditory sensory memory based on the encoding of statistical regularities from the acoustic input (Näätänen et al. 2005), and a complementary view that ascribes deviance detection to a mechanism of release from neural adaptation (May and Tiitinen 2010).

To determine whether MMN is suitable to provide additional information in behavioral auditory central tests, Bauer et al. (2009) assessed 32 children with APD and 13 healthy children. The incidence of MMN was always higher in the healthy children, and they had peak latencies that occurred earlier at frontal, central, and temporal electrode sites.

Another promising use of MMN is the investigation of prosodic abilities of infants as early predictors of specific language impairment (SLI). Weber et al. (2005) based their study on the hypothesis that the prosodic abilities of infants at risk for SLI are less elaborated than those of controls because of the less efficient processing of relevant acoustic cues. In contrast to matched controls, infants with very low word production showed a smaller MMN. This amplitude difference indicates impaired prosodic processing of word stress during early development and thus may be taken as an early marker of risk for SLI.

Rocha-Muniz et al. (2015) investigated speech encoding in the auditory system of children with SLI and compared it with children with APD and children with typical development through the MMN paradigm. The data demonstrated abnormalities in the automatic discrimination of crucial acoustic components of speech sounds in children with SLI and children with APD. This might indicate problems in the physiological processes responsible for ensuring discrimination of acoustic contrasts in pre-attentional and pre-conscious levels, which could contribute to poor perception (also see Kraus et al. 1996). Figure 9.2 shows a normal speech-evoked MMN for a typical-development child and an abnormal speech-evoked MMN for a child with APD.

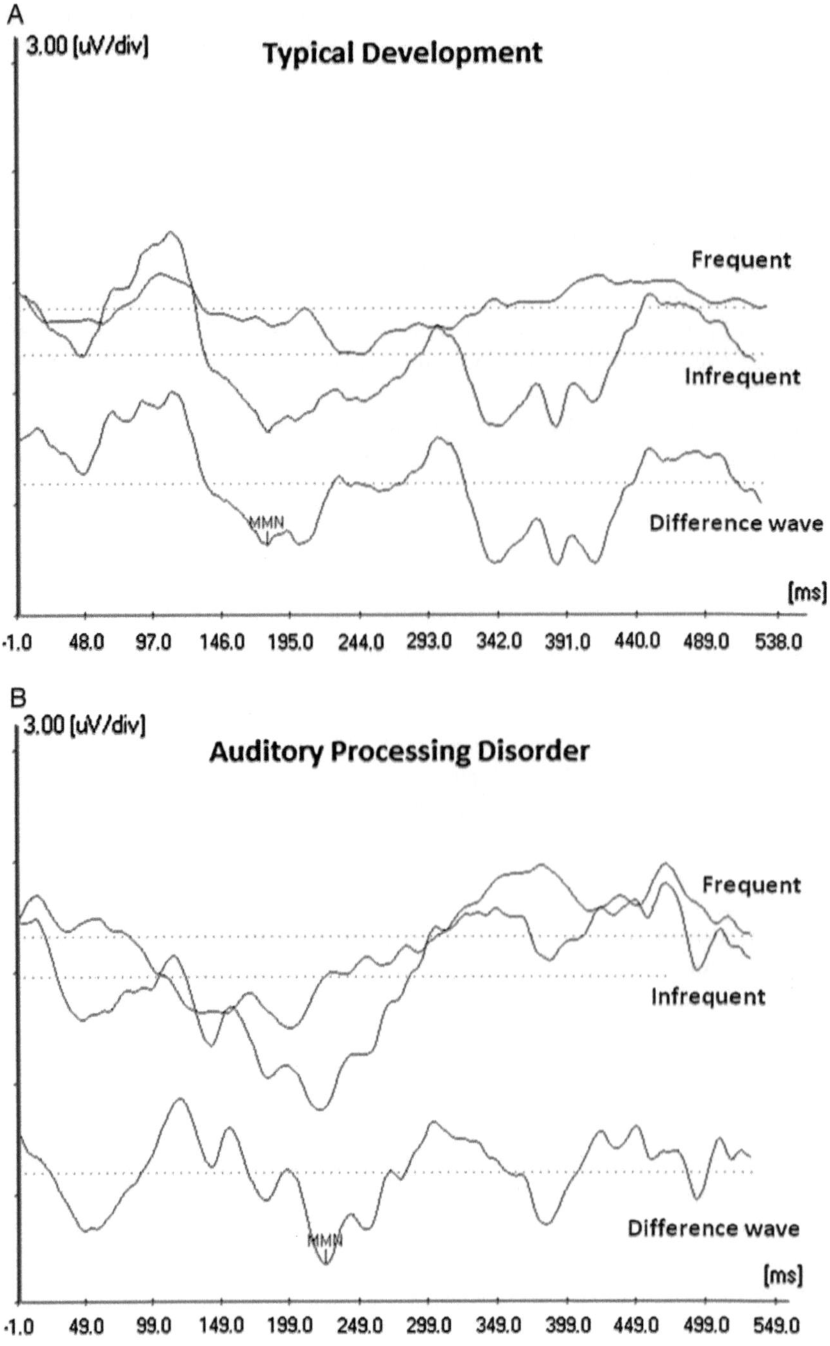
A
3.00 [uV/div]
Typical Development
Frequent
Infrequent
MMN
Difference wave
[ms]
-1.0 48.0 97.0 146.0 195.0 244.0 293.0 342.0 391.0 440.0 489.0 538.0
B
3.00 [uV/div]
Auditory Processing Disorder
Frequent
Infrequent
MMN
Difference wave
[ms]
-1.0 49.0 99.0 149.0 199.0 249.0 299.0 349.0 399.0 449.0 499.0 549.0

◀ **Fig. 9.2** Speech-evoked mismatch negativity (MMN) responses. The grand average *Difference wave* was obtained by subtracting the response to the *Frequent* stimulus from the response to the *Infrequent* stimulus. **(A)** Speech-evoked MMN responses from a child with typical development. **(B)** Speech-evoked MMN responses from a child diagnosed with auditory processing disorder (APD). (Adapted from Rocha-Muniz et al. 2015)

More studies are necessary to determine whether this potential will be useful in clinical settings regarding time and individual results, although it is well known that MMN has good potential to be used for groups (Hall and Johnson 2007; McGee et al. 1997). It is important to note that the AEPs described above are based on the measurement of latency and amplitude, which usually show a major difference in inter-individual comparison as well as large standard deviations in large samples, making it difficult for them to be a clinically useful tool. Additionally, MMNs (and ALPs in general) only offer measures of latency and amplitude, which are blunt and abstract measures of auditory function.

9.4 New Approach to APD Evaluation: Frequency-Following Response

In speech, timing and harmonic cues are important for distinguishing and discriminating consonants and vowels (Delattre et al. 1955; Blumstein and Stevens 1979), pitch cues are important for understanding prosody and intonation, and pitch and the highest harmonics are important for identifying who is speaking (Bachorowski and Owren 1999; Ladefoged 2006). In the last decade, considerable attention has been given to the representation of speech signals in the brainstem auditory pathway. The FFR, also known as the auditory brainstem response to complex sounds (or cABR), is an important method for studying auditory-neurophysiological processing in humans. More specifically, it can measure neural synchrony in response to the crucial phonemic features of speech.

Speech formants are preserved on the discharge periodicities and interspike rate at the auditory nerve fibers (Young and Sachs 1979; Delgutte 1980). Considering these neural properties, the FFR's electrophysiological responses are the synchronized activity (i.e., phase locked) of a population of neurons in the upper brainstem in response to the individual cycles corresponding to the periodicity of the stimulus frequency (Smith et al. 1975; Krishnan 2007). There is evidence to indicate that the FFR emerges from phase-locked activity at the level of the lateral lemniscus (LL) and/or inferior colliculus (IC) (Krishnan 2007; Chandrasekaran and Kraus 2010). Its phase-locking properties refer to a clear and fixed relationship between some aspect of the response and the phase (or time) of the stimulus. However, the location of the specific generator sites remains controversial (Akhoun et al. 2008).

Phase locking, as reflected in the FFR, has been demonstrated for numerous stimuli, such as complex steady-state stimuli (Greenberg et al. 1987; Krishnan 1999, 2002), two-component tones (Greenberg and Marsh 1979), inharmonic tones

(Chambers et al. 1986), pure tones (Moushegian et al. 1973; Ananthanarayan and Durrant 1992), time-variant stimuli (Krishnan and Parkinson 2000; Plyler and Ananthanarayan 2001), synthesized speech (consonant-vowel) stimuli (Cunningham et al. 2001; Russo et al. 2004), emotional stimuli (Strait et al. 2009), and musical stimuli (Musacchia et al. 2007; Parbery-Clark et al. 2011). Furthermore, the FFR has also been recorded in degraded listening conditions such as reverberation (Bidelman and Krishnan 2010) and noise (Cunningham et al. 2001, Russo et al. 2004). This phase-locking activity has been implicated in the temporal encoding of the spectra of steady-state and time-variant speech sounds (Young and Sachs 1979; Blackburn and Sachs 1990).

Numerous studies have used FFR to assess auditory pathway integrity and auditory processing due to an individual's unique ability for phase locking. Poor phase coherence in individuals with speech perception deficits (a form of an APD) has been demonstrated by Ali and Jerger (1992). Using auditory steady state responses (ASSRs), which are 'phase-locking-dependent' and elicited by periodic signals similar to FFR, the authors observed that phase coherence was significantly poorer in the group with disproportionate speech understanding scores. Another study using ASSRs observed significantly increased thresholds in the APD group and dyslexia group compared with typically developed children (Simões 2009). These data suggest that an underlying temporal processing deficit (as a result of poor phase locking) could be contributing to these results.

The brainstem response to a consonant-vowel (CV) speech syllable (FFR in response to speech sounds) was first used by Kraus and colleagues (Cunningham et al. 2001) in a study of children with learning problems and speech sound-perception deficits. FFR in response to speech sounds consists of two separate portions: the transition, corresponding to the CV formant transition (onset); and the sustained stage, corresponding to the relatively unchanging vowel (Russo et al. 2004; Akhoun et al. 2008).

For consonants, the transient onset response marks the beginning component characterized by a harmonic and broadband frication (onset burst) and is followed by a harmonically rich and spectrally dynamic formant transition. The sustained FFR component is synchronized to the periodicity (repeating aspects) of the sound with each cycle faithfully representing its temporal structure, and it has an upper limit of approximately 1000 Hz for neural phase-locking properties (Chandrasekaran and Kraus 2010).

Taken together, the key elements of the brainstem response to speech represent its features with remarkable fidelity. These elements are recorded from the scalp and are presumed to faithfully reflect activity from an ensemble of neural elements within the central auditory pathway (Kraus and Hornickel 2012). The onset, transition, and sustained FFR components reflect the output of brainstem and midbrain structures and encode stimulus-related information with high temporal and spectral accuracy.

Latency measures and the wave morphology of the FFR are apparent and interpretable in infants (Anderson et al. 2015) and preschoolers (White-Schwoch et al. 2015a, b). However, early childhood is marked by a systematic speeding up of

FFR latencies (Skoe et al. 2013). Around 8 years-of-age, latencies are their earliest, and then they slow down and stabilize to adult-like latencies until senescence (Anderson et al. 2012).

The fidelity of the response to the stimulus enables the evaluation of the strength of subcortical encoding of multiple acoustic aspects of complex sounds, including timing (onsets, offsets), pitch (fundamental frequency, F_0), and timbre (integer harmonics of the F_0) (Skoe and Kraus 2010). The analyses of the FFR include measurement of latency and amplitude in the time domain and magnitude of the individual harmonics in the frequency domain (see Sect. 9.4.1). Because of the FFR's remarkable stimulus fidelity, cross-correlation between the stimulus and the response is also a meaningful measure (Krishnan et al. 2005). In addition, responses obtained from two conditions can be cross-correlated to determine the effects of a specific condition (e.g., noise) on a response (Russo et al. 2004).

Temporal and harmonic cues are important for distinguishing consonants, and the CV transition is perceptually vulnerable, particularly in background noise (Nishi et al. 2007), suggesting that temporal perception is important for understanding the linguistic content of speech in noise. The FFR in background noise has shown markedly worse degradation in the portion correspondent to the consonant transition in comparison to the vowel portion (White-Schwoch et al. 2015a). Based on these findings, the authors suggested that in preschoolers the acoustic processing of dynamic speech components may be more susceptible to noise interference than the processing of static features, at least regarding midbrain coding.

The FFR is similar to the evoking stimulus acoustically and visually (Galbraith et al. 1995) so the accuracy of encoding specific speech features, such as timing, pitch, and harmonics, is reasonable to assess to an extent that is not possible when using slower, cortical potentials like MLR and late latency responses. The following section will describe the applications of the FFR for assessment of APD.

9.4.1 Speech-Evoked FFR: General Interpretation

As mentioned in Sect. 9.4, the evaluation of the neural response to speech components (timing, pitch, and harmonics) has been useful in understanding the biological bases of auditory processing in clinical populations (see also Reetzke, Xie, and Chandrasekaran, Chap. 10; Anderson, Chap. 11). Timing has been defined as the latency of each peak and reflects temporal precision of the synchronous neural activity with respect to the onset, periodicity, and offset of the stimulus. Such measures reflect the amount of activity that contributes to the generation of the peak and the temporal synchronization of the response (Russo et al. 2004). Thus, timing measures provide information on the precision with which the brainstem nuclei respond to the acoustic stimuli, whereas amplitude measures provide information on how robust that response is. Abnormalities in these measures might reflect differences in the velocity of signal conduction along dendrites and axonal projections, differences in the neurons' kinetic channels, or differences in the generators' synchronization (Johnson et al. 2007). Response timing is directly related to perceptual abilities.

In addition to the latency measures (i.e., measures in the time domain), it is possible, through a Fourier transformation, to represent the waveform obtained for the sustained portion in the frequency domain, allowing measurement of the response magnitude to a specific frequency or a range of frequencies (Aiken and Picton 2008; Banai et al. 2009). The F_0 provides information regarding pitch, allowing identification of a speaker or emotional intonation of the voice. The amplitude of F_1 (first formant) and HF (high frequency) provides phonetic information, making them important in distinguishing the contrasts between the speech sounds (Russo et al. 2004; Kraus et al. 2009).

Pitch is the perceptual correlate of the periodicity, or repetition rate, of an acoustic waveform (Oxenham 2012). The most important determinant of the pitch of a sound is likely its periodicity. Periodicity is most often quantified by the F_0, which has major contributions to the percept (Cruttenden 1997). The F_0 reveals information regarding voicing and manner of speech and also contributes to prosodic features of speech such as stress and intonation.

The F_0 is preserved in the phase-locked neural activity generating the FFR for steady-state complex tones (Greenberg et al. 1987), time-varying pitch contours of lexical tones (Krishnan et al. 2004), and speech stimuli (consonant-vowel) (Cunningham et al. 2001). Thus, electrophysiological FFR measures should provide an important measure of the neural processing of the F_0.

Pitch is encoded in the early stations of the auditory system by the temporal patterns of spikes. Therefore, difficulties in extracting and correctly using prosodic features of speech are symptoms of APD. Due to neural encoding by temporal patterns and synchrony of neuron spikes, prosodic deficits have been attributed to temporal auditory processing disorders (Gelfand 2010).

Further analysis of the spectral content of the response includes the harmonics, which are shaped by the articulators producing the speech formants (i.e., information about the message or verbal meaning of the utterance). Harmonics refer to the spectral characteristics of an auditory object apart from pitch. The harmonics of a sound are expressed by spectrotemporal properties, that is, the changes in the amplitudes of the sound's constituent frequencies over time (Janata 2015). Peaks in the harmonic spectrum are referred to as formants (e.g., F_1, F_2, F_3, etc.), and they are needed to discern the content of speech.

This measurement of FFR in response to speech sounds was defined as the neural activity that arose due to the harmonics of the fundamental frequency (Banai et al. 2009). The formant structure of the signal, determined by the filtering of the harmonics by the articulators, identifies the speech signal independent of pitch (Hornickel et al. 2012a; Skoe and Kraus 2010). Therefore, the first formant (F_1) and HF measures of the FFR appear to be consistent with phonological processing.

9.5 Auditory Processing Disorder in Light of FFR

Some children with language deficits, either dyslexia or SLI, have deficits in auditory ability compared to typically developing children. According to some authors, these clinical communication disorders may lead to difficulty processing rapidly presented auditory stimuli (Tallal 1976; Wright et al. 1997).

Regarding FFR, studies have shown that clinical populations have distinct patterns of response that may offer converging information about speech-sound coding and, potentially, about weaknesses in the neural processes that are important for everyday communication and auditory processing at large (Kraus and Nicol 2014; Kraus and White-Schwoch 2015). In other words, the evoked response for timing (onsets, offsets), pitch (F_0), and timbre (representation of formants above F_0) are shown to be inefficient or reduced in different ways for distinct clinical populations (Reetzke, Xie, and Chandrasekaran, Chap. 10; Anderson, Chap. 11).

A neural signature of FFR findings has been seen in children with reading and language impairments, that is, they have delayed neural responses and reduced representation of higher speech harmonics (Kraus and Anderson 2016). One reason for these results may be a decrease in the synchrony of neural firing, which can lead to abnormal perceptual abilities.

Complex sounds are susceptible to auditory temporal masking, which is a phenomena in which a brief signal might have its perception influenced by an immediately following sound (backward masking) or immediately preceding sound (forward masking). This type of interaction between temporally close sounds happens frequently in natural speech, and difficulties in overcoming such influence have been linked to language impairments and auditory temporal processing deficits (Wright et al. 1997; Montgomery et al. 2005). Johnson et al. (2007) showed that children with learning impairments who had the slowest response timing to the offset of the speech syllable also had the worst behavioral auditory backward-masking thresholds. In contrast, the children with learning impairments who did not have abnormal offset response timing had backward-masking thresholds within normal limits. The authors suggested that it is not the presence of a learning disorder that results in poor backward-masking performance but rather the inability of the nervous system to faithfully represent the final portion of the stimulus that leads to difficulties in distinguishing a tone from a following masking sound.

Banai et al. (2005) showed that in addition to cortical processing deficits, brainstem responses to speech are abnormal in approximately one-third of children diagnosed with language-based learning problems. In another study, Banai et al. (2009) found poor timing of subcortical auditory encoding and reduced amplitudes for both middle and higher harmonics in the group with reading impairment compared to the group with normal reading ability.

Many other studies have shown that children with a wide range of learning impairments present delayed responses when compared with their typically developing peers (Cunningham et al. 2001; Wible et al. 2004). The delayed responses observed in children with APD and language-based learning disabilities are consistent with a possible interruption of timing representation in the brainstem responses (King et al. 2002; Wible et al. 2005). These responses are vulnerable to degradation in background noise (Nishi et al. 2007), but children with reading and language disorders have difficulty differentiating speech even in quiet conditions (Banai et al. 2005; Rocha-Muniz et al. 2012).

Delayed neural timing—but normal spectral content of the neural response—is also found in children who have been diagnosed specifically with APD, despite the fact that only a few studies have used the FFR to speech sounds to investigate neural speech processing in this specific clinical population. Rocha-Muniz et al. (2012) compared FFR to a 40-ms [da] in three groups of children (ages 6–12) who were typically developing (TD), diagnosed with APD, or diagnosed with SLI. The authors reported abnormal neural encoding of timing measures in the APD and SLI groups. Compared with TD children, the auditory and language impairment groups demonstrated prolonged and less synchronized onset responses (Fig. 9.3). These findings suggest that the neural encoding of acoustic characteristics that vary over time, particularly in the case of rapid changes, may be impaired in these children.

While children with APD and SLI appear to have overlapping abnormal neural processes, they also appear to have distinct neural signatures. Both groups had delayed timing compared to the TD group, but the SLI group presented a more pervasive timing delay than the APD group. In addition, these distinctions are more evident in the spectral content of their response. The SLI group had reduced amplitudes for the higher harmonics compared to either the APD or TD groups.

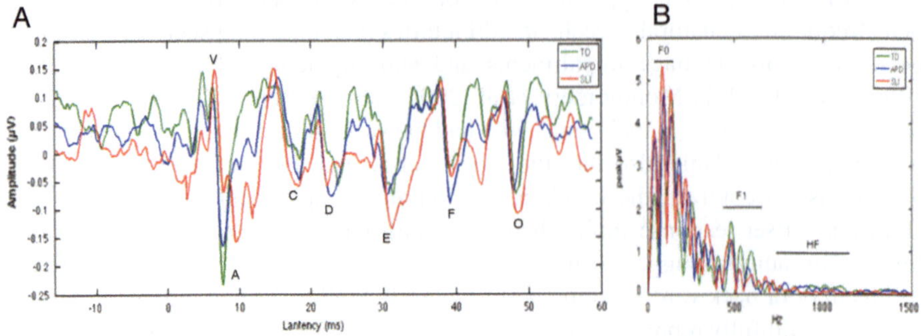

C

	Mean (ms)							Mean (µV)		
	V	A	C	D	E	F	O	F_0	F_1	HF
TD	6.32	7.87	17.57	22.83	30.64	39.37	48.01	5,91	1,32	0,44
APD	6.43	7.93	18.39	22.83	30.64	39.97	48.95	5,46	1,19	0,45
SLI	6.78	8.61	18.76	23.72	31.79	41.14	49.27	6,74	0,96	0,29

Fig. 9.3 **(A)** Grand averages of the time domain response in children with typical development (TD, *solid green lines*), auditory processing disorder (APD, *solid blue lines*), and specific language impairment (SLI, *solid red lines*). Subcortical timing response to speech is delayed in SLI and APD compared with TD. **(B)** Grand-averaged spectra over the formant transition period for three groups. *HF* is the region of the neural responses that differs significantly between children with SLI (lower magnitudes) and the other two groups (TD and APD). **(C)** Mean latency for each peak and mean magnitude of response for each frequency region, according to group. (Reprinted with permission from Rocha-Muniz et al. 2014)

The atypical activity in the high frequencies in the SLI groups (but normal for APD) might reflect group differences in frequency-specific encoding, primarily occurring at higher frequencies that require more precise, rapid activation and recovery mechanisms. This result suggests an underlying weakness in phase locking of the involved neurons and, consequently, poor speech understanding.

Although children with APD or language/reading problems with clinical presentation of APD commonly have temporal auditory processing deficits and abnormal FFR timing, some researchers have demonstrated intact subcortical neural representation of pitch (F_0) using FFR (Hornickel et al. 2012b; Rocha-Muniz et al. 2012). This is in contrast to the more pervasive deficits in children with autism, which include pitch encoding (Russo et al. 2008). Though consistent with behavioral findings (Marshall et al. 2009), this pattern needs to be interpreted with caution, particularly because the pitch trajectory of the stimulus used in these studies was not ecological (i.e., does not occur in natural language). There are some findings showing an FFR advantage in native Mandarin speakers when using an ecological pitch contour (Xu et al. 2006; Chandrasekaran et al. 2007; see Krishnan and Gandour, Chap. 3 for a discussion of how language experience shapes brainstem processing of pitch contours, and Carcagno and Plack, Chap. 4 for a discussion of perceptual learning on pitch tracking).

Abnormalities in FFR have been repeatedly reported in aging populations as well (Anderson et al. 2011, 2012, 2013). Even older adults with normal hearing thresholds report trouble hearing in background noise, echoing one of the primary elements of APD. The FFR in this population is characterized by delayed neural timing (peak latencies) in response to consonant-vowel transition, supporting a deficit in central processing in older adults (Vander Werff and Burns 2011; Anderson et al. 2012). This delay is similar to that seen in children with SLI, dyslexia, APD, and risk for language impairment (White-Schwoch et al. 2015b; Kraus and Anderson 2016).

Despite the influence of peripheral and cognitive factors in both children with language-based learning impairments and older adults, deficits in the FFR are likely to have different etiologies. In older adults, such deficits may arise from peripheral neurodegeneration, cognitive declines, and changes in the balance of neurotransmitters. On the other hand, auditory processing deficits in children are likely to result from a malfunction in making effective sound-to-meaning connections that are necessary for language learning (Hornickel and Kraus 2013). It is important to have in mind that APD may arise from various sources of impairment or delayed development, including auditory nerve, brainstem, auditory cortex, prefrontal cortex, corpus callosum, or other areas (Medwetsky 2011). The FFR may be influenced by impairments in these areas, but the neural patterns may depend upon the specific nature of the impairment.

Although the use of FFR to speech stimuli has exhibited promising results in research settings for more than a decade, its application in clinical practice is still a significant challenge. Rocha-Muniz et al. (2014) aimed to verify the sensitivity, specificity, and efficiency of this AEP. The sensitivity of a clinical test refers to the ability of the test to correctly identify those individuals with the disorder (i.e., true positives). The specificity of a clinical test refers to the ability of the test to correctly identify those individuals in whom the disorder is absent (i.e., true negatives). Efficiency measures the degree of veracity of a diagnostic test on a condition. The study showed approximately 70% efficiency (analyzing the latency of A-peak) in the identification of auditory-language-based disorders (APD and/or SLI). This study is particularly important given the shortcomings of the behavioral APD test battery. Most of the tests widely used in the APD evaluation have been validated with patients with known brain lesions (Musiek et al. 1991), and it is questionable whether this is an appropriate group to presume the effects of APD, particularly in children.

In contrast, Rocha-Muniz et al. (2014) validated the ability of the FFR to identify individual children with APD and SLI diagnoses. Rocha-Muniz et al. (2016) found that 85% of children with an abnormal FFR performed abnormally on the APD tests. Banai et al. (2005) also observed that approximately 80% of poor readers (children) had abnormal brainstem timing measures. Additionally, White-Schwoch et al. (2015b) developed a composite FFR measure that identifies 70% of children with a learning disability and may predict future auditory processing skills. In conclusion, the data presented in this section could be considered evidence that APD might best be viewed as part of a multicomponent characterization of developmental learning/language disorders and, taken as a whole, studies using FFR to speech stimuli have shown patterns of subcomponents of FFR associated with clinical populations.

9.6 Future Directions

As discussed in this chapter, the FFR may be a useful tool for investigating APD and other disorders as well as the occurrence of comorbidities. Future research could include testing individuals with defined lesion sites. Such studies would provide information about neural origins and insight as to its sensitivity to central auditory disorders with a neurological basis.

Skoe et al. (2013) studied a large dataset of FFRs and found that for each set of FFR measures (latency, frequency encoding, response consistency, nonstimulus activity), developmental changes continue well past the age of two. Their data call into question the conventional wisdom that the auditory brainstem is mature by 2 years of age. The authors suggest that future studies should measure neurophysiological and behavioral development in parallel to assess whether the developmental trajectories of the ABR can be followed by specific perceptual or linguistic skills, which is difficult because the many behavioral tests cannot be easily applied to infant, pediatric, adult, and geriatric populations, whereas ABRs use the same testing protocol for all ages.

Additionally, the FFR has the potential to be used to evaluate the nature of the neural representation of speech sounds processed by different strategies employed in hearing aids and cochlear implants (see Anderson, Chap. 11). Processing strategies could be modified to provide the optimal FFR neural representation of acoustic features that are important for the identification and discrimination of speech in individual listeners.

9.7 Summary

The importance of auditory processing skills for successful language learning and later academic achievement was discussed. The need for viable objective and biological measures that are less impacted by nonauditory factors, such as alertness/fatigue and comorbidities, to assess these skills, was also covered.

In addition to providing a biological dimension for assessing the mechanisms that underlie listening disorders, auditory evoked potentials add sensitivity to behavioral assessments. Studies on FFR have shown that these potentials provide a more detailed account of acoustic processing in comparison with other AEPs. How pitch, harmonics and timing may be studied in APD through analysis of FFR was also illustrated.

In summary, this chapter reveals how FFR may provide an objective assessment to enhance the monitoring and understanding of APD and provide insight about the mechanisms of neural encoding that underly auditory processing.

Compliance with Ethics Requirements Eliane Schochat, Caroline Nunes Rocha-Muniz, and Renata Filippini declared that they had no conflicts of interest.

References

AAA (American Academy of Audiology). (2010). American Academy of Audiology clinical practice guidelines: Diagnosis, treatment and management of children and adults with central auditory processing disorder. American Academy of Audiology. http://audiology-web.s3.amazonaws.com/migrated/CAPD%20Guidelines%208-2010.pdf_539952af956c79.73897613.pdf. (Accessed on October 8, 2015)

ASLHA. (2005). *(Central) auditory processing disorders.* Technical Report. American Speech-Language-Hearing Association. Doi:10.1044/policy.PS2005-00114

Aiken, S. J., & Picton, T. W. (2008). Envelope and spectral frequency-following responses to vowel sounds. *Hearing Research, 245*(1), 35–47.

Akhoun, I., Gallego, S., Moulin, A., Ménard, M., et al. (2008). The temporal relationship between speech auditory brain stem responses and the acoustic pattern of the phoneme/ba/ in normal-hearing adults. *Clinical Neurophysiology, 119*(4), 922–933.

Ali, A. A., & Jerger, J. (1992). Phase coherence of the middle-latency response in the elderly. *Scandinavian Audiology, 21*(3), 187–194.

Ananthanarayan, A. K., & Durrant, J. D. (1992). The frequency following response and the onset response: Evaluation of frequency specificity using a forward-masking paradigm. *Ear and Hearing, 13*(4), 228–233.

Anderson, S., Parbery-Clark, A., Yi, H., & Kraus, N. (2011). A neural basis of speech-in-noise perception in older adults. *Ear and Hearing, 32*(6), 750–757.

Anderson, S., Parbery-Clark, A., White-Schwoch, T., & Kraus, N. (2012). Aging affects neural precision of speech encoding. *The Journal of Neuroscience, 32*(41), 14156–14164.

Anderson, S., White-Schwoch, T., Choi, H. J., & Kraus, N. (2013). Training changes processing of speech cues in older adults with hearing loss. *Frontiers in Systems Neuroscience, 7*(97). Doi:10.3389/fnsys.2013.00097

Anderson, S., Parbery-Clark, A., White-Schwoch, T., & Kraus, N. (2015). Development of subcortical speech representation in infant humans. *Journal of the Acoustical Society of America, 137*(6), 3346–3355.

BSA (British Society of Audiology). (2011). Auditory processing disorder. Position statement. British Society of Audiology. http://www.thebsa.org.uk/wpcontent/uploads/2014/04/BSA_APD_PositionPaper_31March11_FINAL.pdf (Accessed October 8, 2015).

Bachorowski, J. A., & Owren, M. J. (1999). Acoustic correlates of talker sex and individual talker identity are present in a short vowel segment produced in running speech. *The Journal of the Acoustical Society of America, 106*(2), 1054–1063.

Banai, K., & Kraus, N. (2014). Auditory processing (disorder): An intersection of cognitive, sensory and reward circuits. In F. E. Musiek & G. D. Chermak (Eds.), *Handbook of central auditory processing disorder* (pp. 191–210). San Diego, CA: Plural Publishing.

Banai, K., Nicol, T., Zecker, S., & Kraus, N. (2005). Brain stem timing: Implications for cortical processing and literacy. *The Journal of Neuroscience, 25*(43), 9850–9857.

Banai, K., Hornickel, J. M., Skoe, E., Nicol, T., et al. (2009). Reading and subcortical auditory function. *Cerebral Cortex, 19*(11), 2699–2707.

Bauer, P., Burger, M., Kummer, P., Lohscheller, J., et al. (2009). Correlation between psychometric tests and mismatch negativity in preschool children. *Folia Phoniatrica et Logopaedica, 61*(4), 206–216.

Bellis, T. J. (2007). Historical foundations and the nature of (central) auditory processing disorder. In G. D. Chermak & F. E. Musiek (Eds.), *Handbook of (central) auditory processing disorder: Auditory neuroscience and clinical diagnosis* (pp. 119–136). San Diego: Plural Publishing.

Bellis, T. J., & Bellis, J. D. (2015). Central auditory processing disorders in children and adults. In G. G. Celesia & G. Hickok (Eds.), *Handbook of clinical neurology. The human auditory system* (pp. 537–556). London: Elsevier.

Bidelman, G. M., & Krishnan, A. (2010). Effects of reverberation on brainstem representation of speech in musicians and non-musicians. *Brain Research, 1355*, 112–125.

Billiet, C. R., & Bellis, T. J. (2011). The relationship between brain stem temporal processing and performance on tests of central auditory function in children with reading disorders. *Journal of Speech, Language, and Hearing Research, 54*(1), 228–242.

Blackburn, C. C., & Sachs, M. B. (1990). The representations of the steady-state vowel sound/e/in the discharge patterns of cat anteroventral cochlear nucleus neurons. *Journal of Neurophysiology, 63*(5), 1191–1212.

Blumstein, S. E., & Stevens, K. N. (1979). Acoustic invariance in speech production: Evidence from the spectral characteristics of stop consonants. *The Journal of the Acoustical Society of America, 66*(4), 1001–1017.

Bocca, E., Calearo, C., & Cassinari, V. A. (1954). New method for testing hearing in temporal lobe tumors: Preliminary report. *Acta Oto-Laryngologica, 44*(3), 219–221.

Boscariol, M., Guimaraes, C. A., Hage, S. R. V., Garcia, V. L., et al. (2011). Auditory processing disorder in patients with language-learning impairment and correlation with malformation of cortical development. *Brain & Development, 33*(10), 824–831.

Chambers, R., Feth, L., & Burns, E. (1986). The relation between the human frequency following response and the low pitch of complex tones. *The Journal of the Acoustical Society of America, 80*, 1673–1680.

Chandrasekaran, B., & Kraus, N. (2010). The scalp-recorded brain stem response to speech: Neural origins and plasticity. *Psychophysiology, 47*(2), 236–246.

Chandrasekaran, B., Krishnan, A., & Gandour, J. T. (2007). Experience-dependent neural plasticity is sensitive to shape of pitch contours. *Neuroreport, 18*(18), 1963–1967.
Chermak, G., & Musiek, F. (1997). *Central auditory processing disorders: New perspectives*. San Diego, CA: Singular.
Chermak, G. D., & Musiek, F. E. (2014). Neurological substrate of central auditory processing disorder. In F. E. Musiek & G. D. Chermak (Eds.), *Handbook of central auditory processing disorder* (pp. 89–112). San Diego, CA: Plural Publishing.
Cruttenden, A. (1997). *Intonation* (2nd ed.). New York: Cambridge University Press.
Cunningham, J., Nicol, T., Zecker, S. G., Bradlow, A., & Kraus, N. (2001). Neurobiologic responses to speech in noise in children with learning problems: Deficits and strategies for improvement. *Clinical Neurophysiology, 112*(5), 758–767.
Davis, P. A. (1939). Effects of acoustic stimuli on the waking human brain. *Journal of Neurophysiology, 2*(6), 494–499.
Dawes, P., Bishop, D. V. M., Sirimanna, T., & Bamiou, D. E. (2008). Profile and aetiology of children diagnosed with auditory processing disorder (APD). *International Journal of Pediatric Otorhinolaryngology, 72*(4), 483–489.
Delattre, P. C., Liberman, A. M., & Cooper, F. S. (1955). Acoustic loci and transitional cues for consonants. *The Journal of the Acoustical Society of America, 27*(4), 769–773.
Delgutte, B. (1980). Representation of speech-like sounds in the discharge patterns of auditory-nerve fibers. *The Journal of the Acoustical Society of America, 68*(3), 843–857.
Escera, C., Yago, E., Polo, M. D., & Grau, C. (2000). The individual replicability of mismatch negativity at short and long inter-stimulus intervals. *Clinical Neurophysiology, 111*(3), 546–511.
Filippini, R., & Schochat, E. (2009). Brainstem evoked auditory potentials with speech stimulus in the auditory processing disorder. *Brazilian Journal of Otorhinolaryngology, 75*(3), 449–455.
Fischer, C., & Luauté, J. (2005). Evoked potentials for the prediction of vegetative state in the acute stage of coma. *Neuropsychological Rehabilitation, 15*(3–4), 372–380.
Galbraith, G. C., Arbagey, P. W., Branski, R., Comerci, N., & Rector, P. M. (1995). Intelligible speech encoded in the human brain stem frequency-following response. *Neuroreport, 6*(17), 2363–2367.
Gelfand, S. A. (2010). Auditory nerve. In S. A. Gelfand (Ed.), *Hearing: An introduction to psychological and physiological acoustics* (5th ed., pp. 103–121). London: Informa Healthcare.
Greenberg, S., & Marsh, J. T. (1979). Spectral basis of human frequency-following response to the missing fundamental. *The Journal of the Acoustical Society of America, 66*, s33. (Abstract)
Greenberg, S., Marsh, J. T., Brown, W. S., & Smith, J. C. (1987). Neural temporal coding of low pitch. I. Human frequency-following responses to complex tones. *Hearing Research, 25*(2), 91–114.
Hall, J. W., III, & Johnson, K. (2007). Electroacoustic and electrophysiologic auditory measures in the assessment of (C)APD. In G. D. Chermak & F. E. Musiek (Eds.), *Handbook of (central) auditory processing disorder: Auditory neuroscience and clinical diagnosis* (pp. 287–315). San Diego, CA: Plural Publishing.
Hall, J. W., III. (2015). e*Handbook of auditory evoked responses: Principles, procedures & protocols.* St. Augustine, FL: Pearson Education, Inc.
He, S., Grose, J. H., & Buchman, C. A. (2012). Auditory discrimination: The relationship between psychophysical and electrophysiological measures. *International Journal of Audiology, 51*(10), 771–782.
Hornickel, J., & Kraus, N. (2013). Unstable representation of sound: A biological marker of dyslexia. *The Journal of Neuroscience, 33*(8), 3500–3504.
Hornickel, J., Knowles, E., & Kraus, N. (2012a). Test-retest consistency of speech-evoked auditory brain stem responses in typically-developing children. *Hearing Research, 284*(1–2), 52–55.
Hornickel, J., Anderson, S., Skoe, E., Yi, H., & Kraus, N. (2012b). Subcortical representation of speech fine structure relates to reading ability. *NeuroReport, 23*(1), 6–9.

Janata, P. (2015). Neural basis of music perception. In G. G. Celesia & G. Hickok, (Eds.), *Handbook of clinical neurology. The human auditory system: Fundamental organization and clinical disorders* (pp. 187–205). Amsterdam: Elsevier.

Jerger, J., & Musiek, F. (2000). Report of the consensus conference on the diagnosis of auditory processing disorders in school-aged children. *Journal of American Academy of Audiology, 11* (9), 467–474.

Jewett, D. L., & Williston, J. S. (1971). Auditory-evoked far fields averaged from the scalp of humans. *Brain, 94*(4), 681–696.

Jirsa, R. E., & Clontz, K. B. (1990). Long latency auditory event-related potentials from children with auditory processing disorders. *Ear and Hearing, 11*(3), 222–232.

Johnson, K., Nicol, T., Zecker, S., & Kraus, N. (2007). Auditory brain stem correlates of perceptual timing deficits. *Journal of Cognitive Neuroscience, 19*(3), 376–385.

Kimura, D. (1961). Some effects of temporal-lobe damage on auditory perception. *Canadian Journal of Psychology, 15*(3), 156–165.

King, C., Warrier, C. M., Hayes, E., & Kraus, N. (2002). Deficits in auditory brain stem pathway encoding of speech sounds in children with learning problems. *Neuroscience Letters, 319*(2), 111–115.

Knight, R. T., Hillyard, S. A., Woods, D. L., & Neville, H. J. (1980). The effects of frontal and temporal-parietal lesions on the auditory evoked potential in man. *Electroencephalography and Clinical Neurophysiology, 50*(1), 112–124.

Knight, R. T., Scabini, D., Woods, D. L., & Clayworth, C. (1988). The effects of lesions of superior temporal gyrus and inferior parietal lobe on temporal and vertex components of the human AEP. *Electroencephalography and Clinical Neurophysiology, 70*(6), 499–509.

Kraus, N., & Anderson, S. (2016). Auditory processing disorder: Biological basis and treatment efficacy. In R. R. Fay & A. N. Popper (Eds.), *Translational research in audiology and the hearing sciences: An essential guide for scientists and clinicians* (pp. 299–318). New York: Springer Science+Business Media.

Kraus, N., & Hornickel, J. (2012). cABR: A biological probe of auditory processing. In D. Geffner & D. Ross-Swain (Eds.), *Auditory processing disorders: Assessment, management and treatment* (pp. 159–183). San Diego: Plural Publishing.

Kraus, N., & Nicol, T. (2014). The cognitive auditory system: The role of learning in shaping the biology of the auditory system. In R. R. Fay & A. N. Popper (Eds.), *Perspectives on auditory research* (pp. 299–319). New York: Springer Science+Business Media.

Kraus, N., & White-Schwoch, T. (2015). Unraveling the biology of auditory learning: A cognitive-sensorimotor-reward framework. *Trends in Cognitive Sciences, 19*(11), 642–654.

Kraus, N., McGee, T., & Comperatore, C. (1989). MLRs in children are consistently present during wakefulness, stage 1, and REM sleep. *Ear and Hearing, 10*, 339–345.

Kraus, N., McGee, T. J., Carrell, T. D., Zecker, S. G., et al. (1996). Auditory neurophysiologic responses and discrimination deficits in children with learning problems. *Science, 273*, 971–973.

Kraus, N., Skoe, E., Parbery-Clark, A., & Ashley, R. (2009). Experience-induced malleability in neural encoding of pitch, timbre, and timing: Implications for language and music. *Annals of the New York Academy of Sciences: Neurosciences and Music III, 1169*, 543–557.

Krishnamurti, S. (2001). P300 auditory event-related potentials in binaural and competing noise conditions in adults with central auditory processing disorders. *Contemporary Issues in Communication Science and Disorders, 28*, 40–47.

Krishnan, A. (1999). Human frequency-following responses to two-tone approximations of steady-state vowels. *Audiology and Neuro-Otology, 4*(2), 95–103.

Krishnan, A. (2002). Human frequency-following responses: Representation of steady-state synthetic vowels. *Hearing Research, 166*(1), 192–201.

Krishnan, A. (2007). Frequency-following response. In R. F. Burkard, J. J. Eggermont, & M. Don (Eds.), *Auditory evoked potentials: Basic principles and clinical application* (pp. 313–334). Philadelphia, Lippincott: Williams & Wilkins.

Krishnan, A., & Parkinson, J. (2000). Human frequency-following responses: Representation of tonal sweeps. *Audiology and Neuro-otology, 5*, 312–321.

Krishnan, A., Xu, Y., Gandour, J. T., & Cariani, P. A. (2004). Human frequency-following response: Representation of pitch contours in Chinese tones. *Hearing Research, 189*(1), 1–12.

Krishnan, A., Xu, Y., Gandour, J., & Cariani, P. (2005). Encoding of pitch in the human brain stem is sensitive to language experience. *Cognitive Brain Research, 25*(1), 161–168.

Kuhl, P. K. (2000). A new view of language acquisition. *Proceedings of the National Academy of Sciences of the USA, 97*(22), 11850–11857.

Kumar, P., & Singh, N. K. (2015). BioMARK as electrophysiological tool for assessing children at risk for (central) auditory processing disorders without reading deficits. *Hearing Research, 324*, 54–58.

Ladefoged, P. (2006). *A course in phonetics* (5th ed.). Boston, MA: Thomson Higher Learning.

Marshall, C. R., Harcourt-Brown, S., Ramus, F., & van der Lely, H. K. J. (2009). The link between prosody and language skills in children with specific language impairment (SLI) and or dyslexia. *International Journal of Language & Communication Disorders, 44*(4), 466–488.

Martin, B. A., Tremblay, K. L., & Stapells, D. R. (2007). Principles and applications of cortical auditory evoked potentials. In R. F. Burkard, M. Don, & J. J. Eggermont (Eds.), *Principles and clinical application of cortical auditory evoked potentials* (pp. 482–507). Baltimore: Lippincott and Williams & Wilkins.

May, P. J., & Tiitinen, H. (2010). Mismatch negativity (MMN), the deviance-elicited auditory deflection, explained. *Psychophysiology, 47*(1), 66–122.

McGee, T., Kraus, N., & Nicol, T. (1997). Is it really a mismatch negativity? An assessment of methods for determining response validity in individual subjects. *Electroencephalography and Clinical Neurophysiology, 104*, 359–368.

McPherson, D. (1996). *Late potentials of the auditory system*. San Diego, CA: Singular Publishing Group.

Medwetsky, L. (2011). Spoken language processing model: Bridging auditory and language processing to guide assessment and intervention. *Language, Speech, and Hearing Services in Schools, 42*(3), 286–296.

Montgomery, C. R., Morris, R. D., Sevcik, R. A., & Clarkson, M. G. (2005). Auditory backward masking deficits in children with reading disabilities. *Brain and Language, 95*, 450–456.

Moore, D. R. (2007). Auditory processing disorders: Acquisition and treatment. *Journal of Communication Disorders, 40*(4), 295–304.

Moore, D. R. (2012). Listening difficulties in children: Bottom-up and top-down contributions. *Journal of Communication Disorders, 45*(6), 411–418.

Moore, J. K., & Linthicum, F. H., Jr. (2007). The human auditory system: A timeline of development. *International Journal of Audiology, 46*(9), 460–478.

Moore, D. R., Rosen, S., Bamiou, D. E., Campbell, N. G., & Sirimanna, T. (2013). Evolving concepts of developmental auditory processing disorder (APD): A British society of audiology APD special interest group 'white paper'. *International Journal of Audiology, 52*(1), 3–13.

Moushegian, G., Rupert, A. L., & Stillman, R. D. (1973). Laboratory note. Scalp-recorded early responses in man to frequencies in the speech range. *Electroencephalography and Clinical Neurophysiology, 35*(6), 665–667.

Musacchia, G., Sams, M., Skoe, E., & Kraus, N. (2007). Musicians have enhanced subcortical auditory and audiovisual processing of speech and music. *Proceedings of the National Academy of Sciences of the USA, 104*(40), 15894–15898.

Musiek, F., & Berge, B. (1998). A neuroscience view of auditory training/stimulation and central auditory processing disorders. In M. Masters, N. Stecker, & J. Katz (Eds.), *Central auditory processing disorders: Mostly management* (pp. 15–32). Boston, MA: Allyn and Bacon.

Musiek, F. E., & Chermak, G. D. (2014). Auditory neuroscience and central auditory processing disorder: An overview. In F. E. Musiek & G. D. Chermak (Eds.), *Handbook of central auditory processing disorder* (pp. 3–15). San Diego, CA: Plural Publishing.

Musiek, F. E., & Lee, W. W. (1995). The auditory brain stem response in patients with brain stem or cochlear pathology. *Ear and Hearing, 16*(6), 631–636.

Musiek, F., Gollegly, K., Kibbe, K., & Verkest-Lenz, S. (1991). Proposed screening test for central auditory processing disorders: Follow-up on the dichotic digits test. *American Journal of Otology, 199, 12*(2), 109–113.
Musiek, F. E., Charette, L., Kelly, T., Lee, W., & Musiek, E. (1999). Hit and false positive rates for middle latency response in patients with central nervous system involvement. *Journal of American Academy of Audiology, 10*(3), 124–132.
Musiek, F. E., Shinn, J. M. S., & Hare, C. M. A. (2002). Plasticity, auditory training and auditory processing disorders. *Seminars in Hearing, 23*(4), 263–275.
Musiek, F. E., Bellis, T. J., & Chermak, G. D. (2005). Nonmodularity of the central auditory nervous system: Implications for (central) auditory processing disorder. *American Journal Audiology, 14*(2), 128–38.
Musiek, F. E., Baran, J., Shinn, J., & Jones, R. (2012). *Disorders of the auditory system* (pp. 317–321). San Diego, CA: Plural Publishing.
Näätänen, R. (1995). The mismatch negativity: A powerful tool for cognitive neuroscience. *Ear and Hearing, 16*(1), 6–18.
Näätänen, R., Jacobsen, T., & Winkler, I. (2005). Memory-based or afferent processes in mismatch negativity (MMN): A review of the evidence. *Psychophysiology, 42*(1), 25–32.
Nishi, K., Lewis, D. E., Hoover, B. M., Choi, S., & Stelmachowicz, P. (2007). Children's recognition of American English consonants in noise. *Journal of the Acoustical Society of America, 127*, 3177–3188.
Oxenham, A. J. (2012). Pitch perception. *The Journal of Neuroscience, 32*(39), 13335–13338.
Özdamar, Ö., & Kraus, N. (1983). Auditory middle-latency responses in humans. *Audiology, 22*, 34–49.
Parbery-Clark, A., Strait, D. L., Anderson, S., Hittner, E., & Kraus, N. (2011). Musical experience and the aging auditory system: Implications for cognitive abilities and hearing speech in noise. *PLoS ONE, 6*(5), e18082.
Plyler, P. N., & Ananthanarayan, A. K. (2001). Human frequency following responses: Representation of second formant transitions in normal-hearing and hearing-impaired listeners. *Journal of the American Academy of Audiology, 12*(10), 423–533.
Rocha-Muniz, C. N., Befi-Lopes, D. M., & Schochat, E. (2012). Investigation of auditory processing disorder and language impairment using the speech-evoked auditory brain stem response. *Hearing Research, 294*, 143–152.
Rocha-Muniz, C. N., Befi-Lopes, D. M., & Schochat, E. (2014). Sensitivity, specificity and efficiency of speech-evoked ABR. *Hearing Research, 317*, 15–22.
Rocha-Muniz, C. N., Befi-Lopes, D. M., & Schochat, E. (2015). Mismatch negativity in children with specific language impairment and auditory processing disorder. *Brazilian Journal of Otorhinolaryngology, 81*(4), 408–415.
Rocha-Muniz, C. N., Filippini, R., Neves-Lobo, I. F., Rabelo, C. M., et al. (2016). Can speech-evoked ABR become a useful tool in clinical practice? *CoDAS, 28*(1), 77–80. doi:10.1590/2317-1782/20162014231.
Russo, N., Nicol, T., Musacchia, G., & Kraus, N. (2004). Brain stem responses to speech syllables. *Clinical Neurophysiology, 115*(9), 2021–2030.
Russo, N., Skoe, E., Trommer, B., Nicol, T., et al. (2008). Deficient brain stem encoding of pitch in children with autism spectrum disorders. *Clinical Neurophysiology, 119*(8), 1720–1731.
Ruusuvirta, T., Huotilainen, M., Fellman, V., & Näätänen, R. (2009). Numerical discrimination in newborn infants as revealed by event-related potentials to tone sequences. *The European Journal of Neuroscience, 30*(8), 620–624.
Sanchez-Longo, L. P., Forster, F. M., & Auth, T. L. (1957). A clinical test for sound localization and its applications. *Neurology, 7*(9), 655–663.
Sanger, D. D., Keith, R. W., & Maher, B. A. (1987). An assessment technique for children with auditory-language processing problems. *Journal of Communication Disorders, 20*(4), 265–279.
Schochat, E., & Musiek, F. (2006). Maturation of outcomes of behavioral and electrophysiologic tests of central auditory function. *Journal of Communication Disorders, 39*(1), 78–92.

Schochat, E., Rabelo, C. M., & Loreti, R. C. A. (2004). Sensitividade e Especificidade do Potencial de Latência Média. *Revista Brasileira de Otorrinolaringologia, 70*(3), 353–358.

Schochat, E., Musiek, F. E., Alonso, R., & Ogata, J. (2010). Effect of auditory training on the middle latency response in children with (central) auditory processing disorder. *Brazilian Journal of Medical and Biological Research, 43*(8), 777–785.

Segalowitz, S. J., Bernstein, D. M., & Lawson, S. (2001). P300 event-related potential decrements in well-functioning university students with mild head injury. *Brain and Cognition, 45*(3), 342–356.

Sharma, A., & Mitchell, T. (2013). The impact of deafness on the human central auditory and visual system. In E. W. Rubel, A. N. Popper, & R. R. Fay (Eds.), *Development of the auditory system*. New York: Springer Science + Business Media.

Sharma, M., Purdy, S. C., & Kelly, A. S. (2009). Comorbity of auditory processing, language, and reading disorders. *Journal of Speech, Language, and Hearing Research, 52*(3), 706–722.

Shehata-Dieler, W., Shimizu, H., Soliman, S. M., & Tusa, R. J. (1991). Middle latency auditory evoked potentials in temporal lobe disorders. *Ear and Hearing, 12*(6), 377–388.

Simões, M. B. (2009). Auditory steady state response in children with dyslexia and with (central) auditory processing disorders. Master's Dissertation. University of São Paulo, Brazil.

Skoe, E., & Kraus, N. (2010). Auditory brain stem response to complex sounds: A tutorial. *Ear and Hearing, 31*(3), 302–324.

Skoe, E., Krizman, J., Anderson, S., & Kraus, K. (2013). Stability and plasticity of auditory brain stem function across the lifespan. *Cerebral Cortex, 25*, 1415–1426.

Small, S. A., & Werker, J. F. (2012). Does the ACC have potential as an index of early speech-discrimination ability? A preliminary study in 4-month-old infants with normal hearing. *Ear and Hearing, 33*(6), 59–69.

Smith, J. C., Marsh, J. T., & Brown, W. S. (1975). Far-field recorded frequency-following responses: Evidence for the locus of brain stem sources. *Electroencephalography and Clinical Neurophysiology, 39*(5), 465–472.

Song, J. H., Banai, K., Russo, N. M., & Kraus, N. (2006). On the relationship between speech- and nonspeech-evoked auditory brain stem responses. *Audiology and Neurotology, 11*(4), 233–241.

Starr, A., & Achor, L. J. (1975). Auditory brain stem responses in neurological disease. *Archives of Neurology, 32*(11), 761–768.

Strait, D. L., Skoe, E., Kraus, N., & Ashley, R. (2009). Musical experience and neural efficiency: Effects of training on subcortical processing of vocal expressions of emotion. *European Journal of Neuroscience, 29*, 661–668.

Tallal, P. (1976). Rapid auditory processing in normal and disordered language development. *Journal of Speech and Hearing Research, 19*(3), 561–571.

Tremblay, K., Kraus, N., McGee, T., Ponton, C. W., & Otis, B. (2001). Central auditory plasticity: Changes in the N1-P2 complex after speech-sound training. *Ear and Hearing, 22*(2), 79–90.

Vander Werff, K. R., & Burns, K. S. (2011). Brain stem responses to speech in younger and older adults. *Ear and Hearing, 32*(2), 168–180.

Xu, Y., Krishnan, A., & Gandour, J. T. (2006). Specificity of experience-dependent pitch representation in the brainstem. *NeuroReport, 17*(15), 1601–1605.

Weber, C., Hahne, A., Friedrich, M., & Friederici, A. D. (2005). Reduced stress pattern discrimination in 5-month-olds as a marker of risk for later language impairment: neurophysiological evidence. *Cognitive Brain Research, 25*(1), 180–187.

Weihing, J., Schochat, E., & Musiek, F. E. (2012). Ear and electrode effects reduce within-group variability in middle latency response amplitude measures. *International Journal of Audiology, 51*(5), 405–412.

White-Schwoch, T., Davies, E. C., Thompson, E. C., Woodruff Carr, K., et al. (2015a). Auditory-neurophysiological responses to speech during early childhood: Effects of background noise. *Hearing Research, 328*, 34–47.

White-Schwoch, T., Woodruff Carr, K., Thompson, E. C., Anderson, S., et al. (2015b). Auditory processing in noise: A preschool biomarker for literacy. *PLoS Biology, 13*(7), e1002196.

Wible, B., Nicol, T., & Kraus, N. (2004). Atypical brainstem representation of onset and formant structure of speech sounds in children with language-based learning problems. *Biological Psychology, 67*, 299–317.

Wible, B., Nicol, T., & Kraus, N. (2005). Correlation between brain stem and cortical auditory processes in normal and language-impaired children. *Brain, 128*, 417–423.

Wijnen, V. J. M., van Boxtel, G. J. M., Eilander, H. J., & de Gelder, B. (2007). Mismatch negativity predicts recovery from the vegetative state. *Clinical Neurophysiology, 118*(3), 597–605.

Witton, C. (2010). Childhood auditory processing disorder as a developmental disorder: The case for a multi-professional approach to diagnosis and management. *International Journal of Audiology, 49*(2), 83–87.

Wright, B. A., Lombardino, L. J., King, W. M., Puranik, C. S., Leonard, C. M., & Merzenich, M. M. (1997). Deficits in auditory temporal and spectral resolution in language-impaired children. *Nature, 387*(6629), 176–178.

Young, E. D., & Sachs, M. B. (1979). Representation of steady-state vowels in the temporal aspects of the discharge patterns of populations of auditory-nerve fibers. *The Journal of the Acoustical Society of America, 66*(5), 1381–1403.

Chapter 10
Neurobiology of Literacy and Reading Disorders

Rachel Reetzke, Zilong Xie, and Bharath Chandrasekaran

Abstract Literacy acquisition is complex and multifactorial. Successful literacy acquisition places extreme demands on sensory and cognitive processes. Individuals with reading disorders demonstrate a range of linguistic, sensory, and cognitive deficits. In this chapter, the relationship between reading ability and the frequency-following response (FFR) is examined. The utility of the FFR in assessment of successful literacy and reading disorders is reviewed along with the use of FFR as an index of remediation. Finally, the chapter concludes with a discussion of current issues and future directions regarding the utility of the FFR as an objective neural metric of deficits in literacy disorders. Throughout these sections the distinct cognitive, linguistic, and experiential influences on the FFR are highlighted to further demonstrate how the FFR to speech may serve as an auditory biomarker to predict literacy disorders.

Keywords Developmental dyslexia · Literacy acquisition · Phonological processing

R. Reetzke · Z. Xie
Department of Communication Sciences and Disorders,
The University of Texas at Austin, 2504A Whitis Avenue,
Austin, TX 78712, USA
e-mail: rreetzke@utexas.edu

Z. Xie
e-mail: xzilong@gmail.com

B. Chandrasekaran (✉)
Department of Communication Sciences and Disorders and Psychology,
Institute for Neuroscience, Center for Perceptual Systems,
The University of Texas at Austin, 2504A Whitis Avenue,
Austin, TX 78712, USA
e-mail: bchandra@utexas.edu

N. Kraus et al. (eds.), *The Frequency-Following Response*,
Springer Handbook of Auditory Research 61,
DOI 10.1007/978-3-319-47944-6_10

10.1 Literacy, Core Deficits, and Theories

An estimated two million children have a learning disability in the United States (U.S. Department of Education, National Center for Education Statistics 2015). Approximately 85% of that population have a diagnosis of developmental dyslexia, a neurological disorder that is characterized by difficulties in the accuracy and fluency of word recognition ability and spelling, likely driven by a phonological deficit (Moats and Dakin 2007). These difficulties cannot be accounted for by low intelligence (IQ), economic disadvantage, insufficient instruction/motivation, or acute neurological injury (Ramus and Ahissar 2012). While developmental dyslexia is a brain-based disorder, the neural mechanisms underlying this disorder are unclear and are the topic of intense investigations that cut across disciplines and theoretical frameworks.

Reading acquisition loads heavily on *phonological recoding,* the ability to translate sound from print information (Goswami 2015). A general consensus is that developmental dyslexia is predominantly phonological in nature (Snowling 1995; Goswami 2015). For example, individuals with a reading disorder may demonstrate difficulties in one or more areas of phonological processing: segmenting and decomposing words into phoneme constituents; rapid phonological retrieval from long-term memory; storing phonological representations for short periods of time; and attending to, discriminating, and manipulating speech sounds (Snowling 1995; Goswami 2015). Thus, it comes as no surprise that phonological ability has been identified as one of the most reliable predictors of eventual reading achievement in preschool-aged children (Ramus 2003; Carroll and Snowling 2004; Papadimitriou and Vlachos 2014). However, there remains ongoing debate regarding the specific nature of the phonological deficit, namely, whether reading impairment is due to degraded phonological representation or impaired access to those phonological representations (Ramus 2014). One possibility is that phonological deficits are secondary to broader sensory deficits (Goswami 2011). Separate theories have been developed under this sensory deficit umbrella, such as impaired processing in the magnocellular visual pathway (Stein 2001; Gori et al. 2015), multimodal perceptual deficits due to attention or short-term memory deficits (Stevens et al. 2013; Harrar et al. 2014), and rapid auditory processing or auditory temporal processing deficits (Tallal 1980; Ahissar et al. 2000).

Instead of focusing on a discussion of causal relationships between core deficits and reading impairments, this chapter concentrates on a neural metric that is able to quantify the encoding of multiple acoustic features of speech sounds as they pertain to literacy: the frequency-following response (FFR). An extant and growing body of literature has demonstrated a relationship between FFRs to speech sounds and literacy skills such as phonological awareness, rapid automatized naming, word reading, and reading fluency (Banai et al. 2009; Hornickel and Kraus 2013). Specifically, FFR has been utilized to uncover the neurophysiological mechanisms related to successful literacy acquisition as well as impaired literacy skills. The FFR reflects phase-locked activity within the subcortical auditory system (but see possible cortical source for the FFR: Coffey et al. 2016) that faithfully captures the acoustic characteristics of speech signals, including timing, fundamental frequency (F_0), and harmonic information (Kraus and Nicol 2005; Krishnan and Gandour 2009).

In addition to providing insight into the biological ingredients that reconstruct speech signals, the FFR is easy to record in difficult-to-test populations (such as children at risk for a reading disorder), shows a high reliability, and reflects auditory experience (Fig. 10.1). It has been argued that FFR properties reflect the collective result of an integrative auditory-cognitive system that is shaped by life experiences (Chandrasekaran et al. 2014; Kraus and White-Schwoch 2015). Therefore, the FFR may serve as a reliable metric of the biological correlates of the diverse array of impairments observed in individuals with reading disorders.

The goal of this chapter is to examine the utility of the FFR in the assessment of reading disorders. Specifically, this chapter assesses the efficacy of the FFR in uncovering mechanistic issues in individuals with reading disorders and in serving as a biological marker during remediation. To achieve these goals, the following areas will be examined: subcortical encoding and plasticity reflected by the FFR as related to phonological processing, an overview of the ability of the FFR to capture predictors of reading disorders, and the FFR as an index of remediation. Finally,

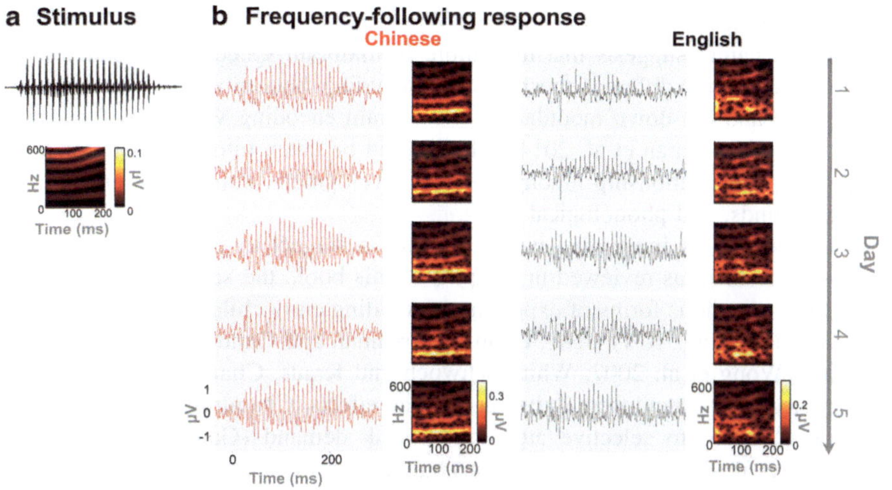

Fig. 10.1 **(a)** The waveform (*top*) and spectrogram (*bottom*) were derived from the stimulus, Mandarin tone 2, which was used to elicit frequency-following responses (FFRs). **(b)** Waveforms (*left*) and spectrograms (*right*) of FFRs elicited by the Mandarin tone 2 across five days of testing (day 1 to day 5, *top to bottom*) in a representative native Chinese participant (*red*) and a representative native English participant (*black*). Notice the FFRs are highly stable across the five test days within each participant. FFRs are additionally malleable to language experience: note that the Chinese participant demonstrates more faithful representation of the stimulus relative to the English participant. (Original figure by authors)

the chapter concludes with a discussion of current issues and future directions regarding the utility of the FFR as an objective neural metric of deficits in literacy.

10.2 Phonological Processing, Brainstem Encoding, and Plasticity

As indicated in Sect. 10.1, literacy acquisition requires the successful development of phonological skills (Wagner and Torgesen 1987; Ramus 2003). The auditory subcortical system processes speech with high fidelity. This fidelity, in turn, provides easier access to stored phonological representations, especially in challenging listening environments. In typical readers, auditory subcortical structures encode speech sounds with high temporal and spectral precision (Chandrasekaran and Kraus 2012). A deficit in the representation of, or access to, speech sounds may impair phonological processing. Scalp-recorded FFR can serve as an objective neuropsychological measure of speech encoding in the auditory pathway and thus has the potential to provide an index of breakdown in phonological encoding.

The inferior colliculus (IC) is posited to be the major neural source of scalp-recorded FFR (Smith et al. 1975; Chandrasekaran and Kraus 2010). Notably, the IC is a convergence center in the auditory system that serves as an essential station for bottom-up sensory input. Additionally, the IC receives massive top-down projections from the cortices (Winer 2005; Malmierca et al. 2015). This integrative nature suggests that the failure in midbrain speech representation can arise from several different mechanisms, including bottom-up and local (to IC) processing and top-down modulation of midbrain encoding via corticofugal pathways (Chandrasekaran et al. 2014). This should be taken into consideration when examining the relationship among literacy development, subcortical encoding of speech sounds, and phonological skills.

Because of this integrative nature, the IC is subject to experience-dependent plasticity. Indeed, as reviewed in Chap. 2 of this book, the scalp-recorded FFR is malleable to various forms of experience, including early childhood experiences in language (e.g., Fig. 10.1) (Jeng, Chap. 2; Krishnan and Gandour, Chap. 3), music training (Wong et al. 2007; White-Schwoch and Kraus, Chap. 6), and short-term experiences that extend from online contextual effects (Chandrasekaran et al. 2009; Escera, Chap. 5) to selective attention or task demand (Galbraith et al. 2003; Varghese et al. 2015) and to modulations on the basis of short-term auditory training (Song et al. 2008; Carcagno and Plack, Chap. 4). According to a recent model (*predictive tuning model*), one mechanism underlying these experience-dependent plastic changes is top-down tuning (from the cortices via corticofugal pathways), which modulates midbrain encoding based on a predictive algorithm that constantly anticipates the incoming signals (Chandrasekaran et al. 2014). A match between the incoming signals and the expectation results in an enhancement in the midbrain representation of the signals, whereas a mismatch leads to poorer signal representation. Based upon the interconnectivity of the midbrain to other neural networks, impairments in cognitive encoding may be reflected in an individual's FFR by virtue of this efferent input (Kraus and White-Schwoch 2015). Thus, the FFR may serve as a metric to capture auditory-cognitive interplay in individuals with reading disorders and, in turn, offer further understanding of the range of linguistic and cognitive impairments observed in this population.

Plasticity observed in subcortical auditory function, reflected by the FFR, suggests that researchers and clinicians may be able to develop training programs to improve auditory function in clinical populations diagnosed with reading disorders. Equally important, these training programs can be used to target vulnerable populations, such as preliterate children identified "at risk" for reading disorders. During training, monitoring the changes in midbrain encoding via scalp-recorded FFR may provide valuable noninvasive information about the progress of neuroplastic changes occurring during the intervention.

In the next two sections, the manifestations of deficits in subcortical encoding, as documented via scalp-recorded FFR, are discussed in the context of reading disorders. Following that section, an intervention study is highlighted with regard to the efficacy of scalp-recorded FFRs as a potential biological marker of remediation.

10.3 FFR Captures Predictors of Literacy and Reading Disorders

Some of the predominant theories of developmental dyslexia have described a phonological deficit as the outcome of a broader deficit in auditory processing (McAnally and Stein 1996; Stefanics et al. 2011). However, the specific nature of the auditory processing deficit remains unresolved. For example, it is unclear whether auditory processing deficit is an impairment in the ability to judge the temporal order of rapid auditory sequences (Tallal 1980), a deficit in frequency discrimination (Mcanally and Stein 1996), or an abnormality in temporal sampling of amplitude modulations (Goswami 2011). Here it is important to note that an auditory processing deficit has not been observed in all individuals with developmental dyslexia, but rather it has been found to define a sizeable subset of individuals with reading disorders (Ramus et al. 2003). For instance, a review of ten studies investigating auditory processing deficits in individuals with dyslexia found that approximately 40% of the population has auditory processing problems (Ramus 2003).

An auditory core-deficit framework fails to account for broader cognitive, linguistic, and speech-in-noise perception deficits that are also found in individuals with reading disorders (Ramus and Ahissar 2012; White-Schwoch et al. 2015a). With such heterogeneity in auditory and cognitive processing abilities observed across individuals with reading disorders, it is all the more important to identify a metric that facilitates reliable prediction of future literacy skills, as well as provides an assessment of impaired systems, such as the FFR.

Why has the FFR been identified as such a metric? First, it is possible that the FFR reflects the complex interplay between the bottom-up (local) and top-down processes (as described in Sect. 10.2); therefore, the FFR may be better able to account for broad cognitive, linguistic, and auditory processing deficits linked to reading impairment (Chandrasekaran et al. 2014). Second, the FFR can be recorded regardless of the developmental age, language, motivation, or attention level of an individual, which facilitates comparisons across different populations (for example, comparing pre-literate children with children who are literate). The FFR is additionally found to be highly reliable across multiple recording sessions (see Fig. 10.1). Finally, specific abnormalities in acoustic features of the FFR have been consistently associated with impaired reading skills, such as a phonological deficit (Fig. 10.2) (White-Schwoch and Kraus 2013; White-Schwoch et al. 2015b). Specific neurophysiological deficits found in individuals with poor reading skills include: abnormal timing and harmonic encoding of speech signals (Banai et al. 2005, 2009), impaired context-dependent modulation of speech signals (Chandrasekaran et al. 2009; Centanni et al. 2014), diminished subcortical and cortical representation of stimulus differences (Banai et al. 2005; Abrams et al. 2006), and greater vulnerability to the deleterious effects of background noise, evidenced by greater delay in subcortical responses to speech sounds (Anderson et al. 2010; White-Schwoch et al. 2015b). Thus, one may be uncertain of the core deficit in reading impairment while still appreciating the potential of the FFR as an objective, biological marker of literacy achievement and, in preliterate children, literacy potential.

In the next four subsections, we will expand on the efficacy of the FFR as a metric in literacy and reading disorders. Specifically, we will assess the FFR as a

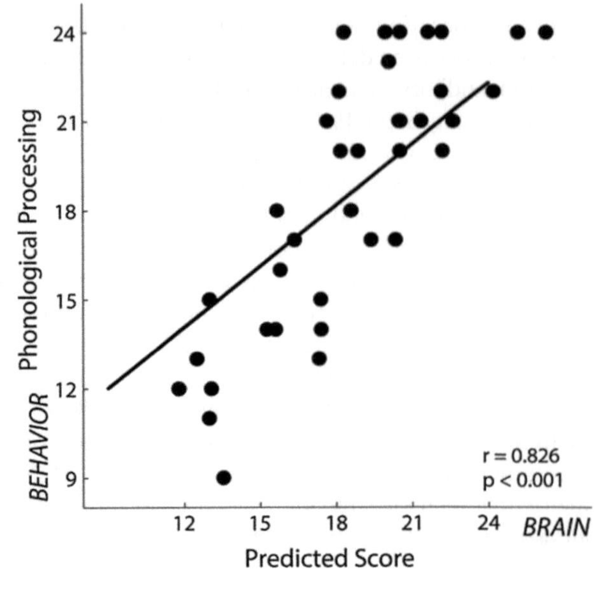

Fig. 10.2 Predictive value of FFR for emerging literacy skills. In this study, neural coding of consonants in noise, as measured by the FFR (*BRAIN*) strongly predicted phonological processing (*BEHAVIOR*), as measured by the Clinical Evaluation of Language Fundamentals Preschool-2: Phonological Awareness subtest. The results suggest that the precision and stability of coding consonants in noise parallels emergent literacy skills. *r*, correlation coefficient (Adapted from White-Schwoch et al. 2015b)

metric for capturing abnormal speech representation, impaired auditory-cognitive interaction, deficits in context-dependent modulation, and impairment in the brainstem-cortical relationship.

10.3.1 FFR Captures Abnormal Speech Representation

For the successful reader, precise representation of the spectral and temporal acoustic features of incoming speech signals are thought to facilitate the transformation of acoustic patterns into phonological representations (Ahissar et al. 2000). The FFR reflects neural encoding of complex acoustic features, including the transient and periodic acoustic events found in speech. Three fundamental acoustic features are intrinsic to speech sounds: timing, fundamental frequency (F_0), and harmonic information. A strong relationship has been demonstrated between these primary features of the FFR to speech syllables (for example, /ba/, /da/, /ga/) and measures of reading achievement. When FFR responses to consonant-vowel syllables are recorded, the timing of FFR reveals information about the onset and offset of the stimuli as well as phonemic differences (e.g., the difference in voice onset time between /da/ and /ta/). Analysis of the F_0 is often performed to assess primary pitch-related acoustic cues in speech signals, which carry important paralinguistic information (e.g., the sex of the talker, and the emotive state of the talker), as well as linguistic information (e.g., signaling lexical meaning in tonal languages). Finally, higher harmonics convey information about the representation of acoustic cues that are directly relevant in differentiating speech sounds (such as formant structure). The FFRs of children with reading disorders have revealed impairment in consonant differentiation (Hornickel et al. 2009), as well as timing and harmonic-encoding deficits associated with phonological processing disorders (Hornickel and Kraus 2013). However, children with reading impairment seem to have intact coding of pitch-related cues (Banai et al. 2009).

One consistent finding across studies has been that individuals with reading disorders have difficulty differentiating and identifying stop consonants (Hornickel et al. 2009). This difficulty may stem from the transient nature and quick changing formants that define stop consonants. For example, phonological awareness, the ability to focus on and manipulate phonemes in words, tracks with better subcortical differentiation of the three stop consonants (Hornickel et al. 2009).

In addition to difficulty with consonant differentiation, individuals with reading disorders may demonstrate abnormal timing and harmonic encoding of speech sounds. The timing of the FFR to a syllable provides information regarding the onset and offset of the consonant-vowel by fractions of milliseconds. Hornickel and Kraus (2013) found that children with poor reading ability have auditory brainstem responses to speech that are significantly more variable than the responses of children with good reading ability, particularly in response to formant transitions (compared to vowels). Therefore, variability in the efficacy of the central nervous system may underlie the disparities in directed attention and the impaired understanding of speech due to inconsistent encoding in the midbrain. This suggests that reading success depends on the stable neural representation of sound and that children who have unstable neural responses are at risk for reading disorders. This body of work motivates further investigation of the dynamic interaction of cognitive processes and subcortical encoding of the acoustic signal (auditory-cognitive interaction) during auditory perception.

10.3.2 FFR Captures Auditory-Cognitive Interaction

As mentioned in Sect. 10.1, a precursor to successful literacy is *phonological recoding*, the ability to derive sounds from visual symbols. Effective reading requires the integration of auditory, visual, motor, and cognitive processes (Goswami 2015). For example, for audiovisual information to be integrated across multiple timescales, relevant information (such as the correct line of text) must be selected, while irrelevant information (such as the incorrect lines of text) must be ignored (Harrar et al. 2014; Stevens et al. 2013). The underlying neural source responsible for generating the FFR is posited to be intricately interconnected with, and influenced by, signals from a number of neural networks, such as those related to multisensory and attentional processing (Huffman and Henson 1990). Therefore, the FFR may be a useful metric for capturing broad auditory-cognitive deficits found in individuals with reading impairment.

One way to measure the interaction of cognitive and sensory processes in humans is through speech-in-noise (SIN) tasks. SIN tasks are specifically useful in the context of reading disorders because individuals with reading impairment often demonstrate perceptual deficits in adverse listening conditions (Sperling et al. 2005; Ziegler et al. 2009). Noisy listening conditions place demands on both auditory and cognitive processes, such as selective attention and working memory (Shinn-Cunningham 2008; White-Schwoch et al. 2015b), which also are impaired in some individuals with reading disorders (Harrar et al. 2014; Stevens et al. 2013). Some argue that individuals with developmental dyslexia exhibit a noise-exclusion deficit, that is, the inability to exclude background noise during signal processing (Sperling et al. 2005).

How can FFR to SIN be used to examine auditory-cognitive interaction? To identify the target speech signal from background noise, the listener must form an auditory object based on the spectrotemporal aspects of the acoustic signal (Shinn-Cunningham 2008). The F_0, timing, and the harmonic features of the signal facilitate the extraction of the target sound; however, noise inhibits access to these salient acoustic cues. For example, noise may overlap with auditory signals in their spectrotemporal profile and cause difficulty in the segregation of the target signal from the features of similar-sounding distracters (Shinn-Cunningham 2008). Therefore, cognitive processes, such as selective attention, must then be engaged in order to segregate the target acoustic signal from the background masker (Shinn-Cunningham 2008). Correlations have been demonstrated among neural processing of speech-in-noise and measures of literacy, such as standardized assessments of phonological awareness (White-Schwoch et al. 2015b).

This observed neurobehavioral association may provide further insight into the underlying mechanisms necessary for successful literacy. Support for this notion comes from evidence demonstrating that stronger neural encoding of consonants in noise have been linked to early literacy skills (White-Schwoch et al. 2015b). For example, a sequence of experiments recorded FFRs to the syllable /da/ in both a quiet and background noise condition from a group of children, ranging in ages from 3 to 14 years old (White-Schwoch et al. 2015b). Significant correlations were found between the reliability of the neural encoding of speech in noise and behavioral measures of phonology, indicating that pre-literacy children with stronger neural encoding demonstrated better literacy skills. In a one year follow-up, this same cohort was identified as stronger emerging readers. In the older children from this study, the same neural measurement was applied, and literacy achievement, as well as diagnosis of a reading impairment, was reliably predicted.

From these results, it is posited that the implementation of experimental conditions with noise may tax the auditory system and extract individual differences in the breakdown of subcortical encoding of speech sounds—observations that otherwise may not be evident in quiet conditions. This may shed light on daily challenges that children encounter when they try to map sounds to meaning in noisy listening environments. The evidence here not only supports the notion that features of phonological processing are associated with the stability of encoding consonants in noise, but most importantly, this body of work points toward an objective measure for emerging literacy skills. Therefore, the FFR may serve as a vital tool for future identification of children who are at-risk for literacy disorders (White-Schwoch et al. 2015b).

As demonstrated through work on SIN perception, top-down cognitive processes play an integral role in successful auditory processing, phonological processing and, ultimately, successful reading. However, the nature of sensory deficits in individuals with reading disorders is currently unclear. That is, it has been posited that literacy acquisition and practicing reading actually assists in the training of sensory and cognitive processes (Goswami 2015). The argument here is that working memory and selective attention processes engaged in successful reading are less practiced in individuals who have a reading disorder. Therefore, impaired sensory and/or cognitive processes observed in individuals with reading disability may be secondary to the reading deficit.

It is imperative that research efforts continue with this line of work to better understand the interactions among cognitive and auditory processes. The FFR to SIN perception tasks can be a useful method for moving this line of investigation forward. In order to determine the extent to which sensory deficits impact literacy, future studies should begin to examine the systematic impact of sensory deficits on other cognitive systems that depend upon perceptual auditory and visual processes. This will not only assist in the identification of individuals with reading deficits but will further understanding of the role this interaction plays in the typical acquisition of literacy.

10.3.3 FFR Captures Deficits in Context-Dependent Modulation

The ability to tune into regularities in the soundscape is fundamental to processing relevant auditory information (Suga 2008); the human auditory system is highly sensitive to stimulus statistics (see Escera, Chap. 5). Experience and expectation have been found to be vital components in the reconstruction process of the auditory signal (Skoe et al. 2015). For example, the subcortical auditory system encodes speech signals more accurately when the signals are behaviorally relevant and in contexts that are predictable (Krishnan and Gandour 2009; Chandrasekaran et al. 2009). Individuals with reading impairment have a reduced ability to modulate the probability with which acoustic signals are presented and reduced ability to adapt to contextual differences (Ahissar et al. 2006; Banai and Ahissar 2006). The inability of individuals with reading disorders to benefit from the repetition of stimulus input has been described as an inability to form "perceptual anchors" (*anchor-deficit hypothesis*) (Ahissar et al. 2006). The anchor-deficit hypothesis suggests that encoding deficits observed in individuals with a reading disorder are the result of deficient encoding of stimulus acoustic properties and are also context dependent.

The FFR has been utilized to capture deficient context-dependent modulation in individuals with poor reading ability. Studies have found that individuals with reading disorders do not adapt to stimulus repetition as well as individuals without a reading disorder. For example, in one study the speech syllable /da/ was presented repetitively in a predictable context and pseudo-randomly within a set of seven other speech syllables in a variable context (Chandrasekaran et al. 2009). In contrast to typically developing readers, children with developmental dyslexia demonstrated reduced FFRs to key auditory features when the syllable was presented in a repetitive, predictable context (Chandrasekaran et al. 2009). The extent to which repetition improved the representation related to performance on a speech-in-noise task may suggest that regularity-detection mechanisms are additionally active during the separation of an auditory object from background noise. The findings of this study indicate that individuals with developmental dyslexia are unable to utilize patterns in sensory streams (such as in statistical learning) in order to improve the representation of incoming perceptual information.

Why are individuals with poor reading ability unable to adapt to stimulus repetition? As mentioned in Sect. 10.2, empirical evidence supports the existence of multiple forms of subcortical plasticity. One argument is that the lack of experience-dependent fine tuning of the FFR in children with dyslexia may reflect a top-down influence of deficient phonological representations at the level of the cortex (Chandrasekaran et al. 2009). That is, a deficit in midbrain encoding may be the result of a failure to establish cortical phonological representations in individuals with developmental dyslexia. Therefore, top-down influence of these deficient phonological representations may then explain why individuals with developmental dyslexia demonstrate less effective use of predictive context to modulate ongoing midbrain activity.

10.3.4 FFR Captures Subcortical-Cortical Relationship

The auditory cortex plays a critical role in rapid acoustic signal processing during speech perception. As described in Sect. 10.2 of this chapter, the corticofugal pathway connects the auditory cortex with the auditory midbrain. While it is well-established that neural transcription of stop consonants, vowels, and pitch variation occur in the auditory midbrain and cortex, more research is needed to better understand the nature of the midbrain-cortical relationship. For individuals with literacy disorders, it is important to understand the role that the midbrain-cortical relationship plays during successful phonological processing.

Recent efforts have focused on investigating the connection between midbrain timing and cortical processing. The question here is whether midbrain timing deficits are associated with abnormal cortical discrimination of acoustic change, specifically in adverse listening conditions. In one study, midbrain timing and cortical sensitivity to acoustic change (mismatch negativity, MMN) were measured in two groups of children with a broad range of learning disabilities (LD), including those with normal and abnormal midbrain response timing, and typical age-matched peers (Banai et al. 2005). The MMNs, literacy, and cognitive abilities were compared among the three groups. The LD children with abnormal midbrain timing were more likely to show reduced processing of acoustic change at the cortical level relative to LD individuals with normal midbrain timing and controls. This group also demonstrated poorer reading, reduced listening comprehension, as well as cognitive impairment. The study suggests that abnormal midbrain timing in learning disabilities is related to higher prevalence of reduced cortical sensitivity to acoustic dynamics and to deficient literacy skills (Wible et al. 2005).

Other studies investigating the midbrain-cortical relationship have focused on the role of the auditory brainstem in cerebral lateralization. For example, cortical event-related potential studies have revealed that auditory midbrain timing significantly predicts cerebral lateralization (Abrams et al. 2006): children who exhibited atypical brainstem timing had less cortical asymmetry as well as poorer reading skills. These findings suggest that subtle impairment at the level of the midbrain is tied to cortical processing of acoustic signals and speech sounds, which is consistent with integrative models of auditory processing and its impairments (Chandrasekaran et al. 2014; Kraus and White-Schwoch 2015).

10.3.5 FFR as an Index of Remediation

Hornickel et al. (2012) reported the use of assistive listening devices (i.e., classroom FM systems) in improving phonological awareness in dyslexic children. The FM system can lead to enhancement in the input acoustic cues such as signal-to-noise ratio of the target talker (e.g., the teacher) (Crandell et al. 2005). One-year of FM-system use improved the consistency of FFRs to speech signals for children with dyslexia, whereas another group of dyslexic children who were educated in the same classrooms, but did not use FM systems, displayed no evidence of improvement. Interestingly, the consistency of FFRs to speech before FM use predicted the amount of improvement in phonological awareness after one-year of FM use, and lower FFR consistency before FM use was linked with the greatest improvement in phonological awareness. Moreover, dyslexic children with the larger improvement in phonological awareness also demonstrated the greater improvement in FFR consistency, in many cases nearing the levels of their typically developing peers. This evidence suggests that the FFR may be used as a biomarker that identifies dyslexic individuals who would benefit most from remediation and also may be used to index the extent of therapeutic effects.

10.4 Current Issues and Future Directions

The FFRs elicited by speech sounds are well suited for clinical application. As discussed in this chapter, the FFR reflects how the brain reconstructs speech signals with high fidelity, the FFR is a metric of auditory-cognitive experience, the FFR is easy to record regardless of age, motivation, or attention level, and additionally, the FFR shows high reliability in individual subjects. FFRs to speech signals provide objective insight into reading ability by reflecting the integrity of the neural encoding of fundamental features of speech, such as pitch, timing, and harmonic information. This metric may be useful clinically to predict and evaluate reading ability in early (White-Schwoch et al. 2015b) and later (Hornickel et al. 2009) childhood. However, there is still much work to be done before this metric can be utilized as a reliable predictor of successful literacy.

More research is specifically needed to better understand the neurobiological basis of FFR application to assessment and treatment of literacy disorders. Future FFR investigations should continue to focus on finding the biological constraints that may be manifested in the nervous system prior to literacy acquisition. These future studies should be conducted with the goal to expand conventional clinical assessment tools for developmental dyslexia with the aim to implement the FFR. In order to do this, a first step should be to collect normative data across the lifespan for clinical reference. This research may result in a vital tool for early detection of developmental dyslexia. With a better understanding of the neural deficits associated with reading disorders, FFR investigations should additionally aim to develop intervention programs in order to target identified neural deficiencies. Finding a way to remediate neural deficits underlying reading disorders may result in a method for improving literacy skills (Hornickel et al. 2012). For example, a program to monitor FFRs of school children could first use the method as a diagnostic marker to assess risk for literacy disorders and then use FFRs systematically as a measure of remediation-related changes.

10.5 Summary

Literacy and disorders of literacy are complex and multifactorial. Successful literacy places extreme demands on sensory and cognitive processes, and individuals with reading disorders demonstrate a range of linguistic, sensory, and cognitive deficits. The studies reviewed in this chapter specifically indicate that reading ability correlates with various midbrain measurements: timing and harmonic encoding of speech signals, speech perception in noise (as evidenced by the greater delay in brainstem responses to speech sounds), context-dependent modulation, and subcortical and cortical representation of stimuli. This relationship may seem puzzling but likely reflects the complexity within the FFR measure as well. The FFR is not simply a reflection of the integrity of stimulus encoding. There are distinct cognitive (Galbraith et al. 2003; Hairston et al. 2013), linguistic (Krishnan and Gandour 2009; Krizman et al. 2012), and experiential influences (Kraus and Chandrasekaran 2010; Kraus and White-Schwoch 2015) on the FFR. While the relative contribution of these influences on the FFR and the relationship to literacy disorders is far from clear, the FFR to speech may serve as a potent biomarker to predict successful literacy acquisition and potential disorders.

Acknowledgments This work was supported by the National Institute on Deafness and Other Communication Disorders of the National Institutes of Health under Award Number R01DC013315 (BC).

Compliance with Ethics Requirements

The authors, Rachel Reetzke, Zilong Xie, and Bharath Chandrasekaran, had no conflict of interest. Data represented in Fig. 10.1 was supported by the National Institute on Deafness and Other Communication Disorders, National Institutes of Health (Grant R01DC013315 to BC). The content is solely the responsibility of the authors and does not necessarily represent the official view of the National Institutes of Health.

References

Abrams, D. A., Nicol, T., Zecker, S. G., & Kraus, N. (2006). Auditory brainstem timing predicts cerebral asymmetry for speech. *The Journal of Neuroscience, 26*(43), 11131–11137.

Ahissar, M., Protopapas, A., Reid, M., & Merzenich, M. M. (2000). Auditory processing parallels reading abilities in adults. *Proceedings of the National Academy of Sciences of the U S A, 97* (12), 6832–6837.

Ahissar, M., Lubin, Y., Putter-Katz, H., & Banai, K. (2006). Dyslexia and the failure to form a perceptual anchor. *Nature Neuroscience, 9*(12), 1558–1564. Doi:10.1038/nn1800

Anderson, S., Skoe, E., Chandrasekaran, B., & Kraus, N. (2010). Neural timing is linked to speech perception in noise. *The Journal of Neuroscience, 30*(14), 4922–4926.

Banai, K., & Ahissar, M. (2006). Auditory processing deficits in dyslexia: Task or stimulus related? *Cerebral Cortex, 16*(12), 1718–1728. Doi:10.1093/cercor/bhj107

Banai, K., Nicol, T., Zecker, S. G., & Kraus, N. (2005). Brainstem timing: Implications for cortical processing and literacy. *The Journal of Neuroscience, 25*(43), 9850–9857.

Banai, K., Hornickel, J., Skoe, E., Nicol, T., et al. (2009). Reading and subcortical auditory function. *Cerebral Cortex, 19*(11), 2699–2707.

Carroll, J. M., & Snowling, M. J. (2004). Language and phonological skills in children at high risk of reading difficulties. *Journal of Child Psychology and Psychiatry, 45*(3), 631–640.

Centanni, T. M., Booker, A. B., Sloan, A. M., Chen, F., et al. (2014). Knockdown of the dyslexia-associated gene Kiaa0319 impairs temporal responses to speech stimuli in rat primary auditory cortex. *Cerebral Cortex, 24*(7), 1753–1766.

Chandrasekaran, B., & Kraus, N. (2010). The scalp-recorded brainstem response to speech: Neural origins and plasticity. *Psychophysiology, 47*(2), 236–246.

Chandrasekaran, B., & Kraus, N. (2012). Biological factors contributing to reading ability: Subcortical auditory function. In A. A. Benasich & R. H. Fitch (Eds.), *Developmental dyslexia: Early precursors, neurobehavioral markers and biological substrates*. Baltimore, MD: Paul H Brookes Publishing.

Chandrasekaran, B., Hornickel, J., Skoe, E., Nicol, T., & Kraus, N. (2009). Context-dependent encoding in the human auditory brainstem relates to hearing speech in noise: Implications for developmental dyslexia. *Neuron, 64*(3), 311–319.

Chandrasekaran, B., Skoe, E., & Kraus, N. (2014). An integrative model of subcortical auditory plasticity. *Brain Topography, 27*(4), 539–552.

Coffey, E. B., Herholz, S. C., Chepesiuk, A. M., Baillet, S., & Zatorre, R. J. (2016). Cortical contributions to the auditory frequency-following response revealed by MEG. *Nature Communications*, 7. Doi:10.1038/ncomms11070

Crandell, C., Smaldino, J., & Flexer, C. (2005). *Sound field amplification: Applications to speech perception and classroom acoustics* (2nd ed.). Clifton Park, NY: Thomson Delmar Learning.

Galbraith, G. C., Olfman, D. M., & Huffman, T. M. (2003). Selective attention affects human brain stem frequency-following response. *NeuroReport, 14*(5), 735–738.

Gori, S., Seitz, A. R., Ronconi, L., Franceschini, S., & Facoetti, A. (2015). Multiple causal links between magnocellular-sorsal pathway deficit and developmental dyslexia. *Cerebral Cortex*. Doi:10.1093/cercor/bhv206

Goswami, U. (2011). A temporal sampling framework for developmental dyslexia. *Trends in Cognitive Sciences, 15*(1), 3–10.

Goswami, U. (2015). Sensory theories of developmental dyslexia: Three challenges for research. *Nature Reviews Neuroscience, 16*(1), 43–54.

Hairston, W. D., Letowski, T. R., & McDowell, K. (2013). Task-related suppression of the brainstem frequency following response. *PLoS ONE, 8*(2), e55215. Doi:10.1371/journal.pone.0055215

Harrar, V., Tammam, J., Pérez-Bellido, A., Pitt, A., et al. (2014). Multisensory integration and attention in developmental dyslexia. *Current Biology, 24*(5), 531–535.

Hornickel, J., & Kraus, N. (2013). Unstable representation of sound: A biological marker of dyslexia. *The Journal of Neuroscience, 33*(8), 3500–3504.

Hornickel, J., Skoe, E., Nicol, T., Zecker, S., & Kraus, N. (2009). Subcortical differentiation of stop consonants relates to reading and speech-in-noise perception. *Proceedings of the National Academy of Sciences of the USA, 106*(31), 13022–13027.

Hornickel, J., Zecker, S. G., Bradlow, A. R., & Kraus, N. (2012). Assistive listening devices drive neuroplasticity in children with dyslexia. *Proceedings of the National Academy of Sciences of the USA, 109*(41), 16731–16736.

Huffman, R. F., & Henson, O. (1990). The descending auditory pathway and acousticomotor systems: Connections with the inferior colliculus. *Brain Research Reviews, 15*(3), 295–323.

Kraus, N., & Chandrasekaran, B. (2010). Music training for the development of auditory skills. *Nature Reviews Neuroscience, 11*(8), 599–605.

Kraus, N., & Nicol, T. (2005). Brainstem origins for cortical 'what'and 'where'pathways in the auditory system. *Trends in Neurosciences, 28*(4), 176–181.

Kraus, N., & White-Schwoch, T. (2015). Unraveling the biology of auditory learning: A cognitive-sensorimotor-reward framework. *Trends in Cognitive Sciences, 19*(11), 642–654.

Krishnan, A., & Gandour, J. T. (2009). The role of the auditory brainstem in processing linguistically relevant pitch patterns. *Brain and Language, 110*(3), 135–148.

Krizman, J., Marian, V., Shook, A., Skoe, E., & Kraus, N. (2012). Subcortical encoding of sound is enhanced in bilinguals and relates to executive function advantages. *Proceedings of the National Academy of Sciences of the USA, 109*(20), 7877–7881.

Malmierca, M. S., Anderson, L. A., & Antunes, F. M. (2015). The cortical modulation of stimulus-specific adaptation in the auditory midbrain and thalamus: A potential neuronal correlate for predictive coding. *Frontiers in Systems Neuroscience, 9*(19). Doi:10.3389/fnsys. 2015.00019

Mcanally, K. I., & Stein, J. F. (1996). Auditory temporal coding in dyslexia. *Proceedings of the Royal Society of London B: Biological Sciences, 263*(1373), 961–965.

Moats, L. C., & Dakin, K. E. (2007). *Basic facts about dyslexia and other reading problems.* Baltimore, MD: International Dyslexia Association.

Papadimitriou, A. M., & Vlachos, F. M. (2014). Which specific skills developing during preschool years predict the reading performance in the first and second grade of primary school? *Early Child Development and Care, 184*(11), 1706–1722.

Ramus, F. (2003). Developmental dyslexia: Specific phonological deficit or general sensorimotor dysfunction? *Current Opinion in Neurobiology, 13*(2), 212–218.

Ramus, F. (2014). Neuroimaging sheds new light on the phonological deficit in dyslexia. *Trends in Cognitive Sciences, 18*(6), 274–275.

Ramus, F., & Ahissar, M. (2012). Developmental dyslexia: The difficulties of interpreting poor performance, and the importance of normal performance. *Cognitive Neuropsychology, 29*(1–2), 104–122.

Ramus, F., Rosen, S., Dakin, S. C., Day, B. L., et al. (2003). Theories of developmental dyslexia: Insights from a multiple case study of dyslexic adults. *Brain, 126*(4), 841–865.

Shinn-Cunningham, B. G. (2008). Object-based auditory and visual attention. *Trends in Cognitive Sciences, 12*(5), 182–186.

Skoe, E., Krizman, J., Spitzer, E., & Kraus, N. (2015). Prior experience biases subcortical sensitivity to sound patterns. *Journal of Cognitive Neuroscience, 27*(1), 124–140.

Smith, J. C., Marsh, J. T., & Brown, W. S. (1975). Far-field recorded frequency-following responses: Evidence for the locus of brainstem sources. *Electroencephalography and Clinical Neurophysiology, 39*(5), 465–472.

Snowling, M. J. (1995). Phonological processing and developmental dyslexia. *Journal of Research in Reading, 18*(2), 132–138.

Song, J. H., Skoe, E., Wong, P. C., & Kraus, N. (2008). Plasticity in the adult human auditory brainstem following short-term linguistic training. *Journal of Cognitive Neuroscience, 20*(10), 1892–1902.

Sperling, A. J., Lu, Z.-L., Manis, F. R., & Seidenberg, M. S. (2005). Deficits in perceptual noise exclusion in developmental dyslexia. *Nature Neuroscience, 8*(7), 862–863.

Stefanics, G., Fosker, T., Huss, M., Mead, N., et al. (2011). Auditory sensory deficits in developmental dyslexia: A longitudinal ERP study. *Neuroimage, 57*(3), 723–732.

Stein, J. (2001). The magnocellular theory of developmental dyslexia. *Dyslexia, 7*(1), 12–36.

Stevens, C., Harn, B., Chard, D. J., Currin, J., et al. (2013). Examining the role of attention and instruction in at–risk kindergarteners electrophysiological measures of selective auditory attention before and after an early literacy intervention. *Journal of Learning Disabilities, 46*(1), 73–86.

Suga, N. (2008). Role of corticofugal feedback in hearing. *Journal of Comparative Physiology A: Neuroethology, Sensory, Neural, and Behavioral Physiology, 194*(2), 169–183.

Tallal, P. (1980). Auditory temporal perception, phonics, and reading disabilities in children. *Brain and Language, 9*(2), 182–198.

Varghese, L., Bharadwaj, H. M., & Shinn-Cunningham, B. G. (2015). Evidence against attentional state modulating scalp-recorded auditory brainstem steady-state responses. *Brain Research, 1626*, 146–164. Doi:10.1016/j.brainres.2015.06.038

Wagner, R. K., & Torgesen, J. K. (1987). The nature of phonological processing and its causal role in the acquisition of reading skills. *Psychological Bulletin, 101*(2), 192–212. Doi:10.1162/jocn_a_0069

Wible, B., Nicol, T., & Kraus, N. (2005). Correlation between brainstem and cortical auditory processes in normal and language-impaired children. *Brain, 128*(2), 417–423.

White-Schwoch, T., & Kraus, N. (2013). Physiologic discrimination of stop consonants relates to phonological skills in pre-readers: A biomarker for subsequent reading ability? *Frontiers in Human Neuroscience, 7*, 899. Doi:10.3389/fnhum.2013.00899

White-Schwoch, T., Davies, E. C., Thompson, E. C., Carr, K. W., et al. (2015a). Auditory-neurophysiological responses to speech during early childhood: Effects of background noise. *Hearing Research, 328*, 34–47.

White-Schwoch, T., Carr, K. W., Thompson, E. C., Anderson, S., et al. (2015b). Auditory processing in noise: A preschool biomarker for literacy. *PLoS Biology, 13*(7), e1002196. Doi:10.1371/journal.pbio.1002196

Winer, J. A. (2005). Decoding the auditory corticofugal systems. *Hearing Research, 207*(1), 1–9.

Wong, P. C., Skoe, E., Russo, N. M., Dees, T., & Kraus, N. (2007). Musical experience shapes human brainstem encoding of linguistic pitch patterns. *Nature Neuroscience, 10*(4), 420–422.

Ziegler, J. C., Pech-Georgel, C., George, F., & Lorenzi, C. (2009). Speech-perception-in-noise deficits in dyslexia. *Developmental Science, 12*(5), 732–745.

Chapter 11
Clinical Translation: Aging, Hearing Loss, and Amplification

Samira Anderson

Abstract The temporal and spectral components of complex stimuli are well represented in the frequency-following response (FFR), making it ideal for assessing speech encoding in clinical populations. Age-related deficits in temporal precision are noted in animals and humans, providing a possible explanation for the auditory temporal processing deficits that contribute to difficulties with speech perception. Hearing loss has variable effects on the FFR; in some cases hearing loss enhances encoding of the stimulus envelope, especially for higher frequency stimuli. However, hearing loss may degrade the envelope for lower frequency, steady stimuli and also appears to degrade representation of the temporal fine structure. A disruption of the balance of neural representation of the envelope and temporal fine structure may provide an explanation for the observation that speech is loud, but unclear, in cases of hearing loss. Aging and hearing loss effects on central auditory processing have ramifications for assessment and management of older adults with hearing loss. Historically, strategies have focused on improving audibility, although interest is rising in the interaction of hearing loss and cognitive function. Along with consideration of cognition function, it is expected that clinicians will soon begin to implement strategies to improve accuracy of central speech encoding, either through changes in hearing aid algorithms or through targeted auditory training.

Keywords Aging · Auditory temporal processing · Auditory training · Frequency-following response · Hearing aids · Hearing loss · Temporal envelope · Temporal fine structure

---

S. Anderson (✉)
Department of Hearing and Speech Sciences, University of Maryland,
0100 Lefrak Hall, College Park, MD 20742, USA
e-mail: sander22@umd.edu

N. Kraus et al. (eds.), *The Frequency-Following Response*,
Springer Handbook of Auditory Research 61,
DOI 10.1007/978-3-319-47944-6_11

11.1 Introduction

According to the Administration of Aging, the number of adults 60 years and older will increase in the next 10 years from 57 million to more than 75 million. One consequence of this increase in the older population is a greater prevalence of communication problems associated with decreased hearing. Decreased hearing in older adults may lead to social isolation, depression, and reduced cognitive function (Heine and Browning 2002; Lin et al. 2011a). For this reason, increased efforts are being directed toward understanding the neural mechanisms that underlie the communication difficulties associated with aging and hearing loss with the aim of implementing appropriate evaluation and management strategies to offset some of these declines.

Although aging is associated with a decline in peripheral hearing thresholds, additional hearing difficulties may arise from declines in central auditory processing. Suprathreshold deficits in temporal processing have been documented for both speech and nonspeech stimuli. Older adults have more difficulty than young adults when detecting changes in the temporal cues of speech that may distinguish one word from another, such as voice-onset time or silence duration (Gordon-Salant et al. 2008). Furthermore, older adults have poorer temporal resolution (e.g., gap detection or duration discrimination) compared to younger adults (Fitzgibbons and Gordon-Salant 1994; Harris et al. 2010). Precise encoding of the temporal features of speech is necessary for accurate perception in noisy or reverberant environments in which the inherent redundancy of speech may be reduced.

Animal models have demonstrated possible neural mechanisms for these temporal processing deficits. A loss of auditory nerve fibers would lead to a reduction in the neural synchrony that is required for precisely timed representation of auditory stimuli. Older animals or humans may experience a degree of auditory neuropathy that has variable effects on speech perception. A selective loss of low–spontaneous–rate auditory nerve fibers is reflected in lower auditory brainstem response Wave I amplitudes in an aging mouse model (Sergeyenko et al. 2013). Other factors, such as delayed neural recovery (Walton et al. 1998) and changes in the balance of excitatory and inhibitory neurotransmission (Caspary et al. 2008), may contribute to the observed age-related changes in temporal resolution.

In both animal and human models, the frequency–following response (FFR), a scalp–recorded farfield response arising primarily from the midbrain, has been used to assess age-related or hearing-related changes in auditory temporal processing. The FFR is well suited to assess temporal encoding as it preserves the temporal and spectral features of the stimulus with remarkable precision. From a clinical perspective, differences on the order of fractions of milliseconds may be clinically significant, indicating possible auditory-based impairments in children with learning disabilities (White-Schwoch et al. 2015), in older adults with speech perception difficulties (Anderson et al. 2013b), and in other populations. These populations may be exhibiting some degree of auditory neuropathy, as even young adults with normal hearing exhibit variability in behavioral and electrophysiological measures of temporal coding (Bharadwaj et al. 2015), suggesting that suprathreshold declines in auditory function may be observed in the presence of normal cochlear function.

Hearing aids are the primary intervention for older adults with hearing difficulties, but increased audibility may not improve auditory temporal processing deficits associated with age–related auditory neuropathy. Increased understanding of the nature of the neural mechanisms underlying these deficits may lead to improved assessment and management strategies.

This chapter first provides a brief summary of the behavioral and physiological literature examining the nature of deficits in auditory function associated with aging and hearing loss. The effects of aging and hearing loss on the FFR are then reviewed. The chapter ends with a discussion of how knowledge of the effects of aging and hearing loss on subcortical neural processing of sound can inform assessment and remediation strategies used in clinical management of older individuals who are experiencing hearing difficulties.

11.2 Perceptual Declines Associated with Aging and Hearing Loss

11.2.1 Aging

Older adults typically report that they can understand what others are saying in quiet settings but that they have difficulty hearing in noisy backgrounds. A similar observation is often made by individuals with auditory neuropathy, suggesting that decreased neural synchrony contributes to age–related decreases in perception. Behavioral evidence of age–related auditory temporal processing deficits has been found for a variety of perceptual tasks. For example, young adults' perception of speech that has been temporally jittered is equivalent to that of older adults' perception of normal speech, suggesting that jitter associated with age-related dysynchrony may account for speech perception difficulties in older adults (Pichora-Fuller et al. 2007). Older adults also exhibit perceptual deficits for time–compressed speech (Wingfield et al. 1999; Gordon-Salant et al. 2007) and for reverberant speech (Halling and Humes 2000).

Poorer performance on nonspeech tasks of temporal processing also supports the idea that decreased temporal resolution contributes to poor speech-in-noise performance in older adults. A signal waveform consists of two temporal components: the temporal envelope corresponds to slow variations in amplitude and the temporal fine structure (TFS) corresponds to the rapid oscillations in the signal that carry the envelope. Older adults exhibit decreased sensitivity to temporal envelope and TFS cues compared to young adults in tasks using tonal stimuli, and this sensitivity to temporal cues relates to identification of consonants and sentences presented in two–talker babble (Füllgrabe et al. 2014). The existence of age-related deficits in speech perception that are distinct from decreases in peripheral hearing or cognition has been debated (Humes et al. 2012), but the deficits found in the Füllgrabe et al. (2014) study were observed in older adults who had hearing thresholds matched to those of the younger adults. Furthermore, the correlation between age and TFS sensitivity remained even after controlling for the effect of cognition, suggesting the existence of central presbycusis that may arise from multiple levels of the auditory system (refer to Gordon-Salant et al. (2010) for a comprehensive review of aging effects on the auditory system).

11.2.2 Hearing Loss

Aging may be a confounding variable when evaluating hearing loss effects on perception of TFS, as individuals with hearing loss are often older than individuals with normal hearing (Lorenzi et al. 2006). To circumvent this problem, studies have used age and hearing loss as continuous variables to evaluate independent effects on TFS sensitivity. For example, King et al. (2014) found orthogonal effects of aging and sensorineural hearing loss on the discrimination of interaural phase differences (IPD) in a group of adults who had a wide range of age and hearing levels. While the TFS thresholds and envelope IPD thresholds increased with age, sensorineural hearing loss appeared to affect the TFS but not the envelope thresholds. In another study, sensitivity to TFS was evaluated using monaural, bilateral, and binaural gap duration discrimination tasks in a group of adults with a wide range of ages, most of whom had fairly normal hearing thresholds (Gallun et al. 2014). They found that age and hearing loss had independent effects on performance across tasks, but age had a greater influence on monaural than binaural performance, while hearing had a greater influence on binaural than monaural performance.

Overall, these studies support the idea that temporal processing deficits in older adults contribute to the difficulties of hearing in noise. Furthermore, declines in either age or hearing may be associated with these deficits, especially for tasks involving sensitivity to TFS. Finally, as noted by Gallun et al. (2014), a substantial amount of variability in performance cannot be predicted by age or hearing loss, suggesting that the remaining variability may be accounted for, at least in part, by neurodegeneration in the auditory nerve and throughout the central auditory system or by cognitive factors.

11.3 Neuroanatomical Changes Associated with Aging and Hearing Loss

Age-related hearing loss (presbycusis) may be associated with a loss of outer hair cells (Schuknecht 1964), a reduction in the endocochlear potential (Schmiedt et al. 2002; Ohlemiller et al. 2006), and a loss of auditory nerve fibers (Felder and

Schrott-Fischer 1995; Lang et al. 2010). While these pathologies may produce an actual loss of audiometric thresholds, recent attention has focused on other age-related neural changes that may result in decreased performance on speech perceptual tasks in the presence of normal thresholds. For example, cochlear synaptic and neural degeneration were found in an aging mouse model (CBA/CaJ) prior to a loss of outer hair cells (Sergeyenko et al. 2013). Decreased auditory brainstem response (ABR) amplitudes in these mice suggest an auditory neuropathy that could lead to speech perception impairments, especially in noise.

Auditory processing deficits may also arise from degeneration at higher levels of the auditory system. Caspary and colleagues have documented changes in the balance of inhibitory and excitatory neurotransmission in the brainstem (Caspary et al. 2006), midbrain (Caspary et al. 1995), and auditory cortex (Hughes et al. 2010). These changes may lead to decreased ability to process rapidly changing temporal speech components and subsequent impairments in perception (as reviewed in Caspary et al. 2008). For example, the ability to detect gaps represents a dimension of temporal resolution that is important for accurate perception of certain speech contrasts, such as consonants that differ in voice-onset time (tie versus die).

In a chinchilla model, noise–induced hearing loss (NIHL) leads to enhanced neural coding of the temporal envelope of sinusoidally amplitude–modulated (SAM) tones presented in quiet in the auditory nerve and inferior colliculus (Kale and Heinz 2010; Zhong et al. 2014). Conversely, in responses to SAM tones presented in noise, coding of the TFS is actually reduced in chinchillas with NIHL (Henry and Heinz 2012). Heinz and colleagues surmise that this pattern of NIHL induced enhancement of the envelope and reduction of the TFS (at least in noise) may arise from a number of different mechanisms, including outer hair cell dysfunction or an increase in the excitability of auditory neurons due to reduced neural input. This change in excitability was also noted in a gerbil model of sensorineural hearing loss (SNHL) that found increased amplitude but decreased frequency of excitatory postsynaptic currents in thalamocortical slices (Kotak et al. 2005). These studies suggest the existence of homeostatic mechanisms in the central auditory system that may serve to increase central gain to offset loss of sensory input (Zhong et al. 2014). This change in central gain may offer an explanation for why older adults with hearing loss often report that the volume is loud enough or too loud but that the clarity is reduced. Exaggerated amplitude fluctuations may lead to a sensation of loudness, but clarity may be diminished due to inadequate representation of the TFS.

In humans, imaging studies using cortical evoked potentials or magnetoencephalography have demonstrated an exaggerated enhancement of responses to auditory stimuli associated with hearing loss (Tremblay et al. 2003; Alain et al. 2014) and with aging (Soros et al. 2009; Alain et al. 2012), further evidence of a central gain mechanism that compensates for a loss of sensory input. Older adults also draw on cognitive resources to compensate for hearing difficulties, especially in noise. In fMRI studies of speech-in-noise perception, older adults show reduced activation of the auditory cortex but increased activation of prefrontal areas related to working memory and attention compared to younger adults (Wong et al. 2009). Furthermore, in older adults, the volume of the left pars triangularis and the thickness of the left superior frontal gyrus predict performance on speech-in-noise tasks (Wong et al. 2010). Although the participants in the Wong et al. studies had clinically normal hearing thresholds (≤25 dB HL), the reduced auditory cortex activation may have been affected by subclinical age-related loss of peripheral sensitivity and by age-related deficits in central processing independent of ear health. Even mild to moderate declines in hearing sensitivity are associated with reduced gray matter volume in the auditory cortices in older adults (Peelle et al. 2011). These results provide a neural basis for the increasing role of cognition in speech perception performance that has been found in behavioral investigations of aging effects on speech understanding (Schneider and Pichora-Fuller 2000; Tun et al. 2002).

11.4 Aging Effects on the FFR

11.4.1 Aging Effects in Animals

Because aging affects temporal precision of neural speech encoding, the FFR is well–suited to evaluate temporal processing deficits associated with aging. Bartlett and colleagues conducted a series of studies on aging effects on the ABR and FFR in Fischer 344 rats, the results of which provide a neurophysiological basis for psychophysical findings in humans. Using farfield recordings, they compared amplitude–modulated following responses (AMFRs) in younger versus older rats and found comparable AMFR amplitudes between the groups in the mid–frequency range (181–512 Hz), but higher amplitudes at low and high modulation frequencies in younger rats compared to older rats (Parthasarathy et al. 2010), consistent with human studies showing age-related declines in envelope detection of modulation frequencies for low (He et al. 2008) and high modulation frequencies (Grose et al. 2009). In the same study, Bartlett and colleagues (Parthasarathy et al. 2010) evaluated the effects of wideband noise on AMFR amplitudes. Interestingly, they found that moderate levels of background noise resulted in significant reductions in AMFR amplitudes in the younger but not the older rats at low frequencies, but at the higher frequencies this pattern was reversed and the older rats had a greater noise–induced decline in AMFR amplitude than the younger rats. A subsequent study compared responses in younger and older rats when amplitude and frequency modulation depths were varied (Parthasarathy and Bartlett 2011) and found an age-related reduction in response amplitudes at the lower, but not the higher, modulation depths (Fig. 11.1). They also found that the older rats had reduced precision of envelope shape coding, suggesting a loss of the ability to sustain neural firing. They surmised that decreased inhibitory neurotransmission associated with aging leads to a reduction in the precision of temporal processing that was demonstrated in these studies.

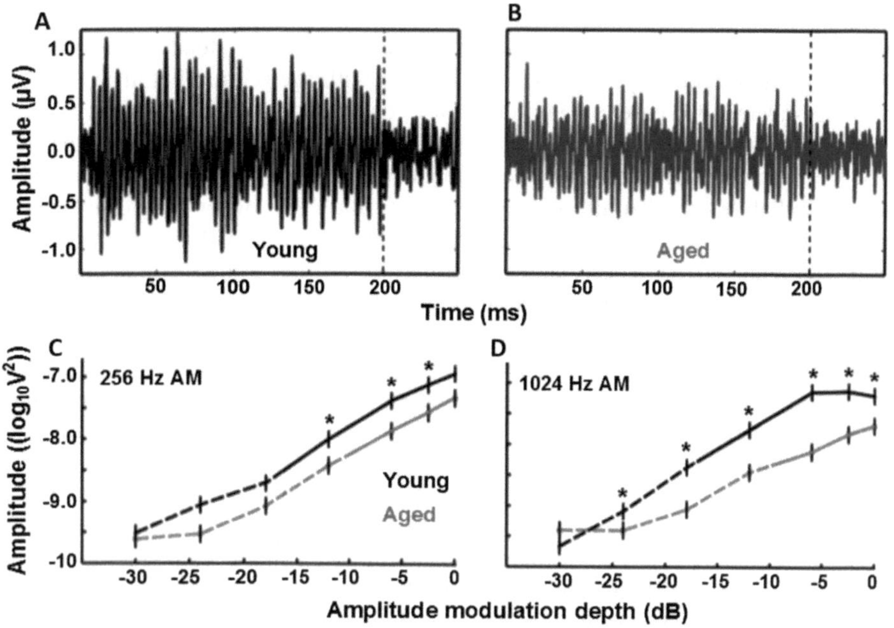

Fig. 11.1 **(A, B)** Responses comparing young rodents (*black graph, A*) and aged rodents (*grey graph, B*) show clear phase locking to amplitude-modulated tones. *Dashed lines* indicate stimulus offset. **(C, D)** Age-related differences in response amplitudes are present for both low frequencies (C) and high frequencies (D), but these differences are more apparent for smaller amplitude modulation depths. *Solid lines* indicate responses above mean modulation detection threshold. $^*p < 0.05$. (Adapted from Parthasarathy and Bartlett 2011, with permission from Elsevier)

Because current clinical testing uses the ABR rather than the FFR, it would be important to establish if the FFR provides information regarding auditory processing beyond what is represented in the ABR. Parthasarathy et al. (2014) compared ABRs and FFRs in a rodent model of aging and found age-related differences in ABR thresholds and amplitudes and in FFR phase locking capability. Interestingly, they found significant correlations between ABR and FFR amplitudes in the young rodents but not in the old rodents, suggesting that these measures provide information about different aspects of neurophysiological sound processing, and that the relationships among these measures change with age.

11.4.2 Aging Effects in Humans

One might expect that the reduced precision of temporal coding in the FFR found in aging animal models would also be found in humans. To evaluate age-related effects on temporal precision in humans, FFRs to steady–state tones and dynamic frequency sweeps were compared in younger and older adults (Clinard et al. 2010; Clinard and Cotter 2015). Response amplitudes elicited by steady–state tones of relatively high frequency ($\sim$1000 Hz) decrease with age (Clinard et al. 2010), and this age-related decrease in amplitude also occurs for lower frequency sweeps (beginning or ending at 400 Hz) that rapidly rise or fall in frequency at rates from 1333 Hz/sec to 6667 Hz/sec (Clinard and Cotter 2015). Response amplitudes to speech syllables are also affected by age, particularly the onset and offset regions after controlling for the effects of hearing (Vander Werff and Burns 2011; Clinard and Tremblay 2013). These studies used a 40-ms [da] syllable that contains a rapidly changing formant transition without a steady-state vowel region. Anderson et al. (2012) recorded responses to a 170-ms [da] in younger and older adults to compare the effects of aging on a speech syllable containing both formant transition and steady-state vowel regions. They found smaller amplitudes and reduced phase locking for both the transition and steady-state regions in the time and frequency domains, but the effects were more pronounced in the steady-state region.

A follow-up study compared FFRs to the vowel [a] and the syllable [da] to determine if age-related delays in peak latencies were due to an inability to phase lock to the rapidly changing formant transition in the syllable [da] (Presacco et al. 2015). They replicated the Anderson et al. (2012) finding of delayed peak latencies specific to the formant transition in the [da]. However, they also found that in young adults, peak latencies were earlier for the [da] than the [a], as expected given that the high frequency stop consonant burst in the [da] would be encoded earlier than the [a] due to cochlear tonotopicity. However, these peak timing differences between syllables were not found in the FFRs of older adults. They concluded that the lack of peak latency differences between syllables in the older adults was likely due to decreased hearing in the high frequencies, even though this group had clinically normal hearing. These findings and those of Vander Werff and Burns (2011) speak to the importance of accounting for group differences in high-frequency thresholds, even when those differences are slight. An important but unexpected finding in Presacco et al. (2015) was a marked reduction in sustained phase locking to the vowel [a] in older adults that was not observed in the younger adults (Fig. 11.2). These results are in line with those of Parthasarathy and Bartlett (2011), which showed age-related changes in the precision of envelope coding. The loss of sustained phase locking may arise from a number of changes associated with aging. For example, a loss of auditory nerve fibers may lead to an inability to sustain neural firing, as may be seen with VIIIth nerve tumors (Lidén and Korsan-Bengtsen 1973). Prolonged response recovery times may also change the shape of the neural response (Walton et al. 1998).

The Anderson et al. (2012) study also assessed trial-to-trial consistency and found that older adults had poorer response consistency than young adults for both transition and steady-state response regions. They surmised that poorer response consistency in older adults may be a neural correlate of temporal jitter that contributes to poorer speech perception in older adults (Pichora-Fuller et al. 2007). Mamo et al. (2015) tested this idea by applying different degrees of jitter to a speech syllable and recording responses to these jittered syllables in younger and older

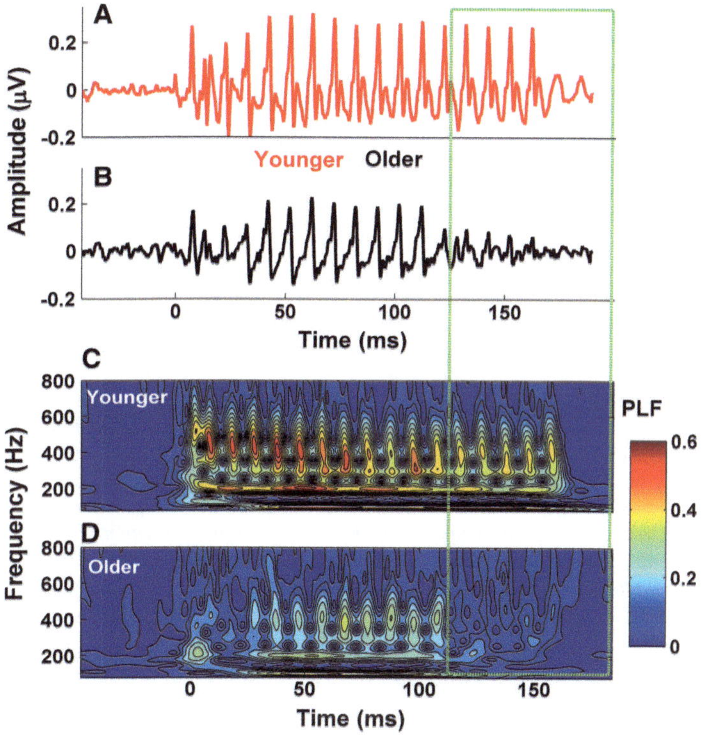

Fig. 11.2 **(A, B)** Average responses to a 170-ms vowel /a/ are displayed for younger (A, *red*, n=15) and older (B, *black*, n=15) human adults. **(C, D)** Phase-locking factor (*PLF*) to the vowel /a/ in the same younger (C) and older adults (D). Note the dramatic reduction in response amplitude and in phase locking after ~110 ms in older adults (this region indicated by the *light green rectangle*. (Adapted from Presacco et al. 2015, with permission from Wolters Kluwer Health, Inc.)

adults. They compared effects of jitter on the envelope and TFS components of speech by presenting the [da] in alternating polarities. Adding responses to the two polarities emphasizes the envelope component and minimizes the fine structure, while subtracting the responses has the opposite effect (Aiken and Picton 2008). Even a mild degree of jitter produced a significant decrease in response amplitudes to the envelope in the younger adults, whereas no reduction was seen in the older adults (Fig. 11.3). In response to the temporal fine structure, the mild jitter condition resulted in a dramatic reduction in harmonic representation in the young adults to the extent that their responses in the mild jitter condition were equivalent to the responses of older adults in the non–jittered condition. Again, older adults' responses did not show a reduction in amplitude with jitter, presumably because a loss of neural synchrony has already introduced jitter into the responses of older adults.

The FFR may also be used to increase understanding of the mechanisms contributing to cognitive functions, such as selective attention. Although they did not

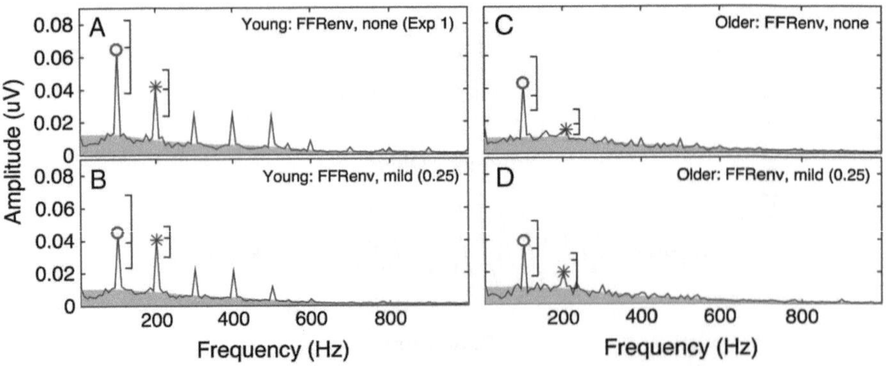

Fig. 11.3 Average responses to a 170-ms [da] syllable are displayed for younger and older human adults. **(A, B)** The spectral amplitude of the F_0 was significantly reduced in the mild jitter condition (B) compared to the no jitter condition (A) in young adults (A, n = 22; B, n = 21). The harmonics were essentially unaffected. **(C, D)** A similar reduction in F_0 amplitude was not seen in the mild jitter (D) compared to the no jitter condition (C) in older adults, presumably because their responses are already affected by neural jitter (C, n = 22; D, n = 7). *open circles*, F_0; *asterisks*, second harmonic; *brackets*, data distribution with *top and bottom dash* indicating top and bottom quartiles and *middle dash* indicating the median; *FFRenv*, frequency-following response to the envelope; *mild (0.25)*, mild jitter; *none*, no jitter. (Adapted from Mamo et al. 2015, with permission from Elsevier)

find age differences in behavioral measures of selective attention, Ruggles et al. (2011) found that the impact of reverberation on selective attention increases with age. They analyzed FFR phase locking to both the stimulus temporal envelope and TFS. When comparing relationships between phase locking and selective attention measures, they found that performance in middle-aged listeners appears to rely on encoding of TFS, whereas performance in young listeners is predicted by encoding of the stimulus envelope. The authors concluded that because effects of reverberation are greater for the TFS than for the envelope, selective attention in younger listeners, who rely primarily on envelope cues, will be affected to a lesser extent than in older listeners, who rely primarily on fine structure cues (see Shinn-Cunningham, Varghese, Wang, and Bharadwaj, Chap. 7 for a more thorough review of the FFR role in spatial hearing and selective attention).

These findings are supported by a recent study relating word intelligibility assessed in different degrees of reverberation to envelope and fine structure components of the FFR in older adults (Fujihira and Shiraishi 2015). This study found that representation of the fine structure (harmonics corresponding to the first formant), but not the envelope, predicted word intelligibility in conditions of mild to moderate reverberation. Although subcortical representation of fine structure degrades to a greater extent with age than the envelope (Anderson et al. 2012; Mamo et al. 2015), these findings support the idea that older adults rely on TFS components across perceptual tasks.

11.4.3 Neural Correlates of Perceptual Deficits

A number of studies have used the FFR to investigate neural correlates of clinical impairments associated with aging. Older adults who have clinically normal hearing thresholds are known to experience more trouble understanding speech in background noise than younger adults (Dubno et al. 1984; Souza et al. 2007), suggesting deficits in central auditory processing or decreased cognitive function (CHABA 1988), but the existence of central presbycusis as an isolated entity remains controversial (Humes et al. 2012). The FFR may be useful for evaluating central presbycusis as it does not place cognitive demands on the participant.

Two recent studies used the FFR to evaluate the neural basis of speech-in-noise impairments in older adults. The first study divided older adults (ages 60–73 years) into groups of higher and lower performance on the Hearing–in–Noise Test (HINT; Nilsson et al. 1994) and compared FFRs in response to a 170-ms [da] syllable presented in quiet and in six-talker babble (Anderson et al. 2011). They found that the group with better HINT scores had larger response amplitudes and more robust representation of the fundamental frequency (F_0) than the group with poorer HINT scores. They cross-correlated responses obtained in quiet with responses obtained in babble noise and found a strong positive correlation between response correlation values and HINT performance, suggesting that the robustness of subcortical speech representation is a factor in successful hearing in background noise (Fig. 11.4).

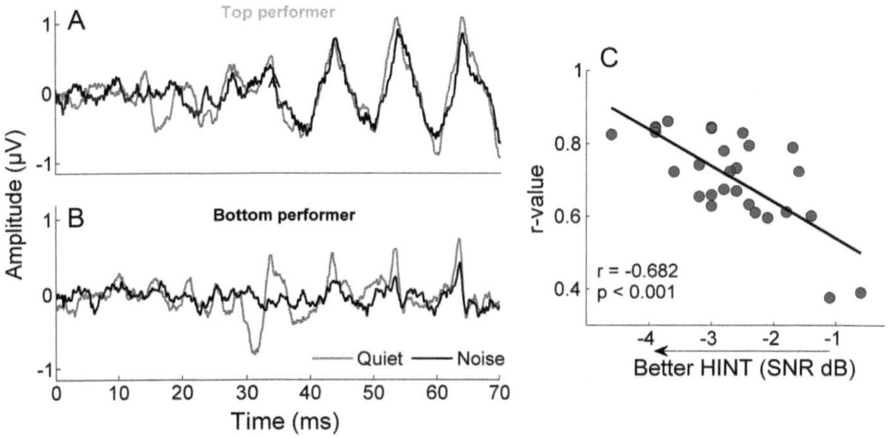

Fig. 11.4 **(A, B)** Individual response waveforms to a 170-ms [da] presented in quiet (*gray*) and in six-talker babble (*black*) from individuals who have good (A) or poor (B) scores on the HINT (Hearing-in-Noise Test). For better visualization, the figure zooms in on the first 70 ms (onset and transition). The consonant transition is degraded by noise to a greater extent in the bottom speech-in-noise performer. **(C)** Responses obtained in quiet were correlated with responses obtained in noise. Higher correlation values related to lower speech-in-noise thresholds. *r*, correlation coefficient; *SNR*, signal-to-noise ratio. (Adapted from Anderson et al. 2011, with permission from Wolters Kluwer Health, Inc.)

In terms of clinical relevance, this information might be useful as a counseling tool to help the patient understand why listening in noise might be so challenging. A follow-up study was performed to determine if the FFR would explain more of the variance in the older adult's own perception of their speech-in-noise ability than traditional clinical measures (Anderson et al. 2013b). This study recorded the FFR using a 40-ms [da] syllable in a group of 111 middle-aged to older-aged adults (ages 45–78 yrs) who had audiometric profiles ranging from normal to mild to moderate sensorineural hearing loss. The protocol was designed to be clinically feasible, and the 40-ms [da] was chosen because the testing time is approximately 20 min. They used the Speech, Spatial, and Qualities of Hearing Scale (SSQ; Gatehouse and Noble 2004) to assess self-reported speech-in-noise performance and the Quick Speech-in-Noise test (*QuickSIN™*; Killion et al. 2004) to assess performance in a clinical setting. Using a step-wise multiple linear regression, they found that hearing thresholds and QuickSIN™ scores predicted 15% of the variance in SSQ scores, and timing measures of the FFR (onset slope, morphology, and offset latency) predicted an additional 15%. They concluded that the FFR provides information about speech-in-noise performance beyond what is obtained using the current audiological protocol, and that it may be useful in the assessment and management of patients presenting with hearing difficulties (see Bidelman, Chap. 8 for more information on the FFR and communication in challenging environments).

Although the previously mentioned studies have found relationships among FFR and clinical speech-in-noise measures, mixed results have been obtained in studies comparing behavioral performance and the FFR using the same stimuli. In studies of frequency discrimination in young adults, periodicity strength of the FFR relates to F_0 difference limens in young adults (Krishnan et al. 2012; Smalt et al. 2012); however, in a study including older adults, Clinard et al. (2010) found age-related deficits in pitch discrimination and FFR representation of the same tones, but these measures were not predictive of each other. Because frequency discrimination performance in older adults is likely to be affected by elevated hearing thresholds, Marmel et al. (2013) investigated the relationship between FFR phase locking and frequency discrimination across a range of age and hearing thresholds to evaluate respective contributions to this relationship. They found that both FFR phase locking and hearing thresholds predicted frequency discrimination performance, while age was not a significant factor. Interestingly, they found that age, but not hearing thresholds, was related to FFR phase locking. Because age-related changes in peripheral hearing thresholds will be seen even in older adults with "clinically normal" hearing thresholds, this study underscores the need to consider the contributions of both age and hearing thresholds when investigating relationships among neural and behavioral measures of auditory performance (Carcagno and Plack, Chap. 4).

11.5 Hearing Loss Effects on the FFR

11.5.1 Hearing Loss Effects in Animals

Early investigations of hearing loss effects on perception have compared young normal hearing individuals with older individuals with hearing loss, thus introducing an aging confound. Animal models of noise-induced hearing loss (NIHL) provide one means of eliminating that confound. Heinz and colleagues have conducted a series of experiments to evaluate effects of NIHL on neural coding of the temporal envelope and fine structure in chinchillas. In the first experiment, responses from auditory nerve fibers were recorded in response to SAM tones or to single-formant stimuli in chinchillas that had normal hearing or had been exposed to narrowband noise levels sufficient to produce a threshold shift of at least 20 dB on the ABR (Kale and Heinz 2010). The strength of envelope coding was enhanced in noise-exposed fibers compared to coding in normal hearing fibers, especially those with higher ABR thresholds, but there was no reduction in the coding of fine structure. This initial study presented stimuli in quiet conditions only. In a follow-up study, Henry and Heinz (2012) recorded responses to spike trains presented in quiet and in three levels of Gaussian noise in chinchillas with and without NIHL to determine if the presence of noise would cause a degradation in the coding of fine structure. They found no differences in vector strength of phase locking to tones in the quiet condition, but with increasing levels of noise, vector strength decreased in the NIHL chinchillas compared to the normal-hearing (NH) chinchillas. Finally, Zhong et al. (2014) used scalp recordings to evaluate noise effects on neural coding in more central structures of the auditory midbrain. They found that noise exposure resulted in an increase in envelope response amplitudes to SAM tones in both quiet and noise conditions (Fig. 11.5).

Zhong et al. (2014) surmised that envelope enhancement associated with hearing impairment may arise from both peripheral and central noise-induced changes. Outer hair cell dysfunction or impairment of high-threshold auditory nerve fibers would lead to steeper input-output functions and enhanced response amplitudes for suprathreshold input levels. Alternatively, results may reflect increased central gain due to homeostatic regulation of excitatory and inhibitory synapses following reduced sensory input (Chambers et al. 2016). Based on these findings, Heinz and colleagues suggest that the enhancement of envelope information at the expense of TFS may contribute to speech perception difficulties in individuals with hearing loss, as the heightened envelope cues may distract the listener from the fine details required for accurate speech discrimination.

Recent attention has been focused on the damage produced by moderate levels of noise exposure that results in cochlear neuropathy—a loss of auditory nerve fibers without concomitant outer hair cell damage (Kujawa and Liberman 2009; Lin, Furman, et al. 2011). Given that this type of auditory dysfunction is not

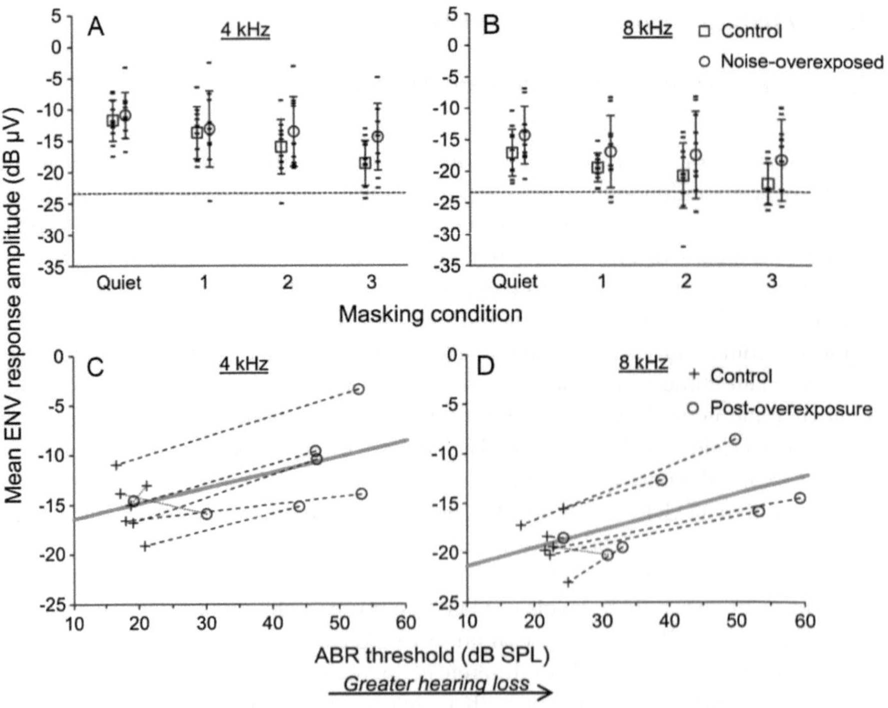

Fig. 11.5 **(A, B)** Considerable overlap is noted in response amplitudes to the temporal envelope of SAM (sinusoidally amplitude-modulated) tones presented in quiet and in three levels of Gaussian noise between chinchillas with and without NIHL (noise-induced hearing loss), possibly reflecting differing degrees of NIHL. *Open symbols* and *error bars* correspond to means and standard deviations, respectively. The *dashed line* represents the noise floor. **(C, D)** Recordings were obtained in seven animals before and after noise exposure and changes in envelope response amplitude were greater in animals that had greater noise-induced shifts in hearing thresholds. *Thick grey lines* represent the predicted relationship between ABR (auditory brainstem response) threshold and response amplitude. Although recordings were also obtained to 1 and 2 kHz carrier frequencies, the greatest effects were obtained to the higher frequencies, which are displayed here. *ENV*, envelope (Adapted from Zhong et al. 2014, with permission from Elsevier Limited)

reflected in audiometric threshold or otoacoustic emission testing, a clinical measure is needed that would be sensitive to cochlear neuropathy. Wave I amplitude of the ABR may reflect a reduction of auditory nerve fibers, but high variability may reduce its clinical efficacy. Shaheen et al. (2015) assessed effects of moderate noise exposure on FFRs to SAM tones and ABRs to tone pips in mice. While ABR amplitude and FFR amplitude and phase locking were reduced in noise-exposed mice, the changes in the FFR were more robust with reduced variability, suggesting that the FFR may serve as an efficacious measure of noise-evoked auditory neuropathy in the clinic.

11.5.2 Hearing Loss Effects in Humans

Anderson et al. (2013a) investigated the effects of hearing loss in humans using FFRs to a 40 ms [da] syllable presented binaurally in quiet and noise. To reduce audibility effects, they created amplified waveforms based on individual hearing loss using the National Acoustics Laboratory-Revised (NAL-R) algorithm (Byrne and Dillon 1986) and presented the [da] syllable in both amplified and unamplified conditions. To minimize effects of aging, they compared two groups of older adults who were matched in age: one group with normal audiometric thresholds and one group with mild to moderate sensorineural hearing loss. Similar to the Kale and Heinz (2010) study, they found that the response amplitude to the envelope was larger in the group with hearing loss than in the group with normal hearing in both aided and unaided conditions, especially in noise.

The initial study did not find differences in fine structure representation, but a follow-up training study comprising a larger number of participants (58 in the follow–up study versus 30 in the initial study) found that spectral amplitudes of the TFS were smaller in the noise condition in hearing-impaired individuals than in normal-hearing individuals (Anderson et al. 2013c) (Fig. 11.6). Because the results of the follow-up study were consistent with Henry and Heinz (2012), the initial lack of significant findings in the first study may have been due to insufficient power. These results support studies demonstrating perceptual deficits for TFS cues associated with hearing loss (King et al. 2014) that may be contributing to deficits in speech perception (Lorenzi et al. 2006; Füllgrabe et al. 2014).

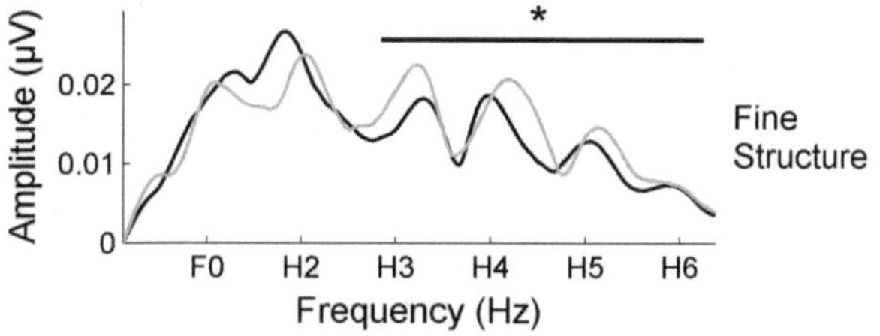

Fig. 11.6 FFRs obtained in human older adults with and without hearing loss (matched in age) to a 40-ms [da] presented in pink noise (+10 dB signal-to-noise ratio). Hearing impaired adults have greater representation of the envelope in the pitch-dominated frequencies (F_0 and second harmonic, H2), whereas normal hearing adults have greater representation of the fine structure in the frequency region corresponding to the first formant. *H3–H6*, harmonics; $^*p < 0.05$, $^{**}p < 0.01$. (Adapted from Anderson et al. 2013, with permission from Frontiers)

The Anderson et al. studies of hearing loss effects used an individually amplified speech stimulus to minimize effects of audibility. Ananthakrishnan et al. (2015) employed a different approach to equate audibility by obtaining FFRs at four different presentation levels and by comparing NH and HI individuals at equal sensation levels. They used a relatively low frequency vowel (/u/) with the first two formants well below 1000 Hz. In contrast with previous findings (Kale and Heinz 2010; Anderson et al. 2013a), they found degradation of both the envelope and the TFS in the HI individuals. They attributed these differences to milder degrees of hearing loss in their study and to differences in compensating for hearing loss. However, it should be noted that the HI group was older than the NH group. Although the effects they found persisted even after they controlled for age, it is important to minimize aging effects by matching groups on this variable to the extent possible in humans. Overall, this study highlights the benefits of using multiple stimulation levels when evaluating the effects of hearing loss.

11.5.3 Neural Correlates of Performance

Using synthesized stop consonants on a /ba/-/da/-ga/ continuum, Plyler and Ananthanarayan (2001) evaluated effects of hearing loss on identification performance and accuracy of FFR encoding of the second formant transition at different presentation levels. They found that although the FFR spectral peak shifts toward the higher frequencies as the second formant transition rises over time in the normal hearing group, this shift was substantially reduced in the group with hearing loss, suggesting that reduced hearing sensitivity may degrade phase locking. Furthermore, wider critical bands and reduced frequency selectivity in the HI group may lead to a broad dispersion of FFR peaks. Although there was no correlation between behavioral performance and FFR representation, the hearing impaired individuals tended to have reduced identification and degraded FFRs, suggesting a relationship in at least some of the HI individuals.

The Plyler and Ananthanarayan (2001) study used a broad range of ages and, therefore, interpretation of their findings is likely to be confounded by aging differences between the NH and HI groups. An alternate approach would be to use hearing level as a continuous variable within age groups. Bidelman et al. (2014) used this approach in a study that evaluated both FFR and cortical-evoked responses to a five-step /u/ to /a/ continuum of synthetic vowels that differed in the first formant frequency, and they compared neural responses to categorical perception of these vowels. Better behavioral performance was related to larger F1 magnitudes in the FFR but to reduced N1-P2 magnitudes in the cortical response across stimuli and groups. When investigating orthogonal effects of aging versus hearing loss, they found that greater levels of hearing loss were related to weaker subcortical pitch salience and larger cortical N1-P2 magnitudes, but age did not correlate with subcortical pitch salience or F1 encoding. However, both hearing loss and aging were associated with stronger cortical responses, an over-enhancement

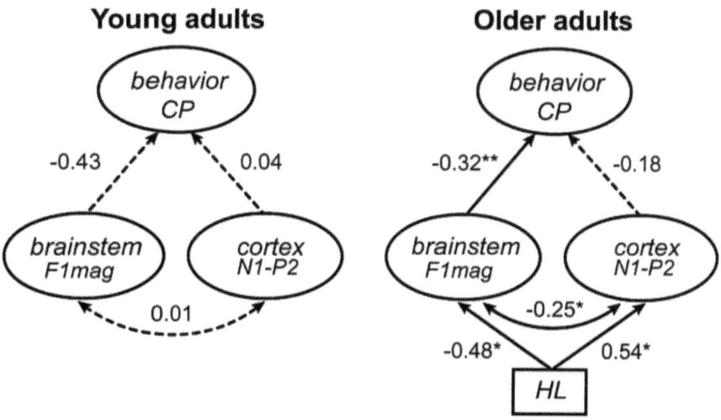

Fig. 11.7 Correlations among categorical perception, brainstem first-formant magnitude *(F1 mag)*, and cortical magnitudes (*N1-P2*) in response to vowels varying on a continuum of the first formant are displayed separately for young and older adults. In older adults only, brainstem magnitudes significantly correlated with cortical magnitudes and with categorical perception. In addition, hearing thresholds were negatively correlated with brainstem magnitudes but positively correlated with cortical magnitudes. *HL*, age-related hearing loss; $^*p < 0.05$, $^{**}p < 0.01$. (Reprinted from Bidelman et al. 2014, with permission from Elsevier Limited)

that has been observed in other studies (Tremblay et al. 2003; Alain et al. 2014). In older adults, smaller FFR magnitudes were related to larger cortical magnitudes, suggesting that weakened subcortical encoding may contribute to exaggerated cortical responses associated with a down regulation of inhibitory neurotransmission (Turner et al. 2005). Because these patterns were not seen in younger adults, the authors concluded that there is greater redundancy between levels of the auditory system in older adults to compensate for deficient encoding associated with aging and hearing loss. This diminished encoding is a factor in impaired perception on the behavioral categorization task (Fig. 11.7).

11.6 Clinical Implications

11.6.1 Amplification

Because the FFR reflects auditory processing, it may prove to be a useful tool for evaluating the benefits of hearing aid amplification. The current recommendation for most individuals with mild to moderate sensorineural hearing loss is the use of hearing aids, but less than 25% of people who would benefit from hearing aids actually use them (Kochkin 2010). The current standard of audiologic care recommends real-ear measurements to verify that hearing aids are providing appropriate levels of amplification for the hearing loss, but this approach does not provide any information about how sound is processed beyond the tympanic membrane.

A number of questionnaires can be used to validate the success of fitting the hearing aid, but these questionnaires may be affected by personality factors and may not reveal the root cause of dissatisfaction with hearing aids. Because digital technology provides a great deal of flexibility in fitting the hearing aid, audiologists often turn to changes in the software to adjust high or low frequency settings without knowing how these changes affect the accuracy of neural speech encoding.

To address some of these issues, a clinical instrument was developed to ensure audibility of speech consonants using cortical evoked potentials (HEARLab™; Munro et al. 2011). This instrument may be useful for assessing infants and individuals who are hard to test, but it may be less useful in a cooperative child or adult who can verify audibility using a behavioral procedure. Furthermore, verification of audibility does not insure that temporal or spectral components of speech components are being accurately encoded.

The traditional ABR to clicks or tone bursts has not been considered a valid approach to the assessment of hearing aids because the transient stimuli that are used for threshold testing would not be compatible with hearing aid time constants. However, the stimuli typically used in FFR testing have durations long enough to exceed the rise and fall times of hearing aid processing. Two studies recently investigated the feasibility of using the FFR to evaluate effects of stimulus level, bandwidth, and amplification in adults with normal hearing and with hearing loss (Easwar et al. 2015a, b). Both studies elicited the FFR with a naturally spoken speech token /susaʃi/ containing low-frequency, mid-frequency, and high-frequency phonemes. To ensure that the protocol was clinically feasible, just 300 sweeps were recorded for each condition and a statistical algorithm was used to determine the probable presence of the response. Bandwidth was evaluated by low–pass filtering the /susaʃi/ token at 1, 2, and 4 kHz. In the initial pilot study with NH adults, increases in level and stimulus bandwidth led to an increase in response amplitudes and in the number of detectable responses. In a follow-up study, experienced hearing-aid users with mild to moderate sensorineural hearing loss underwent the same protocol, but in addition to examining the effects of level and bandwidth, the authors elicited the FFR while the /susaʃi/ token was presented to individually fitted hearing aids through wireless transmission. Again, they found that increases in level and bandwidth and the use of amplification increased the number of detectable responses. Furthermore, speech discrimination scores and sound quality ratings correlated positively with FFR amplitude and detectable responses, suggesting that the FFR might be useful for predicting suprathreshold performance.

Similar to the HEARLab™ system, this previously mentioned protocol was designed to improve verification of the benefits of hearing aids in infants and young children, with a focus on improved audibility. Nevertheless, adult users of hearing aids often report that hearing aids are loud enough to hear conversation, but they have trouble with the clarity of speech. As discussed in Sect. 11.5, loudness may be detrimental to clarity, and it would be worthwhile to understand the factors in subcortical transcription that lead to improved understanding of speech with hearing aids. A better understanding of these factors may lead to adjustments in algorithms for hearing aids or device settings. The feasibility of using the FFR to aid in

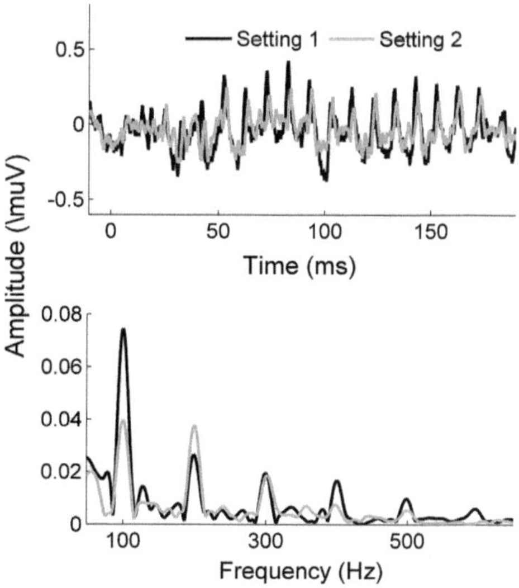

Fig. 11.8 The FFR may reflect changes in hearing aid settings. Responses to a 170-ms [da] syllable were recorded in a sound field in an older individual wearing a hearing aid with one of two settings. The response amplitude in the time and frequency domains was increased with *setting 1* compared to *setting 2*. (Adapted from Anderson and Kraus 2013, with permission from Hindawi Publishing)

adjusting the setting for hearing aids was observed in an individual who was encountering hearing aid difficulties (Fig. 11.8). One factor to consider in these recordings is the stimulus artifact produced by hearing aids. One approach to reducing artifact is to use direct audio input or wireless sound transmission (Bellier et al. 2015). Work is underway to explore the ways that the FFR can be used to maximize successful fitting of hearing aids in both pediatric and adult populations.

11.6.2 Auditory Training

Through the use of digital technology, the benefits of amplification have improved to a considerable extent. Yet, even if a hearing aid is capable of delivering a perfect signal for an individual hearing loss, amplification will not compensate for declines in spectrotemporal processing associated with aging. For this reason, clinicians should consider including auditory training as part of the management protocol. At this time, although there are studies demonstrating the efficacy of auditory training (Song et al. 2012; Ferguson et al. 2014), there is limited understanding of the kinds of protocols that would be most beneficial. The responses and needs of older adults are highly variable; therefore, a "one size fits all approach", such as is used in most commercial training packages, will likely have limited benefits for this heterogeneous population.

Because the FFR represents the temporal and spectral characteristics of the speech signal with precise fidelity, it may provide an appropriate tool for demonstrating training benefits. For example, one of the manifestations of age-related decreases in temporal precision is a delay in FFR peak latencies (Vander Werff and

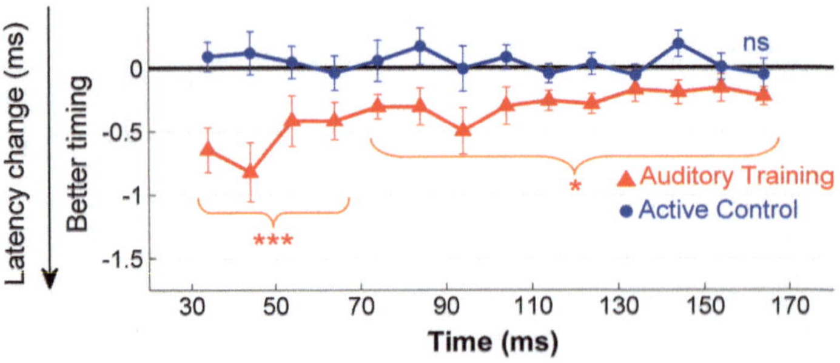

Fig. 11.9 Training-induced changes in FFR peak latencies for a 170-ms [da] syllable recorded in two-talker babble (+10 SNR). In the auditory training group, significant decreases were noted in peak latencies, more so in the region corresponding to the formant transition (30–60 ms) than in the region corresponding to the vowel (60–170 ms). No changes were noted in the active control group (ns = not significant). *Error bars*, ±1 standard error. *$p < 0.05$, ***$p < 0.001$. (Adapted from Anderson et al. 2013d, with permission from Proceedings of the National Academy of Sciences of the USA)

Burns 2011; Anderson et al. 2012). A recent study demonstrated that this aging effect can be partially reversed with training. Adaptive auditory-based cognitive training reduced peak latencies and inter-peak variability in the FFR to a speech syllable presented in quiet and in babble noise, and the greatest effects were seen in noise (Anderson et al. 2013d) (Fig. 11.9). Concomitant improvement was seen in speech-in-noise performance, but the changes between the measures were not related, suggesting that different neural mechanisms contributed to perceptual and neural changes.

11.6.3 Clinical Use

Because the FFR preserves aspects of the speech stimulus so precisely, analysis of specific features that have degraded representation can inform clinical recommendations. For example, an older adult who has difficulty encoding the consonant-vowel transitions of speech may benefit from training that adaptively expands and contracts these transitions. Algorithms for hearing aids may be adjusted depending on the nature of the impairment. Chasin (2011) has recommended adjusting hearing aid parameters based on phoneme-level, word-level, and sentence-level differences in the individual's spoken language. These parameters could be similarly adjusted for processing deficits, such as deficient encoding of the F_0 or inability to sustain phase locking to long-duration vowels.

Another potential clinical use of the FFR would be to predict who would benefit from certain types of clinical management. For example, response consistency in the FFRs of good readers is higher than that of poor readers (Hornickel and Kraus 2013), and response consistency at pretest predicts gain in phonological awareness after a year of using an assistive listening device during school hours in children with reading impairments (Hornickel et al. 2012). School administrators are much more likely to follow up on recommendations that are tailored to an individual rather than to widespread recommendations that are made to everyone in a group.

Audiologists have long been aware that two people with identical audiograms may have vastly different experiences when trying to communicate in a noisy environment (Killion and Niquette 2000). Because the FFR reflects both long-term and short-term experiences (Krishnan and Gandour, Chap. 3; Carcagno and Plack, Chap. 4; White-Schwoch and Kraus, Chap. 6), and the long-term consequences of aging and hearing loss on central auditory processing, its use may provide the clinician with a better understanding of the nature of the deficit that is contributing to the patient's problems with hearing in noise.

11.7 Future Directions

The hearing aid studies cited in Sect. 11.6.1 used the FFR to verify audibility for phonemes containing energy in low to high frequency ranges. Knowledge of the effects of amplification on suprathreshold processing would also be beneficial, both for developers of hearing aid algorithms and for clinicians trying to maximize hearing aid benefits. As mentioned in Sect. 11.5, individuals with hearing loss have an over-representation of the temporal envelope at the expense of the fine structure, especially in noise. It would be useful to determine the specific features of amplification that affect the balance of representation of the envelope and TFS. Modern hearing aids automatically adjust for different environments, but the strategies for this adjustment vary among hearing aid companies. Most hearing aids use some form of nonlinear compression, but time constants and other aspects of compression differ, with some hearing aids having fairly fixed, slow compression time constants, and other hearing aids having an option of slow versus fast speeds. There is evidence supporting the use of slower compression speeds for older individuals or individuals who have reduced cognitive function (Lunner and Sundewall-Thoren 2007; Cox and Xu 2010). It would be useful to determine the effects of varying compression speeds on neural encoding of various speech components in these different populations.

Although evidence suggests that a short training program can improve subcortical encoding of speech, more work is needed to determine the specific features of training that provide the most benefit. Because FFR changes may be specific to the training stimuli (Song et al. 2008; Carcagno and Plack 2011), the use of the FFR may inform the investigator of the aspects of training that can be used to achieve certain perceptual benefits. For example, training on speech-in-noise recognition led to enhancement in the F_0 in young adults (Song et al. 2012). Because the robustness of F_0 appears to be a factor in better speech recognition in noise in young and older adults with normal hearing, perhaps training that adaptively adjusts the signal-to-noise ratio of training stimuli can be particularly effective.

Finally, the FFR is considered a research tool and is not approved for clinical use in the United States. Work is underway to provide normative data and guidance to clinicians regarding the use of the FFR to classify individuals according to specific impairments. In a recent study, a consonant-in-noise score was developed (representing FFR peak latencies, response consistency, and spectral amplitudes) that predicts 68% of the variance in phonological scores in preschool-aged children and correctly classifies school-aged children with or without dyslexia in 69.1% of cases (White-Schwoch et al. 2015). Therefore, the FFR could become a valuable tool in the assessment of children with language-based learning impairments and other populations with communication impairments (Kraus and Anderson, 2016; Schochat, Rocha-Muniz, and Filippini, Chap. 9).

11.8 Summary

Studies have demonstrated the FFR's usefulness in enhancing our understanding of the ways in which aging and hearing loss affect subcortical transcription of speech. Age-related reductions in neural synchrony and subcortical temporal precision are reflected in animal and human FFR studies, with smaller response amplitudes, decreased trial-to-trial consistency, decreased phase locking, and reduced ability to sustain neural firing. These deficits relate to speech perception abilities and may account, in part, for the difficulties older adults experience when trying to understand speech, especially in noisy environments.

Hearing loss effects on the FFR have been more varied, especially in humans, in part due to aging confounds and in part due to differences in strategies for equating audibility. Animal studies of NIHL demonstrate enhanced representation of the temporal envelope but decreased representation of the TFS, especially in noise. These findings were confirmed in human studies but only in response to speech syllables containing a high-frequency transient, stop consonant burst. Degradation in both the envelope and TFS may be found in response to low-frequency vowels. The strength of response magnitude to the first formant of these vowels relates to better categorical perception, suggesting that the FFR may be used as an objective assessment of perception. Disentangling the effects of aging and hearing loss in human studies is problematic, as hearing loss etiologies differ between younger and older individuals. More work is needed to understand the varied effects of hearing loss on the FFR and the ways in which these effects contribute to impaired perception.

Knowledge of changes to the FFR that accompany aging or hearing loss can guide clinical management. Historically, hearing aid algorithms have attempted to compensate for outer hair cell loss by restoring audibility while maintaining comfortable loudness, but recently, the focus has shifted to include cognitive considerations. Knowledge of the specific speech components that are affected by hearing loss or aging, as revealed by the FFR, may also be taken in consideration when developing amplification algorithms. Furthermore, as amplification may not be sufficient to restore degraded temporal processing, auditory training might be used to at least partially restore the deficits revealed by FFR testing in an individual. More work is needed to explore clinical uses and to ascertain the efficacy of FFR use.

Compliance with Ethics Requirements Samira Anderson declared that she had no conflicts of interest.

References

Aiken, S. J., & Picton, T. W. (2008). Envelope and spectral frequency-following responses to vowel sounds. *Hearing Research, 245*(1–2), 35–47.

Alain, C., McDonald, K., & Van Roon, P. (2012). Effects of age and background noise on processing a mistuned harmonic in an otherwise periodic complex sound. *Hearing Research, 283*(1–2), 126–135.

Alain, C., Roye, A., & Salloum, C. (2014). Effects of age-related hearing loss and background noise on neuromagnetic activity from auditory cortex. *Frontiers in Systems Neuroscience*, *8*, 8. http://journal.frontiersin.org/article/10.3389/fnsys.2014.00008. (Accessed August 19, 2016)

Ananthakrishnan, S., Krishnan, A., & Bartlett, E. (2015). Human frequency-following response: Neural representation of envelope and temporal fine structure in listeners with normal hearing and sensorineural hearing loss. *Ear and Hearing, 37*(2), e91–e103. Doi:10.1097/AUD.0000000000000247

Anderson, S., Parbery-Clark, A., Yi, H.-G., & Kraus, N. (2011). A neural basis of speech-in-noise perception in older adults. *Ear and Hearing, 32*(6), 750–757.

Anderson, S., Parbery-Clark, A., White-Schwoch, T., & Kraus, N. (2012). Aging affects neural precision of speech encoding. *The Journal of Neuroscience, 32*(41), 14156–14164.

Anderson, S., Parbery-Clark, A., White-Schwoch, T., Drehobl, S., & Kraus, N. (2013a). Effects of hearing loss on the subcortical representation of speech cues. *Journal of the Acoustical Society of America, 133*(5), 3030–3038.

Anderson, S., Parbery-Clark, A., White-Schwoch, T., & Kraus, N. (2013b). Auditory brainstem response to complex sounds predicts self-reported speech-in-noise performance. *Journal of Speech, Language & Hearing Research, 56*(1), 31–43.

Anderson, S., White-Schwoch, T., Choi, H. J., & Kraus, N. (2013c). Training changes processing of speech cues in older adults with hearing loss. *Frontiers in Systems Neuroscience, 7*(97), 1–9.

Anderson, S., White-Schwoch, T., Parbery-Clark, A., & Kraus, N. (2013d). Reversal of age-related neural timing delays with training. *Proceedings of the National Academy of Sciences of the USA, 110*(11), 4357–4362.

Bellier, L., Veuillet, E., Vesson, J.-F., Bouchet, P., et al. (2015). Speech auditory brainstem response through hearing aid stimulation. *Hearing Research, 325*, 49–54.

Bharadwaj, H. M., Masud, S., Mehraei, G., Verhulst, S., & Shinn-Cunningham, B. G. (2015). Individual differences reveal correlates of hidden hearing deficits. *The Journal of Neuroscience, 35*(5), 2161–2172.

Bidelman, G. M., Villafuerte, J. W., Moreno, S., & Alain, C. (2014). Age-related changes in the subcortical-cortical encoding and categorical perception of speech. *Neurobiology of Aging, 35*(11), 2526–2540.

Byrne, D., & Dillon, H. (1986). The National Acoustic Laboratories' (NAL) new procedure for selecting the gain and frequency response of a hearing aid. *Ear and Hearing, 7*(4), 257–265.

Carcagno, S., & Plack, C. (2011). Subcortical plasticity following perceptual learning in a pitch discrimination task. *Journal of the Association for Research in Otolaryngology, 12*(1), 89–100.

Caspary, D. M., Milbrandt, J. C., & Helfert, R. H. (1995). Central auditory aging: GABA changes in the inferior colliculus. *Experimental Gerontology, 30*(3–4), 349–360.

Caspary, D. M., Hughes, L. F., Schatteman, T. A., & Turner, J. G. (2006). Age-related changes in the response properties of cartwheel cells in rat dorsal cochlear nucleus. *Hearing Research, 216*, 207–215.

Caspary, D. M., Ling, L., Turner, J. G., & Hughes, L. F. (2008). Inhibitory neurotransmission, plasticity and aging in the mammalian central auditory system. *Journal of Experimental Biology, 211*(11), 1781–1791.

Chambers, A. R., Resnik, J., Yuan, Y., Whitton, J. P., et al. (2016). Central gain restores auditory processing following near-complete cochlear denervation. *Neuron, 89*(4), 867–879.

CHABA (Committee on Hearing, Bioacoustics, and Biomechanics). (1988). Speech understanding and aging. *The Journal of the Acoustical Society of America, 83*(3), 859–895.

Chasin, M. (2011). Setting hearing aids differently for different languages. *Seminars in Hearing, 32*(02), 182–188.

Clinard, C., & Tremblay, K. (2013). Aging degrades the neural encoding of simple and complex sounds. *Journal of the American Academy of Audiology, 24*(7), 590–599.

Clinard, C. G., & Cotter, C. M. (2015). Neural representation of dynamic frequency is degraded in older adults. *Hearing Research, 323*, 91–98.

Clinard, C. G., Tremblay, K. L., & Krishnan, A. R. (2010). Aging alters the perception and physiological representation of frequency: Evidence from human frequency-following response recordings. *Hearing Research, 264*(1–2), 48–55.

Cox, R. M., & Xu, J. (2010). Short and long compression release times: Speech understanding, real-world preferences, and association with cognitive ability. *Journal of the American Academy of Audiology, 21*(2), 121–138.

Dubno, J., Dirks, D., & Morgan, D. (1984). Effects of age and mild hearing loss on speech recognition in noise. *The Journal of the Acoustical Society of America, 76*, 87–96.

Easwar, V., Purcell, D. W., Aiken, S. J., Parsa, V., & Scollie, S. D. (2015a). Evaluation of speech-evoked envelope following responses as an objective aided outcome measure: Effect of stimulus level, bandwidth, and amplification in adults with hearing loss. *Ear and Hearing, 36* (6), 635–652.

Easwar, V., Purcell, D. W., Aiken, S. J., Parsa, V., & Scollie, S. D. (2015b). Effect of stimulus level and bandwidth on speech-evoked envelope following responses in adults with normal hearing. *Ear and Hearing, 36*(6), 619–634.

Felder, E., & Schrott-Fischer, A. (1995). Quantitative evaluation of myelinated nerve fibres and hair cells in cochleae of humans with age-related high-tone hearing loss. *Hearing Research, 91*(1–2), 19–32.

Ferguson, M. A., Henshaw, H., Clark, D. P., & Moore, D. R. (2014). Benefits of phoneme discrimination training in a randomized controlled trial of 50- to 74-year-olds with mild hearing loss. *Ear and Hearing, 35*(4), e110–121.

Fitzgibbons, P. J., & Gordon-Salant, S. (1994). Age effects on measures of auditory duration discrimination. *The Journal of Speech and Hearing Research, 37*(3), 662–670.

Fujihira, H., & Shiraishi, K. (2015). Correlations between word intelligibility under reverberation and speech auditory brainstem responses in elderly listeners. *Clinical Neurophysiology, 126*(1), 96–102.

Füllgrabe, C., Moore, B. C. J., & Stone, M. A. (2014). Age-group differences in speech identification despite matched audiometrically normal hearing: Contributions from auditory temporal processing and cognition. *Frontiers in Aging Neuroscience, 6*, Doi:10.3389/fnagi.2014.00347

Gallun, F. J., McMillan, G. P., Molis, M. R., Kampel, S. D., et al. (2014). Relating age and hearing loss to monaural, bilateral, and binaural temporal sensitivity. *Frontiers in Neuroscience, 8*, 172.

Gatehouse, S., & Noble, W. (2004). The speech, spatial and aualities of hearing scale (SSQ). *International Journal of Audiology, 43*(2), 85–99.

Gordon-Salant, S., Fitzgibbons, P. J., & Friedman, S. A. (2007). Recognition of time-compressed and natural speech with selective temporal enhancements by young and elderly listeners. *Journal of Speech, Language and Hearing Research, 50*(5), 1181–1193.

Gordon-Salant, S., Yeni-Komshian, G., & Fitzgibbons, P. (2008). The role of temporal cues in word identification by younger and older adults: Effects of sentence context. *The Journal of the Acoustical Society of America, 124*(5), 3249–3260.

Gordon-Salant, S., Frisina, R. D., Fay, R. R., & Popper, A. N. (Eds.). (2010). *The aging auditory system*. New York: Springer Science+Business Media.

Grose, J. H., Mamo, S. K., & Hall, J. W., 3rd. (2009). Age effects in temporal envelope processing: Speech unmasking and auditory steady state responses. *Ear and Hearing, 30*(5), 568–575.

Halling, D. C., & Humes, L. E. (2000). Factors affecting the recognition of reverberant speech by elderly listeners. *Journal of Speech, Language and Hearing Research, 43*(2), 414–431. Doi:10. 1044/jslhr.4302.414

Harris, K. C., Eckert, M. A., Ahlstrom, J. B., & Dubno, J. R. (2010). Age-related differences in gap detection: Effects of task difficulty and cognitive ability. *Hearing Research, 264*(1–2), 21–29.

He, N.-J., Mills, J. H., Ahlstrom, J. B., & Dubno, J. R. (2008). Age-related differences in the temporal modulation transfer function with pure-tone carriers. *Journal of the Acoustical Society of America, 124*(6), 3841–3849.

Heine, C., & Browning, C. J. (2002). Communication and psychosocial consequences of sensory loss in older adults: Overview and rehabilitation directions. *Disability and Rehabilitation, 24* (15), 763–773.

Henry, K. S., & Heinz, M. G. (2012). Diminished temporal coding with sensorineural hearing loss emerges in background noise. *Nature Neuroscience, 15*(10), 1362–1364.

Hornickel, J., & Kraus, N. (2013). Unstable representation of sound: A biological marker of dyslexia. *The Journal of Neuroscience, 33*(8), 3500–3504.

Hornickel, J., Zecker, S. G., Bradlow, A. R., & Kraus, N. (2012). Assistive listening devices drive neuroplasticity in children with dyslexia. *Proceedings of the National Academy of Sciences of the USA, 109*(41), 16731–16736.

Hughes, L. F., Turner, J. G., Parrish, J. L., & Caspary, D. M. (2010). Processing of broadband stimuli across A1 layers in young and aged rats. *Hearing Research, 264*(1–2), 79–85.

Humes, L. E., Dubno, J. R., Gordon-Salant, S., Lister, J. J., et al. (2012). Central presbycusis: A review and evaluation of the evidence. *Journal of the American Academy of Audiology, 23*(8), 635–666.

Kale, S., & Heinz, M. (2010). Envelope coding in auditory nerve fibers following noise-induced hearing loss. *Journal of the Association for Research in Otolaryngology, 11*(4), 657–673.

Killion, M., & Niquette, P. (2000). What can the pure-tone audiogram tell us about a patient's SNR loss? *Hearing Journal, 53*(3), 46–53.

Killion, M., Niquette, P., Gudmundsen, G., Revit, L., & Banerjee, S. (2004). Development of a quick speech-in-noise test for measuring signal-to-noise ratio loss in normal-hearing and hearing-impaired listeners. *The Journal of the Acoustical Society of America, 116*, 2935–2405.

King, A., Hopkins, K., & Plack, C. J. (2014). The effects of age and hearing loss on interaural phase difference discrimination. *The Journal of the Acoustical Society of America, 135*(1), 342–351.

Kochkin, S. (2010). MarkeTrak VIII: Consumer satisfaction with hearing aids is slowly increasing. *Hearing Journal, 63*(1), 19–32.

Kotak, V. C., Fujisawa, S., Lee, F. A., Karthikeyan, O., et al. (2005). Hearing loss raises excitability in the auditory cortex. *The Journal of Neuroscience, 25*(15), 3908–3918.

Kraus, N., & Anderson, S. (2016). Auditory processing disorder: Biological basis and treatment efficacy. In R. R. Fay & A. N. Popper (Eds.), *Translational research in audiology and the hearing sciences: An essential guide for scientists and clinicians*, (in press). New York: Springer Science+Business Media.

Krishnan, A., Bidelman, G. M., Smalt, C. J., Ananthakrishnan, S., & Gandour, J. T. (2012). Relationship between brainstem, cortical and behavioral measures relevant to pitch salience in humans. *Neuropsychologia, 50*(12), 2849–2859.

Kujawa, S. G., & Liberman, M. C. (2009). Adding insult to injury: Cochlear nerve degeneration after "temporary" noise-induced hearing loss. *The Journal of Neuroscience, 29*(45), 14077–14085.

Lang, H., Jyothi, V., Smythe, N., Dubno, J., et al. (2010). Chronic reduction of endocochlear potential reduces auditory nerve activity: Further confirmation of an animal model of metabolic presbyacusis. *Journal of the Association for Research in Otolaryngology, 11*(3), 419–434.

Lidén, G., & Korsan-Bengtsen, M. (1973). Audiometric manifestations of retrocochlear lesions. *Scandinavian Audiology, 2*(1), 29–40.

Lin, F. R., Ferrucci, L., Metter, E. J., An, Y., et al. (2011a). Hearing loss and cognition in the Baltimore longitudinal study of aging. *Neuropsychology, 25*(6), 763–770.

Lin, H. W., Furman, A. C., Kujawa, S. G., & Liberman, M. C. (2011b). Primary neural degeneration in the Guinea pig cochlea after reversible noise-induced threshold shift. *Journal of the Association for Research in Otolaryngology, 12*(5), 605–616.

Lorenzi, C., Gilbert, G., Carn, H., Garnier, S., & Moore, B. C. J. (2006). Speech perception problems of the hearing impaired reflect inability to use temporal fine structure. *Proceedings of the National Academy of Sciences of the USA, 103*(49), 18866–18869.

Lunner, T., & Sundewall-Thoren, E. (2007). Interactions between cognition, compression, and listening conditions: Effects on speech-in-noise performance in a two-channel hearing aid. *Journal of the American Academy of Audiology, 18*(7), 604–617.

Mamo, S. K., Grose, J. H., & Buss, E. (2015). Speech-evoked ABR: Effects of age and simulated neural temporal jitter. *Hearing Research.* Doi:10.1016/j.heares.2015.09.005

Marmel, F., Linley, D., Carlyon, R. P., Gockel, H. E., et al. (2013). Subcortical neural synchrony and absolute thresholds predict frequency discrimination independently. *Journal of the Association for Research in Otolaryngology, 14*(5), 757–766.

Munro, K. J., Purdy, S. C., Ahmed, S., Begum, R., & Dillon, H. (2011). Obligatory cortical auditory evoked potential waveform detection and differentiation using a commercially available clinical system: HEARLab™. *Ear and Hearing, 32*(6), 782–786.

Nilsson, M., Soli, S., & Sullivan, J. (1994). Development of the hearing in noise test for the measurement of speech reception thresholds in quiet and in noise. *Journal of the Acoustical Society of America, 95*, 1085–1099.

Ohlemiller, K. K., Lett, J. M., & Gagnon, P. M. (2006). Cellular correlates of age-related endocochlear potential reduction in a mouse model. *Hearing Research, 220*(1–2), 10–26.

Parthasarathy, A., & Bartlett, E. L. (2011). Age-related auditory deficits in temporal processing in F-344 rats. *Neuroscience, 192*, 619–630.

Parthasarathy, A., Cunningham, P. A., & Bartlett, E. L. (2010). Age-related differences in auditory processing as assessed by amplitude-modulation following responses in quiet and in noise. *Fronters in Aging Neuroscience*, 2, Doi:10.3389/fnagi.2014.00347

Parthasarathy, A., Datta, J., Torres, J. A., Hopkins, C., & Bartlett, E. L. (2014). Age-related changes in the relationship between auditory brainstem responses and envelope-following responses. *Journal of the Association for Research in Otolaryngology, 15*(4), 649–661.

Peelle, J. E., Troiani, V., Grossman, M., & Wingfield, A. (2011). Hearing loss in older adults affects neural systems supporting speech comprehension. *The Journal of Neuroscience, 31*(35), 12638–12643.

Pichora-Fuller, M. K., Schneider, B. A., MacDonald, E., Pass, H. E., & Brown, S. (2007). Temporal jitter disrupts speech intelligibility: A simulation of auditory aging. *Hearing Research, 223*(1–2), 114–121.

Plyler, P. N., & Ananthanarayan, A. K. (2001). Human frequency-following responses: Representation of second formant transitions in normal-hearing and hearing-impaired listeners. *Journal of the American Academy of Audiology, 12*(10), 523–533.

Presacco, A., Jenkins, K., Lieberman, R., & Anderson, S. (2015). Effects of aging on the encoding of dynamic and static components of speech. *Ear and Hearing, 36*(6), e352–363. Doi:10.1097/AUD.0000000000000193

Ruggles, D., Bharadwaj, H., & Shinn-Cunningham, B. G. (2011). Normal hearing is not enough to guarantee robust encoding of suprathreshold features important in everyday communication. *Proceedings of the National Academy of Sciences of the USA, 108*(37), 15516–15521.

Schmiedt, R. A., Lang, H., Okamura, H.-O., & Schulte, B. A. (2002). Effects of furosemide applied chronically to the round window: A model of metabolic presbycusis. *The Journal of Neuroscience, 22*(21), 9643–9650.

Schneider, B., & Pichora-Fuller, K. (2000). Implications of perceptual deterioration for cognitive aging research. In F. M. I. Craik & T. A. Salthouse (Eds.), *The handbook of aging and cognition* (pp. 155–219). Mahwah, NJ: Lawrence Erlbaum Associates.

Schuknecht, H. F. (1964). Further observations on the pathology of presbycusis. *Archives of Otolaryngology, 80*, 369–382.

Sergeyenko, Y., Lall, K., Liberman, M. C., & Kujawa, S. G. (2013). Age-related cochlear synaptopathy: An early-onset contributor to auditory functional decline. *The Journal of Neuroscience, 33*(34), 13686–13694.

Shaheen, L. A., Valero, M. D., & Liberman, M. C. (2015). Towards a diagnosis of cochlear neuropathy with envelope following responses. *Journal of the Association for Research in Otolaryngology, 16*(6), 727–745.

Smalt, C. J., Krishnan, A., Bidelman, G. M., Ananthakrishnan, S., & Gandour, J. T. (2012). Distortion products and their influence on representation of pitch-relevant information in the human brainstem for unresolved harmonic complex tones. *Hearing Research, 292*(1–2), 26–34.

Song, J. H., Skoe, E., Wong, P. C. M., & Kraus, N. (2008). Plasticity in the adult human auditory brainstem following short-term linguistic training. *Journal of Cognitive Neuroscience, 20*(10), 1892–1902.

Song, J. H., Skoe, E., Banai, K., & Kraus, N. (2012). Training to improve hearing speech in noise: Biological mechanisms. *Cerebral Cortex, 22*(5), 1180–1190.

Soros, P., Teismann, I. K., Manemann, E., & Lutkenhoner, B. (2009). Auditory temporal processing in healthy aging: A magnetoencephalographic study. *BioMed Central Neuroscience, 10*, 34. Doi:10.1186/1471-2202-10-34

Souza, P., Boike, K., Witherell, K., & Tremblay, K. (2007). Prediction of speech recognition from audibility in older listeners with hearing loss: Effects of age, amplification, and background noise. *Journal of the American Academy of Audiology, 18*(1), 54–65.

Tremblay, K., Piskosz, M., & Souza, P. (2003). Effects of age and age-related hearing loss on the neural representation of speech cues. *Clinical Neurophysiology, 114*(7), 1332–1343.

Tun, P., O'Kane, G., & Wingfield, A. (2002). Distraction by competing speech in young and older adult listeners. *Psychology of Aging, 17*, 453–467.

Turner, J. G., Hughes, L. F., & Caspary, D. M. (2005). Affects of aging on receptive fields in rat primary auditory cortex layer V neurons. *Journal of Neurophysiology, 94*(4), 2738–2747.

Vander Werff, K. R., & Burns, K. S. (2011). Brainstem responses to speech in younger and older adults. *Ear and Hearing, 32*(2), 168–180.

Walton, J. P., Frisina, R. D., & O'Neill, W. E. (1998). Age-related alteration in processing of temporal sound features in the auditory midbrain of the CBA mouse. *The Journal of Neuroscience, 18*(7), 2764–2776.

White-Schwoch, T., Woodruff Carr, K., Thompson, E. C., Anderson, S., et al. (2015). Auditory processing in noise: A preschool biomarker for literacy. *PLoS Biology, 13*(7), e1002196.

Wingfield, A., Tun, P. A., Koh, C. K., & Rosen, M. J. (1999). Regaining lost time: Adult aging and the effect of time restoration on recall of time-compressed speech. *Psychology of Aging, 14* (3), 380–389.

Wong, P. C. M., Ettlinger, M., Sheppard, J. P., Gunasekera, G. M., & Dhar, S. (2010). Neuroanatomical characteristics and speech perception in noise in older adults. *Ear and Hearing, 31*(4), 471–479.

Wong, P. C. M., Jin, J. X., Gunasekera, G. M., Abel, R., et al. (2009). Aging and cortical mechanisms of speech perception in noise. *Neuropsychologia, 47*(3), 693–703.

Zhong, Z., Henry, K. S., & Heinz, M. G. (2014). Sensorineural hearing loss amplifies neural coding of envelope information in the central auditory system of chinchillas. *Hearing Research, 309*, 55–62.

MIX
Papier aus verantwortungsvollen Quellen
Paper from responsible sources
FSC® C105338

If you have any concerns about our products,
you can contact us on
ProductSafety@springernature.com

In case Publisher is established outside the EU,
the EU authorized representative is:
Springer Nature Customer Service Center GmbH
Europaplatz 3, 69115 Heidelberg, Germany

Printed by Libri Plureos GmbH
in Hamburg, Germany